LIFE IN THE SEA

Readings from
**SCIENTIFIC
AMERICAN**

LIFE IN THE SEA

With Introductions by
Andrew Todd Newberry
University of California, Santa Cruz

W. H. Freeman and Company
San Francisco

Most of the SCIENTIFIC AMERICAN articles in *Life in the Sea* are available as separate Offprints. For a complete list of articles now available as Offprints, write to W. H. Freeman and Company, 660 Market Street, San Francisco, California 94104.

Library of Congress Cataloging in Publication Data
Main entry under title:

Life in the sea.

    Includes bibliographies and index.
    1. Marine biology—Addresses, essays, lectures.
I. Newberry, Andrew Todd.   II. Scientific American.
QH91.1.L53   574.5′2636   81–17251
ISBN 0–7167–1398–5   AACR2
ISBN 0–7167–1399–3 (pbk.)

Printed in the United States of America

1 2 3 4 5 6 7 8 9 0   KP   0 8 9 8 7 6 5 4 3 2

PREFACE

ife in the Sea is a deliberately ambiguous title for this *Scientific American* Reader. It refers both to the organisms in the sea and to the conditions in which they live. How these two aspects interact is the principle that has guided the selection and arrangement of articles. In doing double duty this way, the articles convey, I hope, how thoroughly organisms are "in conversation with their habitats," that is, embedded in the conditions and selective pressures of their environment.

After an introductory "big picture," the middle three sections constitute the heart of this Reader. These articles reflect some especially effective ways of studying marine biology and describe the results of these investigative methods. The results and data of the articles in these sections inevitably are augmented, dated, and even made obsolete by subsequent studies, but methods retain their cogency. Often elegantly simple experimental manipulations or subtle but crucial observations integrate questions and answers in ways that transcend mere anecdote. Thus, I have selected these articles not only for their facts but also (and sometimes especially) for their methods—not only for what they say about life in the sea but also for how they instruct us in how to study it.

The final section deals with food from the sea and the disposal of wastes there. Considerations of human welfare often seem to slip into marine biology merely as gestures toward "relevance." But, from another profession, Albert Schweitzer put the matter this way: "It seemed to me a matter of course that we all should take our share of the burden of pain that falls upon the world." In this spirit, marine scientists, too, must face squarely the obligations that knowledge carries.

Unavoidably, some essays in this collection appear in other *Scientific American* Readers, but this arrangement creates its own particular sense, different from that of other Readers. Neither selecting nor arranging these articles has been easy. On the one hand, there are bound to be gaps that only future essays in *Scientific American* can fill. On the other hand, from 1948 through 1980 there has been a feast of reports from which I have chosen two dozen for this anthology. I have included the titles of the many others considered in a bibliographic postscript (page 241).

May 1981   Andrew Todd Newberry

CONTENTS

V   THE HUMAN TOUCH

Note on cross-references to SCIENTIFIC AMERICAN articles: Articles included in this book are referred to by title and page number; articles not included in this book but available as Offprints are referred to by title and Offprint number; articles not included in this book and not available as Offprints are referred to by title and date of publication.

THE BIG PICTURE

I THE BIG PICTURE

INTRODUCTION

Ours is the blue planet. Photographs taken from space have brought home the familiar statistic that the seas cover 71 percent of the earth's surface. But the huge volume of the marine realm dominates the earth's biosphere far more than any surface view can convey. We can reckon this volumetric comparison only roughly, but very impressively.

Terrestrial life exploits the land's surface, extending scarcely a few meters into the ground and some dozens of meters into tall trees. If the land were all flat but forested, the volume of the terrestrial realm would be surprisingly small—about 3 million cubic kilometers. Of course, topographic relief greatly increases the land's surface beyond what a two-dimensional map can show. But this wrinkled land is covered mostly with brush, prairies, plains, or desert, not with forests. Taking these modifications into account, we can approximate the volume of the terrestrial realm at 1,250,000 cubic kilometers. Above this thin film of life on land, the air remains largely uninhabited; for whatever reasons, persistent aerial plankton and nekton have not yet evolved.

We know now that the bottom of the sea has as much relief as the continents. And the sea supports living things at all depths, so that the marine realm includes an inhabited medium averaging almost 4 kilometers thick. We take this domain of plankton and nekton very much for granted—it is, after all, "the sea"—but it enormously affects the comparison we are making. The whole marine realm, consequently, comprises some 1.3 *billion* cubic kilometers, easily a thousand times the volume for living that the terrestrial domain offers.

The sea is not only an immense habitat but also a congenial one for biological processes. For example, the very density that makes a wave pack such a punch supports the plankton in midwater, buoys fragile structures of benthic organisms, carries food to sessile feeders, and disperses larval young. And seawater's density, its mix of salts, and its capacity to buffer changes of temperature let marine organisms survive with few of the devices that organisms need to cope with the stresses of living on land.

Marine and terrestrial organisms alike face environmental challenges to their development, persistence, and reproduction. But the conditions of life in the sea—the circumstances through which these challenges assert themselves—differ from those on land. So, too, do the organisms' adaptations for survival: there are characteristically marine combinations of ecological problems and adaptive organismic designs. These are the very combinations that make life in the sea often seem so strange to us.

In his "tour" of the sea, "The Nature of Oceanic Life," John D. Isaacs surveys the diversity not only of inhabitants but also of habitats in the ocean.

Even in the apparent homogeneity of the open ocean, creatures encounter a complex and shifting mosaic of ecological circumstances (see, for example, "The Microstructure of the Ocean," by Michael C. Gregg, SCIENTIFIC AMERICAN, Offprint 905). The sea floor, like the land, presents a bewildering array of conditions that shape the characteristics of any benthic ecosystem or its community of inhabitants. A few among many such factors are the texture of the substrate, floor–water interactions like the sweep of currents or tides or the tug of the surf, depth-related factors like light and pressure that affect midwater and bottom habitats alike, and biological relationships among the benthos.

Afloat or on the bottom, appreciable movement vertically or horizontally implies ecological shifts. With depth, light fades, pressure increases, and salinity and temperature alter as one moves through different water masses, and currents deflect creatures in different directions at different depths. Horizontal movements bring organisms into new water masses, into regions of upwelling and downwelling currents, into habitats of different benthic character (rocky, muddy) or different bathymetric interactions (e.g., moving from the open sea into littoral waters).

Out of this complexity, Isaacs has organized his discussion around the flow of energy and the cycling of materials in the sea. He traces trophic relationships from photosynthetic phytoplankton through herbivorous and carnivorous consumers. In this way, he provides a guide both to general ecological relationships and to their particular patterns in the sea. And along the way, Isaacs pauses to note such peculiarly marine phenomena as the absence of large plants from the high seas, mechanisms of filter feeding, widespread bioluminescence, traits of abyssal life, and many creatures that are profoundly unfamiliar to those biologists whose acquaintance with the diversity of living things and their circumstances stops on the landward side of the water's edge.

James W. Valentine and Eldridge M. Moores conduct another tour of the seas in "Plate Tectonics and the History of Life in the Oceans," an exploration through evolutionary time and geological transformations. The biological phenomena that Isaacs describes persist, but their geographic (and hence ecological) circumstances change. Compared to biological processes, the pace of geological change is very slow, but the changes themselves are of enormous magnitude and consequence: the very rearrangement of continents and seas and especially the redistribution of shallow-water habitats, which harbor the sea's richest biota. The paleoecological story is more speculative than a description of today's seas. In part, this is because the implications of the "conceptual revolution" of plate-tectonic theory are only now being sorted out.

In drawing on plate tectonics, the authors' historical analysis goes beyond a map of contemporary marine biogeographical assemblages to a convincing interpretation of how these patterns came to be arranged and populated. But an intractable problem remains, the same one that confronts evolutionary studies. While the repetitive processes involved lend themselves to testable explanations, their products—biogeographical patterns or phylogenetic lineages—are unique. Statements about the unique are not subject to disproof or confirmation by experimental means. And the methods of comparative biology—lacking as they do, for example, controls—are far less clear-cut than those of experimental biology. Thus, some investigations of clearly scientific topics may lack crucial procedures in "the scientific method." This is not an idle concern. Method is central to confidence in science. Assembling the evidence and critically testing the assertions of the historical natural sciences are different endeavors from their analogs in the experimental sciences. But as the essays in this volume demonstrate, persuasive conclusions about the history of life are attainable.

The Nature of Oceanic Life

1

by John D. Isaacs
September 1969

*The conditions of the marine environment have
given rise to a food web in which the dominant
primary production of organic matter is carried
out by microscopic plants*

I plan to take the reader on a brief tour of marine life from the surface layers of the open sea, down through the intermediate layers to the deep-sea floor, and from there to the living communities on continental shelves and coral reefs. Like Dante, I shall be able to record only a scattered sampling of the races and inhabitants of each region and to point out only the general dominant factors that typify each domain; in particular I shall review some of the conditions, principles and interactions that appear to have molded the forms of life in the sea and to have established their range and compass.

The organisms of the sea are born, live, breathe, feed, excrete, move, grow, mate, reproduce and die within a single interconnected medium. Thus interactions among the marine organisms and interactions of the organisms with the chemical and physical processes of the sea range across the entire spectrum from simple, adamant constraints to complex effects of many subtle interactions.

Far more, of course, is known about the life of the sea than I shall be able even to suggest, and there are yet to be achieved great steps in our knowledge of the living entities of the sea. I shall mention some of these possibilities in my concluding remarks.

A general discussion of a living system should consider the ways in which plants

elaborate basic organic material from inorganic substances and the successive and often highly intricate steps by which organisms then return this material to the inorganic reservoir. The discussion should also show the forms of life by which such processes are conducted. I shall briefly trace these processes through the regions I have indicated, returning later to a more detailed discussion of the living forms and their constraints.

Some organic material is carried to the sea by rivers, and some is manufactured in shallow water by attached plants. More than 90 percent of the basic organic material that fuels and builds the life in the sea, however, is synthesized within the lighted surface layers of open water by the many varieties of phytoplankton. These sunny pastures of plant cells are grazed by the herbivorous zooplankton (small planktonic animals) and by some small fishes. These in turn are prey to various carnivorous creatures, large and small, who have their predators also.

The debris from the activities in the surface layers settles into the dimly lighted and unlighted midlayers of the sea, the twilight mesopelagic zone and the midnight bathypelagic zone, to serve as one source of food for their strange inhabitants. This process depletes the

surface layers of some food and particularly of the vital plant nutrients, or fertilizers, that become trapped below the surface layers, where they are unavailable to the plants. Food and nutrients are also actively carried downward from the surface by vertically migrating animals.

The depleted remnants of this constant "rain" of detritus continue to the sea floor and support those animals that live just above the bottom (epibenthic animals), on the bottom (benthic animals) and burrowed into the bottom. Here filter-feeding and burrowing (deposit-feeding) animals and bacteria rework the remaining refractory particles. The more active animals also find repast in mid-water creatures and in the occasional falls of carcasses and other larger debris. Except in unusual small areas there is an abundance of oxygen in the deep water, and the solid bottom presents advantages that allow the support of a denser population of larger creatures than can exist in deep mid-water.

In shallower water such as banks, atolls, continental shelves and shallow seas conditions associated with a solid bottom and other regional modifications of the general regime enable rich populations to develop. Such areas constitute about 7 percent of the total area of the ocean. In some of these regions added food results from the growth of larger fixed plants and from land drainage.

With the above bare recitation for general orientation, I shall now discuss these matters in more detail.

The cycle of life in the sea, like that on land, is fueled by the sun's visible light acting on green plants. Of every million photons of sunlight reaching the earth's surface, some 90 enter into the net production of basic food. Perhaps 50

NEW EVIDENCE that an abundance of large active fishes inhabit the deep-sea floor was obtained recently by the author and his colleagues in the form of photographs such as the one on the opposite page. The photograph was made by a camera hovering over a five-gallon bait can at a depth of 1,400 meters off Lower California. The diagonal of the bait can measures a foot. The larger fish are mostly rat-tailed grenadiers and sablefish. The fact that large numbers of such fish are attracted almost immediately to the bait suggests that two rather independent branches of the marine food web coexist in support of the deep-bottom creatures by dead material: the rain of fine detritus, which supports a variety of attached filter-feeding and burrowing organisms, and rare, widely separated falls of large food fragments, which support active creatures adapted to the discovery and utilization of such food.

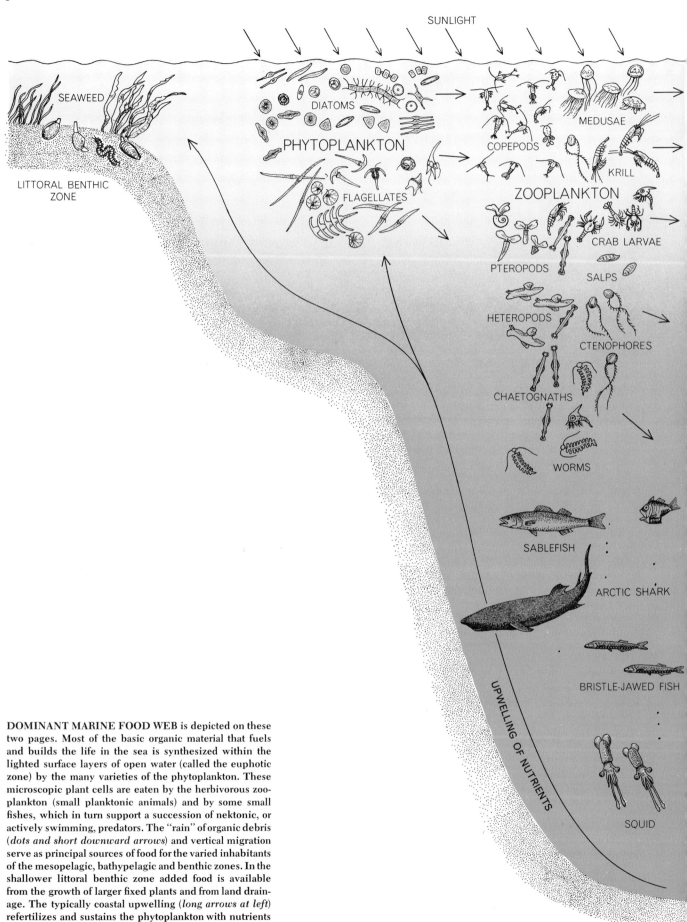

SUNLIGHT

SEAWEED

DIATOMS

PHYTOPLANKTON

FLAGELLATES

LITTORAL BENTHIC ZONE

MEDUSAE

COPEPODS

KRILL

ZOOPLANKTON

CRAB LARVAE

PTEROPODS

SALPS

HETEROPODS

CTENOPHORES

CHAETOGNATHS

WORMS

SABLEFISH

ARCTIC SHARK

BRISTLE-JAWED FISH

UPWELLING OF NUTRIENTS

SQUID

DOMINANT MARINE FOOD WEB is depicted on these two pages. Most of the basic organic material that fuels and builds the life in the sea is synthesized within the lighted surface layers of open water (called the euphotic zone) by the many varieties of the phytoplankton. These microscopic plant cells are eaten by the herbivorous zooplankton (small planktonic animals) and by some small fishes, which in turn support a succession of nektonic, or actively swimming, predators. The "rain" of organic debris (*dots and short downward arrows*) and vertical migration serve as principal sources of food for the varied inhabitants of the mesopelagic, bathypelagic and benthic zones. In the shallower littoral benthic zone added food is available from the growth of larger fixed plants and from land drainage. The typically coastal upwelling (*long arrows at left*) refertilizes and sustains the phytoplankton with nutrients released by bacterial decomposition of organic detritus on the bottom. The organisms are not drawn to same scale.

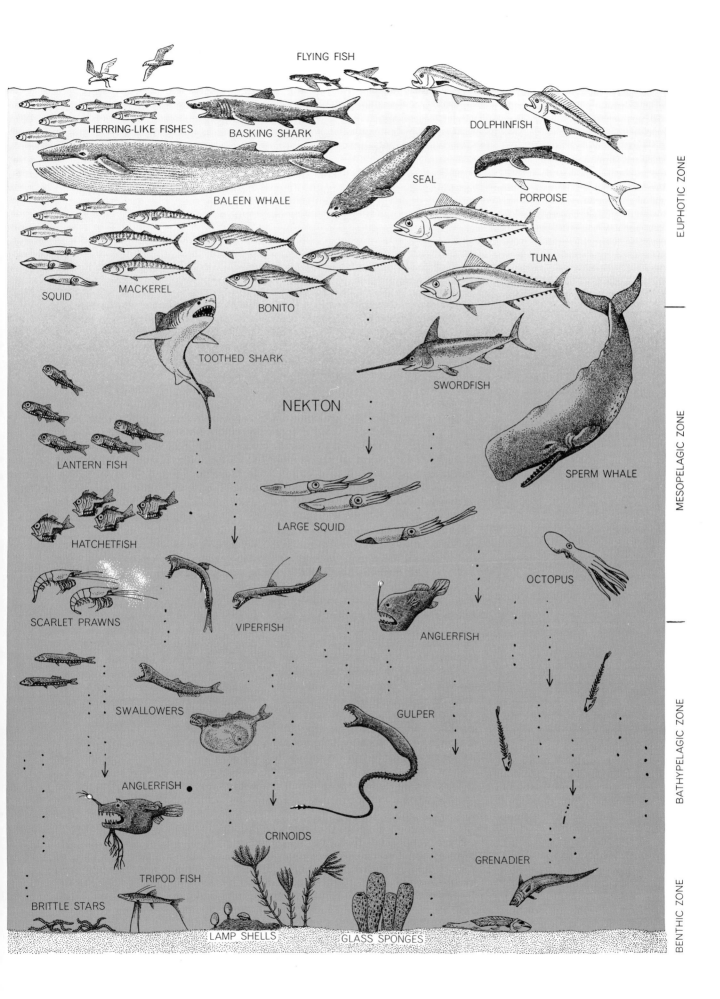

FLYING FISH

HERRING-LIKE FISHES

BASKING SHARK

DOLPHINFISH

BALEEN WHALE

SEAL

PORPOISE

TUNA

SQUID

MACKEREL

BONITO

TOOTHED SHARK

NEKTON

SWORDFISH

SPERM WHALE

LANTERN FISH

HATCHETFISH

LARGE SQUID

OCTOPUS

SCARLET PRAWNS

VIPERFISH

ANGLERFISH

SWALLOWERS

GULPER

ANGLERFISH

CRINOIDS

GRENADIER

TRIPOD FISH

BRITTLE STARS

LAMP SHELLS

GLASS SPONGES

EUPHOTIC ZONE

MESOPELAGIC ZONE

BATHYPELAGIC ZONE

BENTHIC ZONE

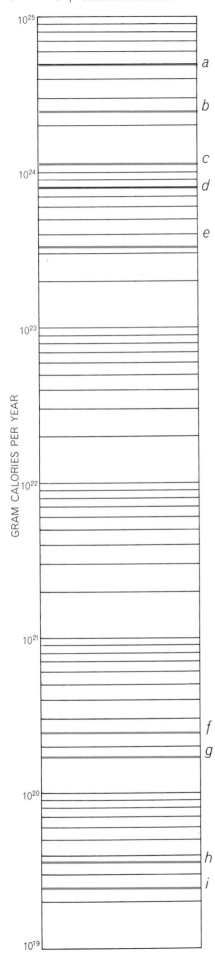

GRAM CALORIES PER YEAR

of the 90 contribute to the growth of land plants and about 40 to the growth of the single-celled green plants of the sea, the phytoplankton [*see illustration at left*]. It is this minute fraction of the sun's radiant energy that supplies the living organisms of this planet not only with their food but also with a breathable atmosphere.

The terrestrial and marine plants and animals arose from the same sources, through similar evolutionary sequences and by the action of the same natural laws. Yet these two living systems differ greatly at the stage in which we now view them. Were we to imagine a terrestrial food web that had developed in a form limited to that of the open sea, we would envision the land populated predominantly by short-lived simple plant cells grazed by small insects, worms and snails, which in turn would support a sparse predaceous population of larger insects, birds, frogs and lizards. The population of still larger carnivores would be a small fraction of the populations of large creatures that the existing land food web can nurture, because organisms in each of these steps pass on not more than 15 percent of the organic substance.

In some important respects this imaginary condition is not unlike that of the dominant food web of the sea, where almost all marine life is sustained by microscopic plants and near-microscopic herbivores and carnivores, which pass on only a greatly diminished supply of food to sustain the larger, more active and more complex creatures. In other respects the analogy is substantially inaccurate, because the primary marine food production is carried out by cells dispersed widely in a dense fluid medium.

This fact of an initial dispersal imposes a set of profound general conditions on all forms of life in the sea. For comparison, the concentration of plant food in a moderately rich grassland is of the order of a thousandth of the volume of the gross space it occupies and of the order of half of the mass of the air in which it is immersed. In moderately rich areas of the sea, on the other hand, food

is hundreds of times more dilute in volume and hundreds of thousands of times more dilute in relative mass. To crop this meager broth a blind herbivore or a simple pore in a filtering structure would need to process a weight of water hundreds of thousands of times the weight of the cell it eventually captures. In even the densest concentrations the factor exceeds several thousands, and with each further step in the food web dilution increases. Thus from the beginnings of the marine food web we see many adaptations accommodating to this dilution: eyes in microscopic herbivorous animals, filters of exquisite design, mechanisms and behavior for discovering local concentrations, complex search gear and, on the bottom, attachments to elicit the aid of moving water in carrying out the task of filtration. All these adaptations stem from the conditions that limit plant life in the open sea to microscopic dimensions.

It is in the sunlit near-surface of the open sea that the unique nature of the dominant system of marine life is irrevocably molded. The near-surface, or mixed, layer of the sea varies in thickness from tens of feet to hundreds depending on the nature of the general circulation, mixing by winds and heating [see "The Atmosphere and the Ocean," by R. W. Stewart, SCIENTIFIC AMERICAN Offprint 881. Here the basic food production of the sea is accomplished by single-celled plants. One common group of small phytoplankton are the coccolithophores, with calcareous plates, a swimming ability and often an oil droplet for food storage and buoyancy. The larger microscopic phytoplankton are composed of many species belonging to several groups: naked algal cells, diatoms with complex shells of silica and actively swimming and rotating flagellates. Very small forms of many groups are also abundant and collectively are called nannoplankton.

The species composition of the phytoplankton is everywhere complex and varies from place to place, season to season and year to year. The various regions of the ocean are typified, however, by

PRODUCTIVITY of the land and the sea are compared in terms of the net amount of energy that is converted from sunlight to organic matter by the green cells of land and sea plants. Colored lines denote total energy reaching the earth's upper atmosphere (*a*), total energy reaching earth's surface (*b*), total energy usable for photosynthesis (*c*), total energy usable for photosynthesis at sea (*d*), total energy usable for photosynthesis on land (*e*), net energy used for photosynthesis on land (*f*), net energy used for photosynthesis at sea (*g*), net energy used by land herbivores (*h*) and net energy used by sea herbivores (*i*). Although more sunlight falls on the sea than on the land (by virtue of the sea's larger surface area), the total land area is estimated to outproduce the total sea area by 25 to 50 percent. This is primarily due to low nutrient concentrations in the euphotic zone and high metabolism in marine plants. The data are from Walter R. Schmitt of Scripps Institution of Oceanography.

dominant major groups and particular species. Seasonal effects are often strong, with dense blooms of phytoplankton occurring when high levels of plant nutrients suddenly become usable or available, such as in high latitudes in spring or along coasts at the onset of upwelling. The concentration of phytoplankton varies on all dimensional scales, even down to small patches.

It is not immediately obvious why the dominant primary production of organic matter in the sea is carried out by microscopic single-celled plants instead of free-floating higher plants or other intermediate plant forms. The question arises: Why are there no pelagic "trees" in the ocean? One can easily compute the advantages such a tree would enjoy, with its canopy near the surface in the lighted levels and its trunk and roots extending down to the nutrient-rich waters under the mixed layer. The answer to this fundamental question probably has several parts. The evolution of plants in the pelagic realm favored smallness rather than expansion because the mixed layer in which these plants live is quite homogeneous; hence small incremental extensions from a plant cell cannot aid it in bridging to richer sources in order to satisfy its several needs.

On land, light is immediately above the soil and nutrients are immediately below; thus any extension is of immediate benefit, and the development of single cells into higher erect plants is able to follow in a stepwise evolutionary sequence. At sea the same richer sources exist but are so far apart that only a very large ready-made plant could act as a bridge between them. Although such plants could develop in some other environment and then adapt to the pelagic conditions, this has not come about. It is difficult to see how such a plant would propagate anyway; certainly it could not propagate in the open sea, because the young plants there would be at a severe disadvantage. In the sea small-scale differential motions of water are rapidly damped out, and any free-floating plant must often depend on molecular diffusion in the water for the uptake of nutrients and excretion of wastes. Smallness and self-motion are then advantageous, and a gross structure of cells cannot exchange nutrients or wastes as well as the same cells can separately or in open aggregations.

In addition the large-scale circulation of the ocean continuously sweeps the pelagic plants out of the region to which they are best adapted. It is essential that some individuals be returned to renew

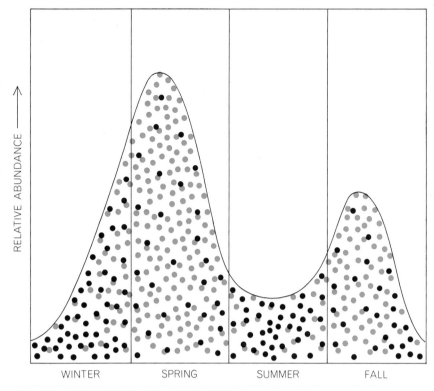

RELATIVE ABUNDANCE ⟶

WINTER   SPRING   SUMMER   FALL

SPECIES COMPOSITION AND ABUNDANCE of the phytoplankton varies from season to season, particularly at high latitudes. During the winter the turbulence caused by storms replenishes the supply of nutrients in the surface layers. During this period flagellates (*black dots*) tend to dominate. In early spring the increase in the amount of sunlight reaching the surface stimulates plant growth, and diatoms (*colored dots*) are stimulated to grow. Later in spring grazing by zooplankton and a decrease in the supply of nutrients caused by calmer weather result in a general reduction in the phytoplankton population, which reaches a secondary minimum in midsummer, during which time flagellates again dominate. The increased mixing caused by early autumn storms causes a rise in the supply of nutrients and a corresponding minor surge in the population of diatoms. The decreasing sunlight of late fall and grazing by zooplankton again reduce the general level of the plant population.

the populations. More mechanisms for this essential return exist for single-celled plants than exist for large plants, or even for any conventional spores, seeds or juveniles. Any of these can be carried by oceanic gyres or diffused by large-scale motions of surface eddies and periodic counterflow, but single-celled plants can also ride submerged countercurrents while temporarily feeding on food particles or perhaps on dissolved organic material. Other mechanisms of distribution undoubtedly are also occasionally important. For example, living marine plant cells are carried by storm-borne spray, in bird feathers and by well-fed fish and birds in their undigested food.

No large plant has solved the many problems of development, dispersal and reproduction. There *are* no pelagic trees, and these several factors in concert therefore restrict the open sea in a profound way. They confine it to an initial food web composed of microscopic forms, whereas larger plants live attached only to shallow bottoms (which

comprise some 2 percent of the ocean area). Attached plants, unlike free-floating plants, are not subject to the aforementioned limitations. For attached plants all degrees of water motion enhance the exchange of nutrients and wastes. Moreover, their normal population does not drift, much of their reproduction is by budding, and their spores are adapted for rapid development and settlement. Larger plants too are sometimes found in nonreproducing terminal accumulations of drifting shore plants in a few special convergent deep-sea areas such as the Sargasso Sea.

Although species of phytoplankton will populate only regions with conditions to which they are adapted, factors other than temperature, nutrients and light levels undoubtedly are important in determining the species composition of phytoplankton populations. Little is understood of the mechanisms that give rise to an abundance of particular species under certain conditions. Grazing herbivores may consume only a part of the size range of cells, allowing certain sizes and types to dominate temporarily. Little is understood of the mechanisms that give rise to an abundance of particular species under certain conditions. Chemical by-products of certain species probably exclude certain other species. Often details of individual cell behavior are probably also important in the introduction and success of a species in a particular area. In some cases we can glimpse what these mechanisms are.

For example, both the larger diatoms and the larger flagellates can move at appreciable velocities through the water. The diatoms commonly sink downward, whereas the flagellates actively swim upward toward light. These are probably patterns of behavior primarily for increasing exchange, but the interaction of such unidirectional motions with random turbulence or systematic convective motion is not simple, as it is with an inactive particle. Rather, we would expect diatoms to be statistically abundant in upward-moving water and to sink out of the near-surface layers when turbulence or upward convection is low.

Conversely, flagellates should be statistically more abundant in downwelling water and should concentrate near the surface in low turbulence and slow downward water motions. These effects seem to exist. Off some continental coasts in summer flagellates may eventually collect in high concentrations. As they begin to shade one another from the light, each individual struggles closer to the lighted surface, producing such a high density that large areas of the water are turned red or brown by their pigments. The concentration of flagellates in these "red tides" sometimes becomes too great for their own survival. Several species of flagellates also become highly toxic as they grow older. Thus they sometimes both produce and participate in a mass death of fish and invertebrates that has been known to give rise to such a high yield of hydrogen sulfide as to blacken the white houses of coastal cities.

Large diatom cells, on the other hand, spend a disproportionately greater time in upward-moving regions of the water and an unlimited time in any region where the upward motion about equals their own downward motion. (The support of unidirectionally moving objects by contrary environmental motion is observed in other phenomena, such as the production of rain and hail.) Diatom cells are thus statistically abundant in upwelling water, and the distribution of diatoms probably is often a reflection of the turbulent-convective regime of the water. Sinking and the dependence of

the larger diatoms on upward convection and turbulence for support aids them in reaching upwelling regions, where nutrients are high; it helps to explain their dominance in such regions and such other features of their distribution as their high proportion in rich ocean regions and their frequent inverse occurrence with flagellates. Differences in adaptations to the physical and chemical conditions, and the release of chemical products, probably reinforce such relations.

In some areas, such as parts of the equatorial current system and shallow seas, where lateral and vertical circulation is rapid, the species composition of phytoplankton is perhaps more simply

a result of the inherent ability of the species to grow, survive and reproduce under the local conditions of temperature, light, nutrients, competitors and herbivores. Elsewhere second-order effects of the detailed cell behavior often dominate. Those details of behavior that give rise to concentrations on any dimensional scale are particularly important to all subsequent steps in the food chain.

All phytoplankton cells eventually settle from the surface layers. The depletion of nutrients and food from the surface layers takes place continuously through the loss of organic material, plant cells, molts, bodies of animals, fecal pellets and so forth, which release their content of chemical nutrients at

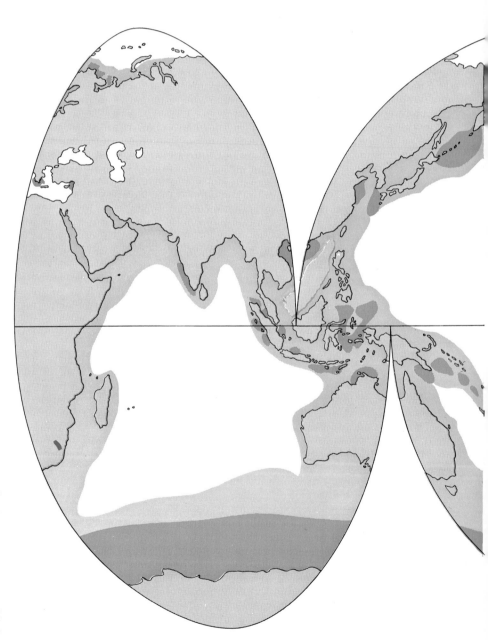

FAVORABLE CONDITIONS for the growth of phytoplankton occur wherever upwelling or mixing tends to bring subsurface nutrients up to the euphotic layer of the ocean. This map,

various depths through the action of bacteria and other organisms. The periodic downward migration of zooplankton further contributes to this loss.

These nutrients are "trapped" below the level of light adequate to sustain photosynthesis, and therefore the water in which plants must grow generally contains very low concentrations of such vital substances. It is this condition that is principally responsible for the comparatively low total net productivity of the sea compared with that of the land. The regions where trapping is broken down or does not exist—where there is upwelling of nutrient-rich water along coasts, in parts of the equatorial regions, in the wakes of islands and banks and in high

latitudes, and where there is rapid recirculation of nutrients over shallow shelves and seas—locally bear the sea's richest fund of life.

The initial factors discussed so far have placed an inescapable stamp on the form of all life in the open sea, as irrevocably no doubt as the properties and distribution of hydrogen have dictated the form of the universe. These factors have limited the dominant form of life in the sea to an initial microscopic sequence that is relatively unproductive, is stimulated by upwelling and mixing and is otherwise altered in species composition and distribution by physical, chemical and biological processes on all dimensional scales. The same factors also

limit the populations of higher animals and have led to unexpectedly simple adaptations, such as the sinking of the larger diatoms as a tactic to solve the manifold problems of enhancing nutrient and waste exchange, finding nutrients, remaining in the surface waters and repopulating.

The grazing of the phytoplankton is principally conducted by the herbivorous members of the zooplankton, a heterogeneous group of small animals that carry out several steps in the food web as herbivores, carnivores and detrital (debris-eating) feeders. Among the important members of the zooplankton are the arthropods, animals with external

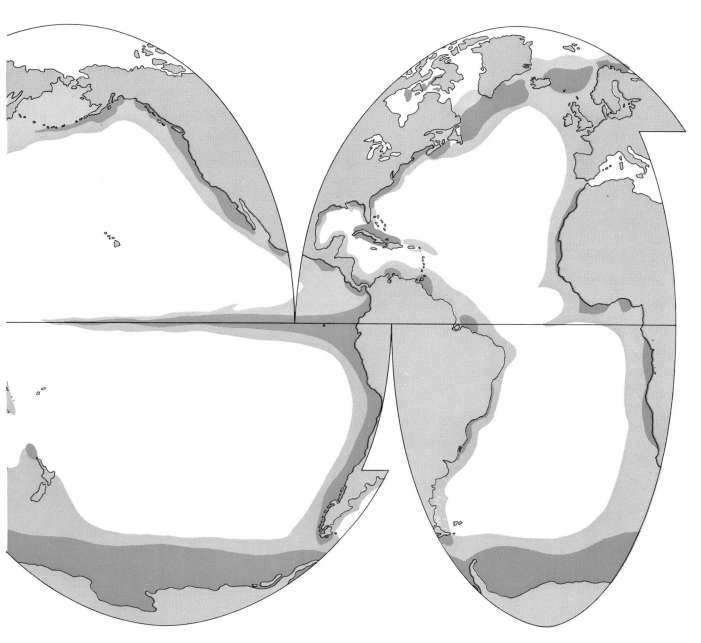

which is adapted from one compiled by the Norwegian oceanographer Harald U. Sverdrup, shows the global distribution of such waters, in which the productivity of marine life would be expected to be very high (*dark color*) and moderately high (*light color*).

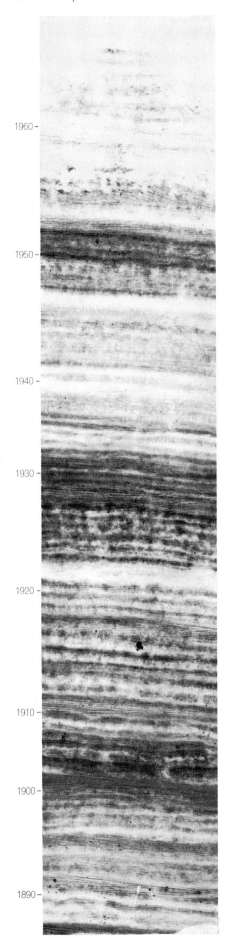

1960 –

1950 –

1940 –

1930 –

1920 –

1910 –

1900 –

1890 –

RARE SEDIMENTARY RECORD of the recent annual oceanographic, meteorological and biological history of part of a major oceanic system is revealed in this radiograph of a section of an ocean-bottom core obtained by Andrew Soutar of the Scripps Institution of Oceanography in the Santa Barbara Basin off the California coast. In some near-shore basins such as this one the absence of oxygen causes refractory parts of the organic debris to be left undecomposed and the sediment to remain undisturbed in the annual layers called varves. The dark layers are the densest and represent winter sedimentation. The lighter and less dense layers are composed mostly of diatoms and represent spring and summer sedimentation.

skeletons that belong to the same broad group as insects, crabs and shrimps. The planktonic arthropods include the abundant copepods, which are in a sense the marine equivalent of insects. Copepods are represented in the sea by some 10,-000 or more species that act not only as herbivores, carnivores or detrital feeders but also as external or even internal parasites! Two or three thousand of these species live in the open sea. Other important arthropods are the shrimplike euphausiids, the strongest vertical migrators of the zooplankton. They compose the vast shoals of krill that occur in high latitudes and that constitute one of the principal foods of the baleen whales. The zooplankton also include the strange bristle-jawed chaetognaths, or arrowworms, carnivores of mysterious origin and affinities known only in the marine environment. Widely distributed and abundant, the chaetognaths are represented by a surprisingly small number of species, perhaps fewer than 50. Larvae of many types, worms, medusae (jellyfish), ctenophores (comb jellies), gastropods (snails), pteropods and heteropods (other pelagic mollusks), salps, unpigmented flagellates and many others are also important components of this milieu, each with its own remarkably complex and often bizarre life history, behavior and form.

The larger zooplankton are mainly carnivores, and those of herbivorous habit are restricted to feeding on the larger plant cells. Much of the food supply, however, exists in the form of very small particles such as the nannoplankton, and these appear to be available almost solely to microscopic creatures. The immense distances between plant cells, many thousands of times their diameter, place a great premium on the development of feeding mechanisms that avoid the simple filtering of water

through fine pores. The power necessary to maintain a certain rate of flow through pores or nets increases inversely at an exponential rate with respect to the pore or mesh diameter, and the small planktonic herbivores, detrital feeders and carnivores show many adaptations to avoid this energy loss. Eyesight has developed in many minute animals to make possible selective capture. A variety of webs, bristles, rakes, combs, cilia and other structures are found, and they are often sticky. Stickiness allows the capture of food that is finer than the interspaces in the filtering structures, and it greatly reduces the expenditure of energy.

A few groups have developed extremely fine and apparently quite effective nets. One group that has accomplished this is the Larvacea. A larvacian produces and inhabits a complex external "house," much larger than its owner, that contains a system of very finely constructed nets through which the creature maintains a gentle flow [*see illustration on page 14*]. The Larvacea have apparently solved the problem of energy loss in filtering by having proportionately large nets, fine strong threads and a low rate of flow.

The composition of the zooplankton differs from place to place, day to night, season to season and year to year, yet most species are limited in distribution, and the members of the planktonic communities commonly show a rather stable representation of the modes of life.

The zooplankton are, of course, faced with the necessity of maintaining breeding assemblages and, like the phytoplankton, with the necessity of establishing a reinoculation of parent waters. In addition, their behavior must lead to a correspondence with their food and to the pattern of large-scale and small-scale spottiness already imposed on the marine realm by the phytoplankton. The swimming powers of the larger zooplankton are quite adequate for finding local small-scale patches of food. That this task is accomplished on a large scale is indirectly demonstrated by the observed correspondence between the quantities of zooplankton and the plant nutrients in the surface waters. How this large-scale task is accomplished is understood for some groups. For example, some zooplankton species have been shown to descend near the end of suitable conditions at the surface and to take temporary residence in a submerged countercurrent that returns them upstream.

There are many large and small puz-

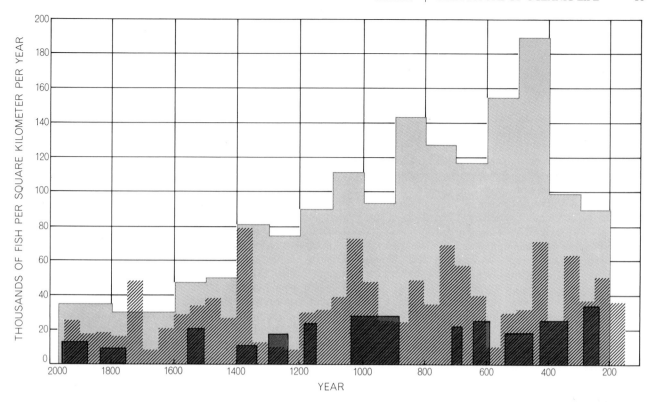

ESTIMATED FISH POPULATIONS in the Santa Barbara Basin over the past 1,800 years were obtained for three species by counting the average number of scales of each species in the varves of the core shown on the opposite page. Minimum population estimates for fish one year old and older are given for Pacific sardines (*gray*), northern sardines (*colored areas*) and Pacific hake (*hatched*).

zles in the distribution of zooplankton. As an example, dense concentrations of phytoplankton are often associated with low populations of zooplankton. These are probably rapidly growing blooms that zooplankton have not yet invaded and grazed on, but it is not completely clear that this is so. Chemical repulsion may be involved.

The concentration of larger zooplankton and small fish in the surface layers is much greater at night than during the day, because of a group of strongly swimming members that share their time between the surface and the mesopelagic region. This behavior is probably primarily a tactic to enjoy the best of two worlds: to crop the richer food developing in the surface layers and to minimize mortality from predation by remaining always in the dark, like timid rabbits emerging from the thicket to graze the nighttime fields, although still in the presence of foxes and ferrets. Many small zooplankton organisms also make a daily migration of some vertical extent.

In addition to its primary purpose daily vertical migration undoubtedly serves the migrating organisms in a number of other ways. It enables the creatures to adjust their mean temperature, so that by spending the days in cooler water the amount of food used during

rest is reduced. Perhaps such processes as the rate of egg development are also controlled by these tactics. Many land animals employ hiding behavior for similar kinds of adjustment. Convincing arguments have also been presented to show that vertical migration serves to maintain a wide range of tolerance in the migrating species, so that they will be more successful under many more conditions than if they lived solely in the surface layers. This migration must also play an important part in the distribution of many species. Interaction of the daily migrants with the water motion produced by daily land-sea breeze alternation can hold the migrants offshore by a kind of "rectification" of the oscillating water motion. More generally, descent into the lower layers increases the influence of submerged countercurrents, thereby enhancing the opportunity to return upstream, to enter upwelling regions and hence to find high nutrient levels and associated high phytoplankton productivity.

Even minor details of behavior may strongly contribute to success. Migrants spend the day at a depth corresponding to relatively constant low light levels, where the movement of the water commonly is different from that at the surface. Most of the members rise some-

what even at the passage of a cloud shadow. Should they be carried under the shadow of an area rich in phytoplankton, they migrate to shallower depths, thereby often decreasing or even halting their drift with respect to this rich region to which they will ascend at night. Conversely, when the surface waters are clear and lean, they will migrate deeper and most often drift relatively faster.

We might simplistically view the distribution of zooplankton, and phytoplankton for that matter, as the consequence of a broad inoculation of the oceans with a spectrum of species, each with a certain adaptive range of tolerances and a certain variable range of feeding, reproducing and migrating behavior. At some places and at some times the behavior of a species, interacting even in detailed secondary ways with the variable conditions of the ocean and its other inhabitants, results in temporary, seasonal or persistent success.

There are a few exceptions to the microscopic dimensions of the herbivores in the pelagic food web. Among these the herrings and herring-like fishes are able to consume phytoplankton as a substantial component of their diet. Such an adaptation gives these fishes access to many times the food supply of the more carnivorous groups. It is therefore no surprise that the partly herbivorous fishes comprise the bulk of the world's fisheries [see the article "The Food Resources of the Ocean," by S. J. Holt, beginning on page 201].

The principal food supplies of the pelagic populations are passed on in incremental steps and rapidly depleted quantity to the larger carnivorous zooplankton, then to small fishes and squids, and ultimately to the wide range of larger carnivores of the pelagic realm. In this region without refuge, either powerful static defenses, such as the stinging cells of the medusae and men-o'-war, or increasing size, acuity, alertness, speed and strength are the requirements for survival at each step. Streamlining of form here reaches a high point of development, and in tropical waters it is conspicuous even in small fishes, since the lower viscosity of the warmer waters will enable a highly streamlined small prey to escape a poorly streamlined predator, an effect that exists only for fishes of twice the length in cold, viscous, arctic or deep waters.

The pelagic region contains some of the largest and most superbly designed creatures ever to inhabit this earth: the exquisitely constructed pelagic tunas; the multicolored dolphinfishes, capturers of flying fishes; the conversational porpoises; the shallow- and deep-feeding swordfishes and toothed whales, and the greatest carnivores of all, the baleen whales and some plankton-eating sharks, whose prey are entire schools of krill or small fishes. Seals and sea lions feed far into the pelagic realm. In concert with

these great predators, large carnivorous sharks await injured prey. Marine birds, some adapted to almost continuous pelagic life, consume surprising quantities of ocean food, diving, plunging, skimming and gulping in pursuit. Creatures of this region have developed such faculties as advanced sonar, unexplained senses of orientation and homing, and extreme olfactory sensitivity.

These larger creatures of the sea commonly move in schools, shoals and herds. In addition to meeting the needs of mating such grouping is advantageous in both defensive and predatory strategy, much like the cargo-ship convoy and submarine "wolf pack" of World War II. Both defensive and predatory assemblages are often complex. Small fishes of several species commonly school together. Diverse predators also form loosely cooperative groups, and many species of marine birds depend almost wholly on prey driven to the surface by submerged predators.

At night, schools of prey and predators are almost always spectacularly illuminated by bioluminescence produced by the microscopic and larger plankton. The reason for the ubiquitous production of light by the microorganisms of the sea remains obscure, and suggested explanations are controversial. It has been suggested that light is a kind of inadvertent by-product of life in transparent organisms. It has also been hypothesized that the emission of light on disturbance is advantageous to the plankton in making the predators of the plankton conspicuous to *their* predators! Unquestionably it does act this way. Indeed, some fisheries base the detection of their prey on

the bioluminescence that the fish excite. It is difficult, however, to defend the thesis that this effect was the direct factor in the original development of bioluminescence, since the effect was of no advantage to the individual microorganism that first developed it. Perhaps the luminescence of a microorganism also discourages attack by the light-avoiding zooplankton and is of initial survival benefit to the individual. As it then became general in the population, the effect of revealing plankton predators to their predators would also become important.

The fallout of organic material into the deep, dimly lighted mid-water supports a sparse population of fishes and invertebrates. Within the mesopelagic and bathypelagic zones are found some of the most curious and bizarre creatures of this earth. These range from the highly developed and powerfully predaceous intruders, toothed whales and swordfishes, at the climax of the food chain, to the remarkable squids, octopuses, euphausiids, lantern fishes, gulpers and anglerfishes that inhabit the bathypelagic region.

In the mesopelagic region, where some sunlight penetrates, fishes are often countershaded, that is, they are darker above and lighter below, as are surface fishes. Many of the creatures of this dimly lighted region participate in the daily migration, swimming to the upper layers at evening like bats emerging from their caves. At greater depths, over a half-mile or so, the common inhabitants are often darkly pigmented, weak-bodied and frequently adapted to unusual feeding techniques. Attraction of prey by luminescent lures or by mimicry of small prey, greatly extensible jaws and expansible abdomens are common. It is, however, a region of Lilliputian monsters, usually not more than six inches in length, with most larger fishes greatly reduced in musculature and weakly constructed.

There are some much larger, stronger and more active fishes and squids in this region, although they are not taken in trawls or seen from submersibles. Knowledge of their existence comes mainly from specimens found in the stomach of sperm whales and swordfish. They must be rare, however, since the slow, conservative creatures that are taken in trawls could hardly coexist with large numbers of active predators. Nevertheless, populations must be sufficiently large to attract the sperm whales and swordfish. There is evidence that the sperm whales possess highly developed long-range hunting sonar. They may lo-

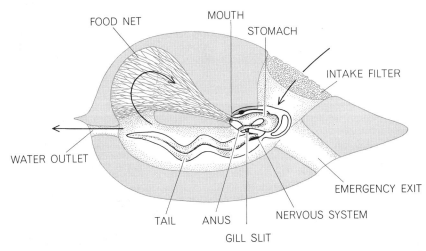

LARVACIAN is representative of a group of small planktonic herbivores that has solved the problem of energy loss in filtering, apparently without utilizing "stickiness," by having proportionately large nets, strong fine threads and a low rate of water flow. The larvacian (*black*) produces and inhabits a complex external "house" (*color*), much larger than its owner, which contains a system of nets through which the organism maintains a gentle flow. In almost all other groups simple filters are employed only to exclude large particles.

cate their prey over relatively great distances, perhaps miles, from just such an extremely sparse population of active bathypelagic animals.

Although many near-surface organisms are luminescent, it is in the bathypelagic region that bioluminescence has reached a surprising level of development, with at least two-thirds of the species producing light. Were we truly marine-oriented, we would perhaps be more surprised by the almost complete absence of biological light in the land environment, with its few rare cases of fireflies, glowworms and luminous bacteria. Clearly bioluminescence can be valuable to higher organisms, and the creatures of the bathypelagic realm have developed light-producing organs and structures to a high degree. In many cases the organs have obvious functions. Some fishes, squids and euphausiids possess searchlights with reflector, lens and iris almost as complex as the eye. Others have complex patterns of small lights that may serve the functions of recognition, schooling control and even mimicry of a small group of luminous plankton. Strong flashes may confuse predators by "target alteration" effects, or by producing residual images in the predators' vision. Some squids and shrimps are more direct and discharge luminous clouds to cover their escape. The luminous organs are arranged on some fishes so that they can be used to countershade their silhouettes against faint light coming from the surface. Luminous baits are well developed. Lights may also be used for locating a mate, a problem of this vast, sparsely populated domain that has been solved by some anglerfishes by the development of tiny males that live parasitically attached to their relatively huge mates.

It has been shown that the vertebrate eye has been adapted to detect objects in the lowest light level on the earth's surface—a moonless, overcast night under a dense forest canopy—but not lower. Light levels in the bathypelagic region can be much lower. This is most probably the primary difference that accounts for the absence of bioluminescence in higher land animals and the richness of its development in the ocean forms.

The densest populations of bathypelagic creatures lie below the most productive surface regions, except at high latitudes, where the dearth of winter food probably would exhaust the meager reserves of these creatures. All the bathypelagic populations are sparse, and in this region living creatures are less than one hundred-millionth of the water volume. Nevertheless, the zone is of immense dimensions and the total populations may be large. Some genera, such as the feeble, tiny bristle-jawed fishes, are probably the most numerous fishes in the world and constitute a gigantic total biomass. There are some 2,000 species of fishes and as many species of the larger invertebrates known to inhabit the bathypelagic zone, but only a few of these species appear to be widespread. The barriers to distribution in this widely interconnected mid-water region are not obvious.

The floor of the deep sea constitutes an environment quite unlike the mid-water and surface environments. Here are sites for the attachment of the larger

CHAMPION FILTER FEEDER of the world ocean in terms of volume is the blue whale, a mature specimen of which lies freshly butchered on the deck of a whaling vessel in this photograph. The whale's stomach has been cut open with a flensing knife to reveal its last meal: an immense quantity of euphausiids, or krill, each measuring about three inches in length. The baleen whales are not plankton-filterers in the ordinary sense but rather are great carnivores that seek out and engulf entire schools of small fish or invertebrates. The photograph was made by Robert Clarke of the National Institute of Oceanography in Wormley, England.

invertebrates that filter detritus from the water. Among these animals are representatives of some of the earliest multicelled creatures to exist on the earth, glass sponges, sea lilies (crinoids)—once thought to have been long extinct—and lamp shells (brachiopods).

At one time it was also thought that the abyssal floor was sparsely inhabited and that the populations of the deep-ocean floor were supplied with food only by the slow, meager rain of terminal detrital food material that has passed through the surface and bathypelagic populations. Such refractory material requires further passage into filter feeders or through slow bacterial action in the sediment, followed by consumption by larger burrowing organisms, before it becomes available to active free-living animals. This remnant portion of the food web could support only a very small active population.

Recent exploration of the abyssal realm with a baited camera throws doubt on the view that this is the exclusive mechanism of food transfer to the deep bottom. Large numbers of active fishes and other creatures are attracted to the bait almost immediately [see illustration on page 6]. It is probably true that several rather independent branches of the food web coexist in support of the deep-bottom creatures: one the familiar rain of fine detritus, and the other the rare, widely separated falls of large food particles that are in excess of the local feeding capacity of the broadly diffuse bathypelagic population. Such falls would include dead whales, large sharks or other large fishes and fragments of these, the multitude of remnants that are left when predators attack a school of surface fish and now, undoubtedly, garbage from ships and kills from underwater explosions. These sources result in an influx of high-grade food to the sea floor, and we would expect to find a population of active creatures adapted to its prompt discovery and utilization. The baited cameras have demonstrated that this is so.

Other sources of food materials are braided into these two extremes of the abyssal food web. There is the rather subtle downward diffusion of living and dead food that results initially from the daily vertical migration of small fish and zooplankton near the surface. This migration appears to impress a sympathetic daily migration on the mid-water populations down to great depths, far below the levels that light penetrates. Not only may such vertical migration bring feeble bathypelagic creatures near the bottom but also it accelerates in itself the flux of dead food material to the bottom of the deep sea.

There must also be some unassignable flux of food to the abyssal population resulting from the return of juveniles to their habitat. The larvae and young of many abyssal creatures develop at much shallower levels. To the extent that the biomass of juveniles returning to the deep regions exceeds the biomass of spawn released from it, this process, which might be called "Faginism," constitutes an input of food.

Benthic animals are much more abundant in the shallower waters off continents, particularly offshore from large rivers. Here there is often not only a richer near-surface production and a less hazardous journey of food to the sea floor but also a considerable input of food conveyed by rivers to the bottom. The deep slopes of river sediment wedges are typified by a comparatively rich population of burrowing and filtering animals that utilize this fine organic material. All the great rivers of the world save one, the Congo, have built sedimentary wedges along broad reaches of their coast, and in many instances these wedges extend into deep water. The shallow regions of such wedges are highly productive of active and often valuable marine organisms. At all depths the wedges bear larger populations than are common at similar depths elsewhere. Thus one wonders what inhabits the fan of the Congo. That great river, because of a strange invasion of a submarine canyon into its month, has built no wedge but rather is depositing a vast alluvial fan in the two-mile depths of the Angola Basin. This great deep region of the sea floor may harbor an unexplored population that is wholly unique.

In itself the pressure of the water at great depths appears to constitute no insurmountable barrier to water-breathing animal life. The depth limitations of many creatures are the associated conditions of low temperature, darkness, sparse food and so on. It should perhaps come as no surprise, therefore, that some of the fishes of high latitudes, which are of course adapted to cold dark waters, extend far into the deep cold waters in much more southern latitudes. Off the coast of Lower California, in water 1,200 to 6,000 feet deep, baited cameras have found an abundance of several species of fishes that are known at the near surface only far to the north. These include giant arctic sharks, sablefish and others. It appears that some of the fishes that have been called arctic species are actually fishes of the dark cold waters of the seas, which only "outcrop" in the Arctic, where cold water is at the surface.

I have discussed several of the benthic and epibenthic environments without pointing out some of the unique features the presence of a solid interface entails. The bottom is much more variable than the mid-water zone is. There are as a result more environmental niches for an organism to occupy, and hence we see organisms that are of a wider range of form and habit. Adaptations develop for hiding and ambuscade, for mimicry and controlled patterns. Nests and burrows can be built, lairs occupied and defended and booby traps set.

Aside from the wide range of form and function the benthic environment elicits from its inhabitants, there are more fundamental conditions that influence the nature and form of life there. For example, the dispersed food material settling from the upper layers becomes much concentrated against the sea floor. Indeed, it may become further concentrated by lateral currents moving it into depressions or the troughs of ripples.

In the mid-water environment most creatures must move by their own energies to seek food, using their own food stores for this motion. On the bottom, however, substantial water currents are present at all depths, and creatures can await the passage of their food. Although this saving only amounts to an added effectiveness for predators, it is of critical importance to those organisms that filter water for the fine food material it contains, and it is against the bottom interface that a major bypass to the microscopic steps of the dominant food web is achieved. Here large organisms can grow by consuming microscopic or even submicroscopic food particles. Clams, scallops, mussels, tube worms, barnacles and a host of other creatures that inhabit this zone have developed a wide range of extremely effective filtering mechanisms. In one step, aided by their attachment, the constant currents and the concentration of detritus against the interface, they perform the feat, most unusual in the sea, of growing large organisms directly from microscopic food.

Although the benthic environment enables the creatures of the sea to develop a major branch of the food web that is emancipated from successive microscopic steps, this makes little difference to the food economy of the sea. The sea is quite content with a large population of tiny organisms. From man's standpoint, however, the shallow benthic environment is an unusually effective producer of larger creatures for his food, and he widely utilizes these resources.

Man may not have created an ideal environment for himself, but of all the environments of the sea it is difficult to conceive of one better for its inhabitants than the one marine creatures have created almost exclusively for themselves: the coral islands and coral reefs. In these exquisite, immense and well-nigh unbelievable structures form and adaptation reach a zenith.

An adequate description of the coral reef and coral atoll structure, environments and living communities is beyond the scope of this article. The general history and structure of atolls is well known, not only because of an inherent fascination with the magic and beauty of coral islands but also because of the wide admiration and publicity given to the prescient deductions on the origin of atolls by Charles Darwin, who foresaw much of what modern exploration has affirmed.

From their slowly sinking foundations of ancient volcanic mountains, the creatures of the coral shoals have erected the greatest organic structures that exist. Even the smallest atoll far surpasses any of man's greatest building feats, and a large atoll structure in actual mass approaches the total of all man's building that now exists.

These are living monuments to the success of an extremely intricate but balanced society of fish, invertebrates and plants, capitalizing on the basic advantages of benthic populations already discussed. Here, however, each of the reef structures acts almost like a single great isolated and complex benthic organism that has extended itself from the deep poor waters to the sunlit richer surface. The trapping of the advected food from the surface currents enriches the entire community. Attached plants further add to the economy, and there is considerable direct consumption of plant life by large invertebrates and fish. Some of the creatures and relationships that have developed in this environment are among the most highly adapted found on the earth. For example, a number of the important reef-building animals, the corals, the great tridacna clams and others not only feed but also harbor within their tissues dense populations of single-celled green plants. These plants photosynthesize food that is then directly available within the bodies of the animals; the plants in turn depend on the animal waste products within the body fluids, with which they are bathed, to derive their basic nutrients. Thus within the small environment of these plant-animal composites both the entire laborious nutrient cycle and the microscopic food web of the sea appear to be substantially bypassed.

There is much unknown and much to be discovered in the structure and ecology of coral atolls. Besides the task of unraveling the complex relationships of its inhabitants there are many questions such as: Why have many potential atolls never initiated effective growth and remained submerged almost a mile below the surface? Why have others lost the race with submergence in recent times and now become shallowly submerged, dying banks? Can the nature of the circulation of the ancient ocean be deduced from the distribution of successful and unsuccessful atolls? Is there circulation within the coral limestone structure that adds to the nutrient supply, and is this related to the curious development of coral knolls, or coral heads, within the lagoons? Finally, what is the potential of cultivation within these vast, shallow-water bodies of the deep open sea?

There is, of course, much to learn about all marine life: the basic processes of the food web, productivity, populations, distributions and the mechanisms of reinoculation, and the effects of intervention into these processes, such as pollution, artificial upwelling, transplantation, cultivation and fisheries. To learn of these processes and effects we must understand the nature not only of strong simple actions but also of weak complex interactions, since the forms of life or the success of a species may be determined by extremely small second- and third-order effects. In natural affairs, unlike human codes, *de minimis curat lex—* the law *is* concerned with trivia!

Little is understood of the manner in which speciation (that is, the evolution of new species) occurs in the broadly intercommunicating pelagic environment with so few obvious barriers. Important yet unexpected environmental niches may exist in which temporary isolation may enable a new pelagic species to evolve. For example, the top few millimeters of the open sea have recently been shown to constitute a demanding environment with unique inhabitants. Further knowledge of such microcosms may well yield insight into speciation.

As it has in the past, further exploration of the abyssal realm will undoubtedly reveal undescribed creatures including members of groups thought long extinct, as well as commercially valuable populations. As we learn more of the conditions that control the distribution of species of pelagic organisms, we shall become increasingly competent to read the pages of the earth's marine-biological, oceanographic and meteorological history that are recorded in the sediments by organic remains. We shall know more of primordial history, the early production of a breathable atmosphere and petroleum production. Some of these deposits of sediment cover even the period of man's recorded history with a fine time resolution. From such great records we should eventually be able to increase greatly our understanding of the range and interrelations of weather, ocean conditions and biology for sophisticated and enlightened guidance of a broad spectrum of man's activities extending from meteorology and hydrology to oceanography and fisheries.

Learning and guidance of a more specific nature can also be of great practical importance. The diving physiology of marine mammals throws much light on the same physiological processes in land animals in oxygen stress (during birth, for example). The higher flowering plants that inhabit the marine salt marshes are able to tolerate salt at high concentration, desalinating seawater with the sun's energy. Perhaps the tiny molecule of DNA that commands this process is the most precious of marine-life resources for man's uses. Bred into existing crop plants, it may bring salt-water agriculture to reality and nullify the creeping scourge of salinization of agricultural soils.

Routine upstream reinoculation of preferred species of phytoplankton and zooplankton might stabilize some pelagic marine populations at high effectiveness. Transplanted marine plants and animals may also animate the dead saline lakes of continental interiors, as they have the Salton Sea of California.

The possible benefits of broad marine-biological understanding are endless. Man's aesthetic, adventurous, recreational and practical proclivities can be richly served. Most important, undoubtedly, is the intellectual promise: to learn how to approach and understand a complex system of strongly interacting biological, physical and chemical entities that is vastly more than the sum of its parts, and thus how better to understand complex man and his interactions with his complex planet, and to explore with intelligence and open eyes a huge portion of this earth, which continuously teaches that when understanding and insight are sought for their own sake, the rewards are more substantial and enduring than when they are sought for more limited goals.

2

Plate Tectonics and the History of Life in the Oceans

by James W. Valentine and Eldridge M. Moores
April 1974

The breakup of the ancient supercontinent of Pangaea triggered a long-term evolutionary trend that has led to the unprecedented variety of the present biosphere

During the 1960's a conceptual revolution swept the earth sciences. The new world view fundamentally altered long established notions about the permanency of the continents and the ocean basins and provided fresh perceptions of the underlying causes and significance of many major features of the earth's mantle and crust. As a consequence of this revolution it is now generally accepted that the continents have greatly altered their geographic position, their pattern of dispersal and even their size and number. These processes of continental drift, fragmentation and assembly have been going on for at least 700 million years and perhaps for more than two billion years.

Changes of such magnitude in the relative configuration of the continents and the oceans must have had far-reaching effects on the environment, repatterning the world's climate and influencing the composition and distribution of life in the biosphere. These more or less continual changes in the environment must also have had profound effects on the course of evolution and accordingly on the history of life.

Natural selection, the chief mechanism by which evolution proceeds, is a very complex process. Although it is constrained by the machinery of inheritance, natural selection is chiefly an ecological process based on the relation between organisms and their environment. For any species certain heritable variations are favored because they are particularly well suited to survive and to reproduce in their prevailing environment. To answer the question of why any given group of organisms has evolved, then, one needs to understand two main factors. First, it is necessary to

know what the ancestral organisms were that formed the "raw material" on which selection worked. And second, one must have some idea of the sequence of environmental conditions that led the ancestral stock to evolve along a particular pathway to a descendant group. Given these factors, one can then infer the organism-environment interactions that gave rise to the evolutionary events. The study of the relations between ancient organisms and their environment is called paleoecology.

The new ideas of continental drift that came into prominence in the 1960's revolve around the theory of plate tectonics. According to this theory, new sea floor and underlying mantle are currently being added to the crust of the earth at spreading centers under deep-sea ridges and in small ocean basins at rates of up to 10 centimeters per year. The sea floor spreads laterally away from these centers and eventually sinks into the earth's interior at subduction zones, which are marked by deep-sea trenches. Volcanoes are created by the consumption process and flank the trenches. The lithosphere, or rocky outer shell of the earth, therefore comprises several major plates that are generated at spreading centers and consumed at subduction zones. Most lithospheric plates bear one continent or more, which passively move with the plate on which they rest. Because the continents are too light to sink into the trenches they remain on the surface. Continents can fragment at new ridges, however, and hence oceans may appear across them. Conversely, continents can be welded together when they collide at the site of a trench. Thus continents may be assembled into supercontinents, fragmented into small continents

and generally moved about the earth's surface as passive riders on plates. In tens or hundreds of millions of years entire oceans may be created or destroyed, and the number, size and dispersal pattern of continents may be vastly altered.

The record of such continental fragmentation and reassembly is evident as deformed regions in the earth's mountain belts, particularly those mountain belts that contain the rock formations known as ophiolites. These formations are characterized by a certain sequence of rocks consisting (from bottom to top) of ultramafic rock (a magnesium-rich rock composed mostly of olivine), gabbro (a coarse-grained basaltic rock), volcanic rocks and sedimentary rocks. The major ophiolite belts of the earth are believed to represent preserved fragments of vanished ocean basins [*see illustration on pages 20 and 21*]. The existence of such a belt within a continent (for example the Uralian belt in the U.S.S.R.) is evidence for the former presence there of an ocean basin separating two continental fragments that at some time in the past collided with each other and were welded into the single larger continent. The timing of such events as the opening of ocean basins, the dispersal of continents and the closing of oceans by continental collisions can accordingly be "read" from the geology of a given mountain system.

Of course, the biological environment is constantly being altered as well. For example, the changes in continental configuration will greatly affect the ocean currents, the temperature, the nature of seasonal fluctuations, the distribution of nutrients, the patterns of productivity and many other factors of fundamental importance to living organisms. Therefore evolutionary trends in marine animals must have varied through geologic time in response to the major environmental changes, as natural selection acted to adapt organisms to the new conditions.

It should in principle be possible to detect these changes in the fossil record. Indeed, paleontologists have long recognized that vast changes in the composition, distribution and diversity of marine life are well documented by the fossil record. Now for the first time, however, it is possible to reconstruct the sequence of environmental changes based on the theory of plate tectonics, to determine their environmental consequences and to attempt to correlate them with the sequence of faunal changes that is seen in the fossil record. Such a thorough reconstruction ultimately may explain many of the enigmatic faunal changes known for many years. Even at this early stage paleontologists have succeeded in shedding much new light on a number of major extinctions and diversifications of the past.

As a first step toward understanding the relation between plate tectonics and the history of life it is helpful to investigate the relations that exist today between marine life, the present pattern of continental drift and plate-tectonic theory. The vast majority of marine species (about 90 percent) live on the continental shelves or on shallow-water portions of islands or subsurface "rises" at depths of less than about 200 meters

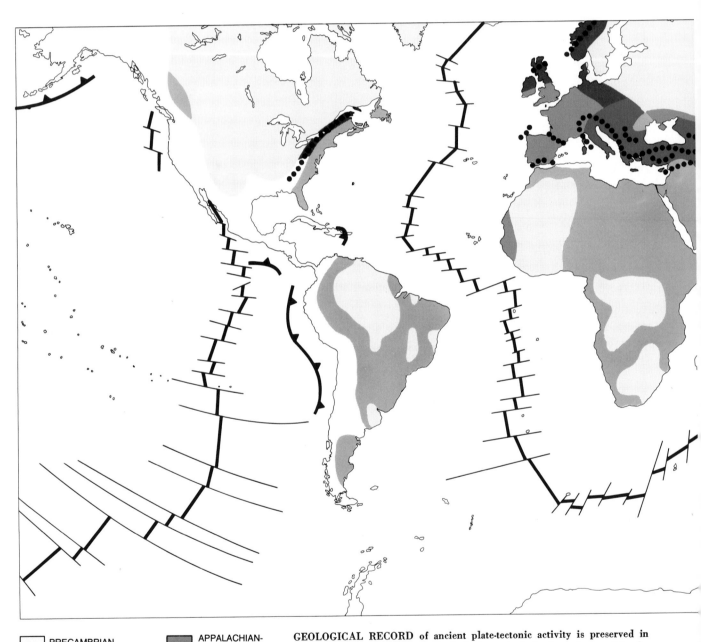

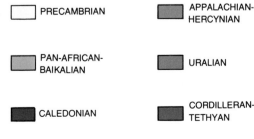

PRECAMBRIAN

PAN-AFRICAN-BAIKALIAN

CALEDONIAN

APPALACHIAN-HERCYNIAN

URALIAN

CORDILLERAN-TETHYAN

GEOLOGICAL RECORD of ancient plate-tectonic activity is preserved in certain deformed mountain belts (*color*), particularly those that contain the characteristic rock sequences known as ophiolites (*black dots*). The Pan-African–Baikalian belt, for example, is made up of rocks dating from 873 to 450 million years ago and may represent the assembly of all or nearly all the landmasses near the beginning of Phanerozoic time. This supercontinent may then have fragmented into four or more smaller continents, sometime just before and during the Cambrian period. The Caledonian mountain system may represent the collision of two continents at about late Silurian or early De-

(660 feet); most of the fossil record also consists of these faunas. Therefore it is the pattern of shallow-water sea-floor animal life that is of particular interest here.

The richest shallow-water faunas are found today at low latitudes in the Tropics, where communities are packed with vast numbers of highly specialized species. Proceeding to higher latitudes, diversity gradually falls; in the Arctic or Antarctic regions less than a tenth as many animals are living as in the Trop-

ics, when comparable regions are considered [*see illustration on pages 22 and 23*]. The diversity gradient correlates well with a gradient in the stability of food supplies; as the seasons become more pronounced, fluctuations in primary productivity become greater. Although this strong latitudinal gradient dominates the earth's overall diversity pattern, there are important longitudinal diversity trends as well. In regions of similar latitude, for example, diversity is lower where there are sharp seasonal

changes (such as variations in the surface-current pattern or in the upwelling of cold water) that affect the nutrient supply by causing large fluctuations in productivity.

At any given latitude, therefore, diversity is highest off the shores of small islands or small continents in large oceans, where fluctuations in nutrient supplies are least affected by the seasonal effects of landmasses, whereas diversity is lowest off large continents, particularly when they face small oceans, where shallow-water seasonal variations are greatest. In short, whereas latitudinal diversity increases generally from high latitudes to low, longitudinal diversity increases generally with distance from large continental landmasses. In both of these trends the increase in diversity is correlated with increasing stability of food resources. The resource-stability pattern depends largely on the shape of the continents and should also be sensitive to the extent of inland seas and to the presence of coastal mountains. Seas lying on continental platforms are particularly important: not only do extensive shallow seas provide much habitat area for shallow-water faunas but also such seas tend to damp seasonal climatic changes and to have an ameliorating influence on the local environment.

Today shallow marine faunas are highly provincial, that is, the species living in different oceans or on opposite sides of the same ocean tend to be quite different. Even along continuous coastlines there are major changes in species composition from place to place that generally correspond to climatic changes. The deep-sea floor, generated at oceanic ridges, forms a significant barrier to the dispersal of shallow-water organisms, and latitudinal climatic changes clearly form other barriers. The present dominantly north-south series of ridges forms a pattern of longitudinally alternating oceans and continents, thereby creating a series of barriers to shallow-water marine organisms. The steep latitudinal climatic gradient, on the other hand, creates chains of provinces along north-south coastlines. As a result the marine faunas today are partitioned into more than 30 provinces, among which there is in general only a low percentage of common species [*see illustration on pages 24 and 25*]. It is estimated that the shallow-water marine fauna represents more than 10 times as many species today as would be present in a world with only a single province, even a highly diverse one.

The volcanic arcs that appear over subduction zones form fairly continuous

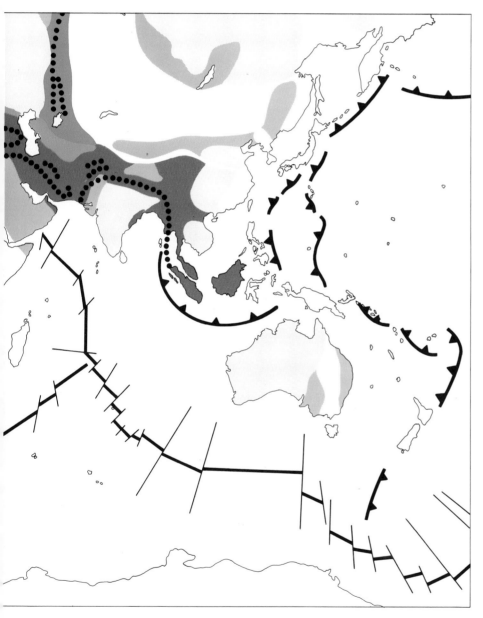

vonian time (approximately 400 million years ago). The Appalachian-Hercynian system may represent a two-continent collision during the late Carboniferous period (300 million years ago). The Uralian mountains may represent a similar collision at about Permo-Triassic time (220 million years ago). The Cordilleran-Tethyan system represents regions of Mesozoic mountain-building and includes the continental collisions that resulted in the Alpine-Himalayan mountain system. The ophiolite belts shown are the preserved remnants of ocean floor exposed in the mountain systems in question. Spreading ridges such as the Mid-Atlantic Ridge are indicated by heavy lines cut by lighter lines, which correspond to transform faults. Subduction zones are marked by heavy black curved lines with triangles.

island chains and provide excellent dispersal routes. When long island chains are arranged in an east-west pattern so as to lie within the same climatic zone, they are inhabited by wide-ranging faunas that are highly diverse for their latitude. Indeed, the widest ranging marine province, and also by far the most diverse, is the Indo-Pacific province, which is based on island arcs in its central regions. The faunal life of this province spills from these arcs onto tropical continental shelves in the west (India and East Africa) and also onto tropical intraplate volcanoes (the Polynesian and Micronesian islands) that are reasonably close to them. This vast tropical biota is cut off from the western American mainland by the East Pacific Barrier, a zoogeographic obstruction formed by a spreading ridge.

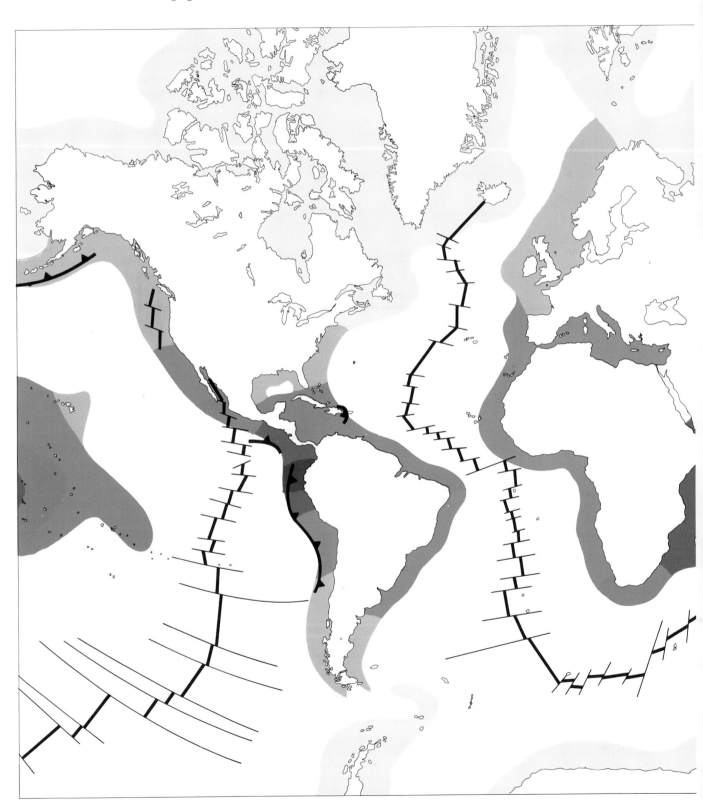

RELATIVE DIVERSITY of shallow-water, bottom-dwelling species in the present oceans is suggested by the colored patterns in this world map. The diversity classes are not based on absolute counts but are inferred from the diversity patterns of the best-

Since current patterns of marine provinciality and diversity fit closely with the present oceanic and continental geography and the resulting environmental patterns, one would expect ancient provinces and ancient diversity patterns also to fit past geographies. One of the best-established of ancient geographies is the one that existed near the beginning of the Triassic period, about 225 million years ago. The continents were then assembled into a single supercontinent named Pangaea, which must have had a continuous shallow-water margin running all the way around it, with no major physical barriers to the dispersal of shallow-water marine animals [see illustration on page 26]. Therefore provinciality must have been low compared with today, and it must have been attributable entirely to climatic effects. It is likely that the marine climate was quite mild and that even in high latitudes water temperatures were much warmer than they are today. As a result climatic provinciality must have been greatly reduced also. Furthermore, the seas at that time were largely confined to the ocean basins and did not extend significantly over the continental shelves. Thus the habitat area for shallow-water marine organisms was greatly reduced, first by the diminution of coastline that accompanies the creation of a supercontinent from smaller continents, and second by the general withdrawal of seas from continental platforms. The reduced habitat area would make for low species diversity. Finally, the extreme emergence of such a supercontinent would provide unstable nearshore conditions, with the result that food resources would have been very unstable compared with those of today. All these factors tend to reduce species diversity; hence one would expect to find that Triassic biotas were widespread and were made up of comparatively few species. That is precisely what the fossil record indicates.

Prior to the Triassic period, during the late Paleozoic, diversity appears to have been much higher [see top illustration on page 28]. It was sharply reduced again near the close of the Permian period during a vast wave of extinction that on balance is the most severe known to have been suffered by the marine fauna. The late Paleozoic species that were the more elaborately adapted specialists became virtually extinct, whereas the surviving descendants tended to have simple skeletons. A high proportion of these survivors appear to have been detritus feeders or suspension feeders that harvested the water layers just above the sea floor. These successful types seem to be ecologically similar to the populations found today in unstable environments, for instance in high latitudes; the unsuccessful specialists, on the other hand, seem ecologically similar to the populations found in stable environments, for instance in the Tropics. Thus the extinctions appear to have been caused by the reduced potential for diversity of the shallow seas, a trend associated with less provinciality, less habitat area and less stable environmental conditions.

In the period following the great extinction, as Pangaea broke up and the

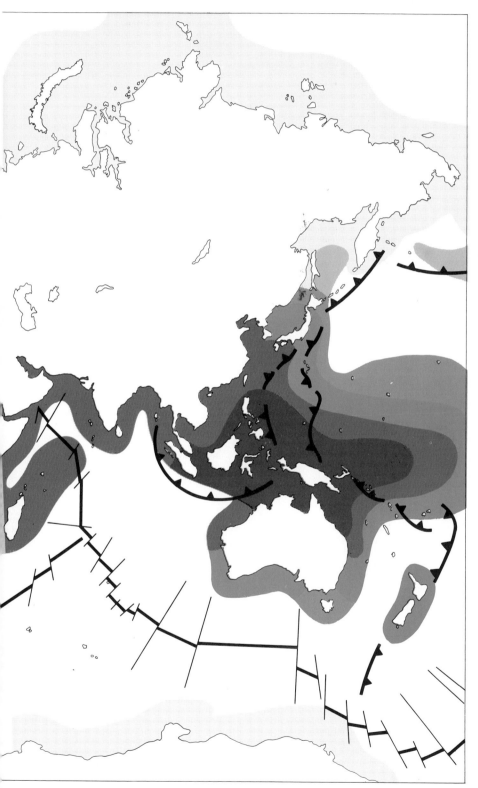

known skeletonized groups, chiefly the bivalves, gastropods, echinoids and corals. The highest class (darkest color) is about 20 times as diverse as the lowest (lightest color).

resulting continents themselves gradually fragmented and migrated to their
present positions, provinciality increased, communities in stabilized regions became filled with numerous specialized animals and the overall diversity
of species in the world ocean rose to unprecedented heights, even though occasional waves of extinctions interrupted
this long-range trend.

There is another time in the past besides the early Triassic period when
low provinciality and low diversity were
coupled with the presence of a high proportion of detritus feeders and near-
bottom suspension feeders. That is in the
late Precambrian and Cambrian periods,
when a widespread, soft-bodied fauna of
low diversity gave way to a slightly pro-
vincialized, skeletonized fauna of some-

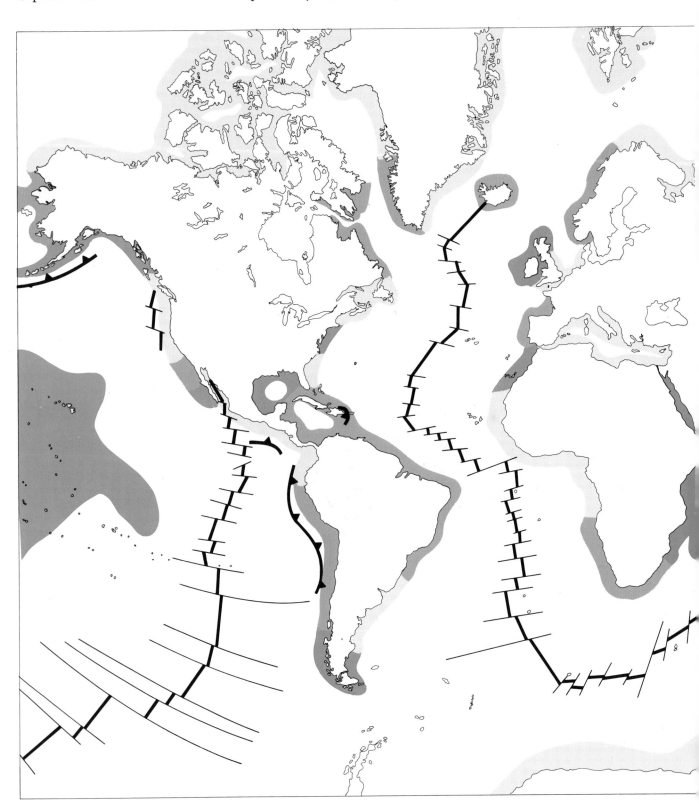

PRINCIPAL SHALLOW-WATER MARINE PROVINCES at present are indicated by the colored areas. The dominant north-
south chains of provinces along the continental coastlines are created by the present high latitudinal gradient in ocean temperature

what higher diversity. It seems likely that the late Precambrian environment was quite unstable and that there may well have been a supercontinent in existence, or at least that the continents then were collected into a more compact assemblage than at present. In the late Precambrian period one finds the first unequivocal records of invertebrate life, including burrowing forms that were probably coelomic, or hollow-bodied, worms. In the Cambrian four continents may have existed although they were not arranged in the present pattern. During the Cambrian a skeletonized fauna appears that is at first almost entirely surface-dwelling and that includes chiefly detritus-feeding and suspension-feeding forms, probably with some browsers.

It seems possible, therefore, that the late Precambrian species were adapted to highly unstable conditions and became diversified chiefly as a bottom-living, detritus-feeding assemblage. The coelomic body cavity, evidently a primitive adaptation for burrowing, was developed and diversified into a variety of forms, perhaps as many as five basic ones: highly segmented worms that lived under the ocean floor and were detritus feeders; slightly segmented worms that lived attached to the ocean floor and were suspension feeders; slightly segmented worms that lived attached to the ocean floor and were detritus feeders; "pseudosegmented" worms that lived on the ocean floor and were detritus feeders or browsers, and nonsegmented worms that lived under the ocean floor and fed by means of an "introvert." In addition to these coelomates there were a number of coelenterate stocks (such as corals, sea anemones and jellyfishes) and probably also flatworms and other noncoelomate worms.

From the chiefly wormlike coelomate stocks higher forms of animal life have originated; many of them appear in the Cambrian period, when they evidently first became organized into the groups that characterize them today. Animals with skeletons appeared in the fossil record at that time. Presumably the invasion of the sea-floor surface by coelomates and the origin of numerous skeletonized species accompanied a general amelioration of environmental conditions as the continents became dispersed; the skeletons themselves can be viewed as adaptations required for worms to lead various modes of life on the surface of the sea floor rather than under it. The sudden appearance of skeletons in the fossil record therefore is associated with a generalized elaboration of the bottom-dwelling members of the marine ecosystem. Later, free-swimming and underground lineages developed from the skeletonized ocean-floor dwellers, with the result that skeletons became general in all marine environments.

The correlation of major events in the history of life with major environmental changes inferred from plate-tectonic processes is certainly striking. Even though details of the interpretation are still provisional, it seems certain that further work on this relation will prove fruitful. Indeed, the ability of geologists

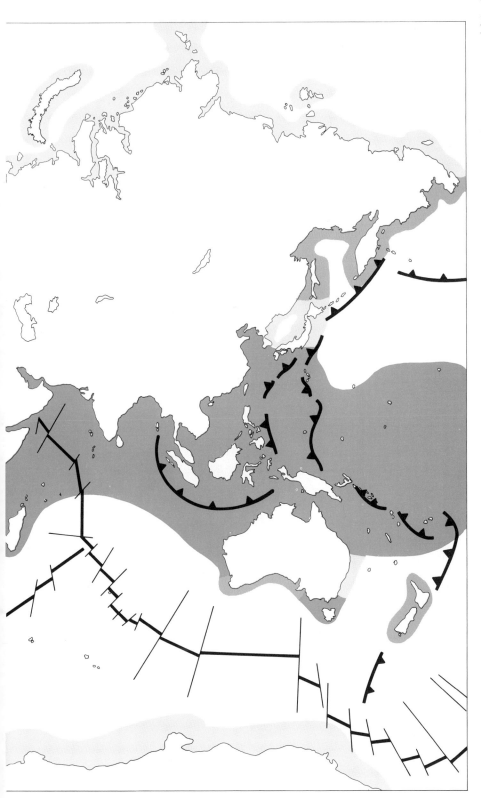

and by the undersea barriers formed by spreading ridges. The vast Indo-Pacific province (**darkest color**) spills out onto scattered islands as indicated. There are 31 provinces shown.

to determine past continental geographies should provide the basis for reconstructing the historical sequence of global environmental conditions for the first time. That sequence can then be compared with the sequence of organisms revealed in the fossil record. The following tentative account of such a comparison, on the broadest scale and without detail, will indicate the kind of history that is emerging; it is based on the examples reviewed above and on similar considerations.

Before about 700 million years ago bottom-dwelling, multicellular animals had developed that somewhat resembled flatworms. As yet no fossil evidence for their evolutionary pathways exists, but evidence from embryology and comparative anatomy suggests that they arose from swimming forms, possibly larval jellyfish, which in turn evolved from primitive single-celled animals.

Approximately 700 million years ago, perhaps in response to the onset of fluctuating environmental conditions

brought about by continental clustering, a true coelomic body cavity was evolved to act as a hydrostatic skeleton in roundworms; this adaptation allowed burrowing in soft sea floors and led to the diversification of a host of worm architectures as that mode of life was explored. Burrows of this type are still preserved in some late Precambrian rocks. As the environment later became more stable, several of the worm lineages evolved more varied modes of life. The changes in body plan necessary to adapt to such

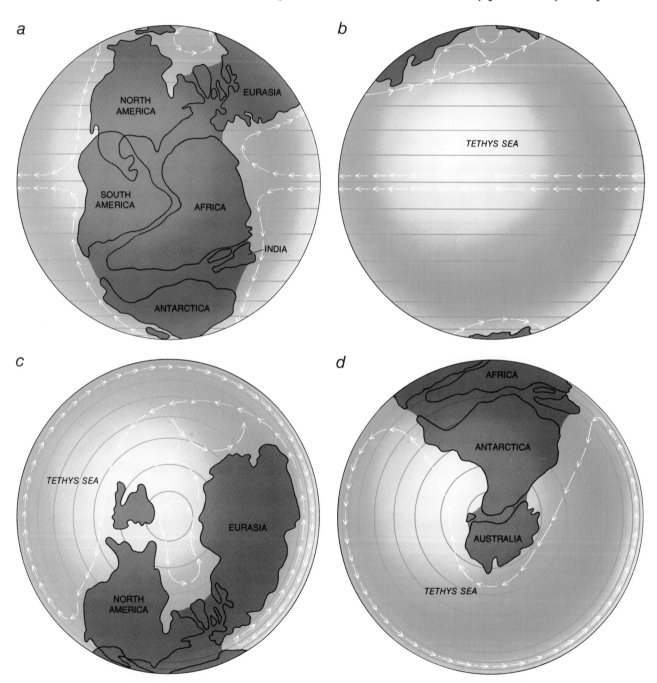

ANCIENT OCEAN CURRENTS in the vicinity of Pangaea, the single "supercontinent" that is believed to have existed near the beginning of the Triassic period some 225 million years ago, are indicated here in two equatorial views (a, b) and two polar views (c, d). Owing to a combination of geographic and environmental factors, including the predominantly warm-water currents shown, one would expect the continuous shallow-water margin that surrounded Pangaea to have been populated by comparatively few but widespread species. Such low species diversity combined with low provinciality is precisely what the fossil record indicates.

a life commonly involved the development of a skeleton. There were evidently three or four main types of worms that are represented by skeletonized descendants today. One type was highly segmented like earthworms, and presumably burrowed incessantly for detrital food; these were represented in the Cambrian period by the trilobites and related species. A second type was segmented into two or three coelomic compartments and burrowed weakly for domicile, afterward filtering suspended food from the seawater just above the ocean floor; these evolved into such forms as brachiopods and bryozoans. A third type consisted of long-bodied creepers with a series of internal organs but without true segmentation; from these the classes of mollusks (such as snails, clams and cephalopods) have descended. Probably a fourth type consisted of unsegmented burrowers that fed on surface detritus and gave rise to the modern sipunculid worms. These may also have given rise to the echinoderms (which include the sea cucumber and the spiny sea urchin), and eventually to the chordates and to man. Although the lines of descent are still uncertain among these primitive and poorly known groups, the adaptive steps are becoming clearer.

The major Cambrian radiation of the underground species into sea-floor surface habitats established the basic evolutionary lineages and occupied the major marine environments. Further evolutionary episodes tended to modify these basic animals into more elaborate structures. After the Cambrian period shallow-water marine animals became more highly specialized and richer in species, suggesting a continued trend toward resource stabilization. Suspension feeders proliferated and exploited higher parts of the water column, and predators also became more diversified. This trend seems to have reached a peak (or perhaps a plateau) in the Devonian period, some 375 million years ago. The characteristic Paleozoic fauna was finally swept away during the reduction in diversity that accompanied the great Permian-Triassic extinctions. Thus the rise of the Paleozoic fauna accompanied an amelioration in environmental conditions and increased provinciality, whereas the decline of the fauna accompanied a reestablishment of severe, unstable conditions and decreased provinciality. The subsequent breakup and dispersal of the continents has led to the present biosphere.

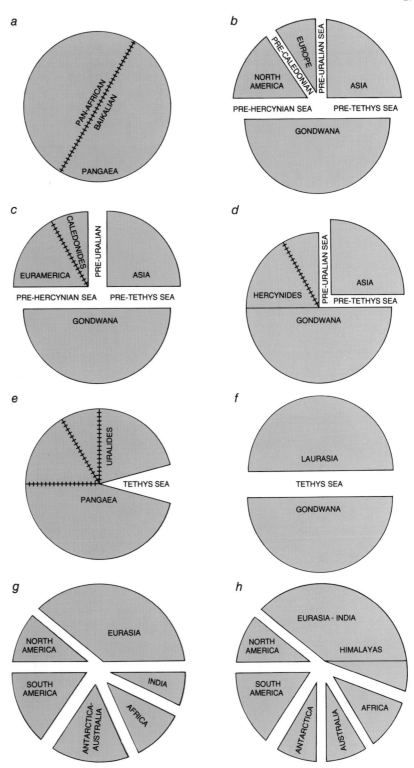

SIMPLIFIED DIAGRAMS are employed to suggest the relative configuration of the continents and the oceans during the past 700 million years. The late Precambrian supercontinent (a), which probably existed some 700 million years ago, may have been formed from previously separate continents. The Cambrian world (b) of about 570 million years ago consisted of four continents. The Devonian period (c) of about 390 million years ago was distinguished by three continents following the collapse of the pre-Caledonian Ocean and the collision of ancient Europe and North America. In the late Carboniferous period (d), about 300 million years ago, Euramerica became welded to Gondwana along the Hercynian belt. In the late Permian period (e), about 225 million years ago, Asia was welded to the remaining continents along the Uralian belt to form Pangaea. In early Mesozoic time (f), about 190 million years ago, Laurasia and Gondwana were more or less separate. In the late Cretaceous period (g), about 70 million years ago, Gondwana was highly fragmented and Laurasia partially so. The present continental pattern (h) shows India welded to Eurasia.

Today we live in a highly diverse world, probably harboring as many species as have ever lived at any time, associated in a rich variety of communities and a large number of provinces, probably the richest and largest ever to have existed at one time. We have been furnished with an enviably diverse and interesting biosphere; it would be a tragedy if we were to so perturb the environment as to return the biosphere to a low-diversity state, with the concomitant extinction of vast arrays of species. Of course, natural processes might eventually recoup the lost diversity, if we waited patiently for perhaps a few tens of millions of years. Alternatively we can work to preserve the environment in its present state and therefore to preserve the richness and variety of nature.

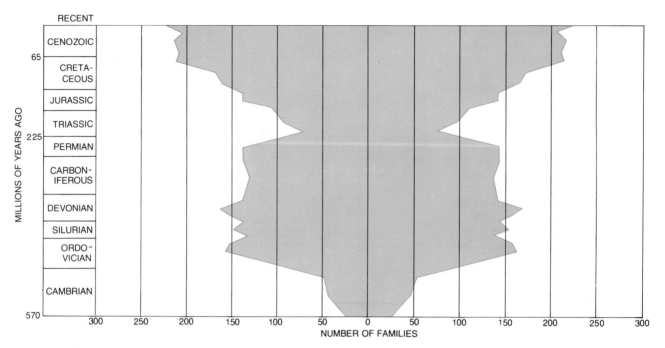

FLUCTUATIONS in the number of families, and hence in the level of diversity, of well-skeletonized invertebrates living on the world's continental shelves during the past 570 million years are plotted by geologic epoch in this graph. Time proceeds upward.

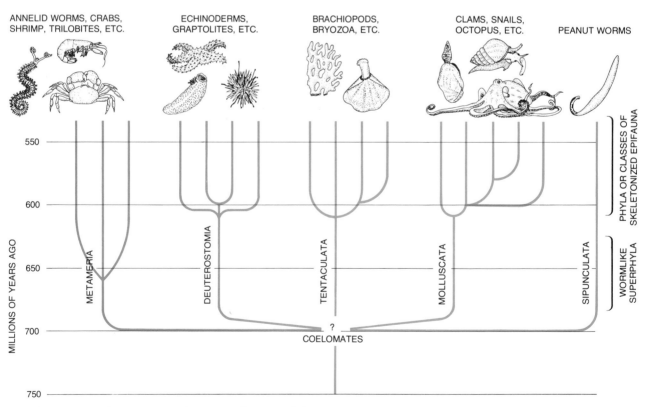

PHYLOGENETIC MODEL of the evolution of coelomate, or hollow-bodied, marine organisms is based on inferred adaptive pathways. The late Precambrian lineages were chiefly worms, which gave rise to epifaunal (bottom-dwelling), skeletonized phyla during the Cambrian period. The organisms depicted in the drawings at top are modern descendants of the major Cambrian lineages.

II

HABITATS
AND INHABITANTS

HABITATS AND INHABITANTS II

INTRODUCTION

These next three sections include essays about marine phenomena (such as buoyancy, wax metabolism, and certain behavioral patterns) along with ones about particular kinds of marine organisms (such as corals, dolphins, appendicularians, and shad). In trying to understand the conditions of life in the sea, we study the attributes of organisms that live under these conditions. Conversely, to make sense of the particular traits of an organism, we probe its activities in that organism's particular ecological circumstances. The former approach is exemplified by Benson and Lee's study of wax in the sea; copepods illuminated this research. The latter approach appears in Alldredge's review of appendicularians, which she found to be major consumers of nannoplankton. Denton turns to fish and squid to study adaptations for buoyancy; Clarke adduces buoyancy as a reason for the features of the spermaceti organ in the sperm whale's head. This shift back and forth between habitat and organism, this twofold approach, repeatedly appears in modern natural history's most ingenious and successful research.

Understanding depends on observing. Observation itself can present an extraordinary challenge. Two articles in this section suggest the difficulties involved. In Robert S. Dietz's "The Sea's Deep Scattering Layers," the challenge is vertical exploration of the ocean to see what is there. Seeing for oneself—going and looking directly—provides evidence that is somehow in a league by itself. If seeing is not all there is to believing, it is often surely a crucial part.

Sonic echograms first revealed a great deal about the daily movements and geographic distribution of deep scattering layers in the sea. Trawlings at appropriate depths during the day and complementary collections from surface waters at night suggested the composition of these layers. But the first direct observations from deep submersible vessels put our knowledge about deep scattering layers on an entirely new footing of confidence and thoroughness. The organisms at last were seen on their own terms. They could be counted without being caught. They could be watched in their habitat. It is this access to nature—what Niko Tinbergen has called "watching and wondering"—that field biology repeatedly seeks as its firmest basis of preparing and testing its analyses.

In space we have depended on photographic and television cameras to "see for ourselves" beyond the moon. On the deep-sea floor, too, these proxy eyes have served us with impressive immediacy. John D. Isaacs and Richard A. Schwartzlose recount the pioneering work in "Active Animals of the Deep-Sea Floor." Increasingly, deep submersibles now visit the bottom of the oceans. But cameras remain less intrusive and have taken invaluable pictures of the place and its activities. In both outer and inner space, the duration of the visit is often crucial. On the sea floor as much as on the moon or Mars, the camera has been our long-term witness, efficiently recording the objects and events of habitats we otherwise would know by far less convincing means. These accounts of exploration reveal how strongly vision shapes inquiry. They carry the force of Goya's most compelling of statements: "I saw this."

Three articles in this section deal with the problems that animals have staying afloat, or at least maneuverably buoyant, in the sea. Eric Denton compares two groups of fairly large animals, fish and cephalopods, in his essay, "The Buoyancy of Marine Animals." Various means of changing depth without harm or maintaining constant depth and ambient pressure reveal both the potentials and the restraints of different anatomical and physiological designs. The myriad devices of fish, especially, suggest that many evolutionary lineages have solved problems of buoyancy independently of one another. Denton stresses the active nature of buoyancy mechanisms. Swim bladders, cuttlebones, and, of course, swimming itself are continuously adjusted to the habitat's conditions and the animal's activities; these structures' functions derive from their controlled use, not merely from their inert presence in the body. Adaptive traits are traits in action.

Whereas Eric Denton compares alternative achievements of buoyancy in markedly different animals, "The Buoyancy of the Chambered Nautilus," by Peter Ward, Lewis Greenwald, and Olive E. Greenwald, examines one such mechanism in detail. Beginning with the animal in its habitat and the role of buoyancy in its natural history, these investigators carry their scrutiny to the cellular level, because mechanisms of homeostatic adjustment must be clarified at the cellular level, where they are effected and controlled. The puzzle in this research was how the nautilus withdraws cameral fluid from its shell compartments, through the wall of the siphuncle, against hydrostatic pressures that prevail in the animal's habitat. Seizing "the right animal to answer the question," these researchers draw on an analogous phenomenon in the gall bladder of the rabbit. The rabbit moves fluid against osmotic and hydrostatic gradients there; it does so at the cellular level by "local osmosis." Cytological and experimental evidence suggests that a similar mechanism empties fluid from the chambers of the nautilus. As it grows larger and heavier, the nautilus can adjust the rate at which it slowly drains its newest chambers to retain neutral buoyancy. With a workable and tested model of the cellular events that control buoyancy, the investigators draw in the development, morphology, behavior, and ecology of the nautilus to complete the explanation at its observational point of departure—observation of the nautilus "in conversation with its habitat."

The nautilus usually inhabits moderately deep waters around coral reefs and apparently does not wander far. In striking contrast, the sperm whale roams the seas from polar waters to the equator and from the surface to depths of thousands of meters. Both animals can achieve virtually neutral buoyancy. But the nautilus does so at the slow pace of developmental compensation for its body's growth, while the sperm whale must compensate rapidly and reversibly for the different water densities of the habitats through which it swims and dives. Malcolm R. Clarke's investigation of the spermaceti organ, which he reviews in "The Head of the Sperm Whale," must deal with buoyancy mechanisms in a different way than is possible with the nautilus. The size of the sperm whale, dissections so complex that one's "sense of orientation is easily lost," the impossibility so far of experimentation with living animals, the sheer rarity of adequate access—these obstacles thwart much close scrutiny and test.

Instead, Clarke has approached the functions of the spermaceti organ and its oil more as an engineer might, by way of various models. In recognition of this indirection, he often states his case in the conditional "could" and "would." But in this cautious way he develops a compelling argument based on the behavior of the whale, the anatomy of its snout and especially the arrangement of the right nasal passage, the physical properties of spermaceti oil at the temperatures that prevail in the whale's head, and the physics of water density in the sea. Alternative hypotheses based on calculations from these various aspects argue that the spermaceti organ is principally a device for buoying the sperm whale; the spermaceti organ could well be constantly and quickly fine-tuned to adjust the whale's buoyancy to the different habitats it encounters in its active life. The result is not the full story. But it is one that sets forth persuasive models and hypotheses, thereby providing direction for whatever tests may be undertaken when practical circumstances someday permit.

Andrew A. Benson and Richard F. Lee examine two main questions in "The Role of Wax in Oceanic Food Chains." The first is what use organisms may make of the waxes they synthesize—whether the materials have demonstrable functions that enhance survival. The second is what the materials provide as food to carnivores. Wax is widely and abundantly distributed among marine animals. In copepods, for example, it serves particularly as a medium to store energy in "deep reserve" against times of starvation. In corals, the wax in the mucus exuded around polyps remains a substance of perplexing significance. As an energy-rich food source for predators, wax contributes importantly to the diets of animals feeding on copepods, corals, and other prey that make wax in substantial amounts. But the predators must synthesize wax-digesting enzymes to take advantage of this food. Ecological relationships thus depend on physiological and biochemical attributes; conversely, metabolic traits imply ecological ones. And for these researchers, the pieces of a biological puzzle fit as suitable organisms are chosen to help solve it.

If unexpected substances can be immensely important in the ecology of the sea, so can unfamiliar organisms. Alice Alldredge's article, "Appendicularians," drives this lesson home. Appendicularians are bizarre in virtually all their anatomical and behavioral traits—the elaborate "house" and improbably designed net, the apparently wasteful way in which these structures are regularly discarded and made anew, the body's twists and distortions. But these parts combine into a planktonic herbivore of extraordinary consequence. On the one hand, appendicularians filter seawater with exceptional thoroughness to remove phytoplankton; on the other hand, the discarded houses add to the slowly sinking pelagic surfaces of "marine snow," drastically affecting the midwater habitat of bacterial and protistan nannoplankton. Appendicularians are natural curiosities, if any creatures are. But their impact on the ecology of the sea should warn us against facile equation of odd appearance with peripheral significance in ecosystems we understand in only rough and rudimentary ways.

In "Stomatopods," their account of those "formidable predators," Roy L. Caldwell and Hugh Dingle show how we can work from the anatomy of particular structures toward the whole natural history of the animals that possess them. Their report is grounded in a thorough description of the stomatopods' raptorial limbs, how they work mechanically, and especially how these creatures use them to spear or smash prey—in other words, in sound and solid functional anatomy. From this, the authors develop their larger account of stomatopod biology. Thus, other anatomical and behavioral traits and even behavioral bases of stomatopod biogeography are carefully drawn into relationship with these findings about the raptorial limbs. The authors' experiments are simple and ingenious, principally testing the links by which often disparate studies are joined to interpret the stomatopod's life and world. The result has a biographical rather than an analytical quality; it is a synthesis rather than a dissection of observations. The procedures, tests, and goals of natural history are different from those of analytical biology. Natural history must contend with peculiarly risky interpretations of data that often resist well-controlled test and confirmation. The research described in this article suggests how convincing synthesis may nonetheless emerge from particularly careful concern with the juxtaposition of observation, and from utmost care and thoroughness in those observations themselves.

3

The Sea's Deep Scattering Layers

by Robert S. Dietz
August 1962

The sound pulses of devices used to measure the depth of the ocean are often scattered by several "phantom bottoms" that rise by night and sink by day. The animals that make up these layers are now being identified

Nautical charts display hundreds of shoals rising from the deep sea and marked "ED"—existence doubtful. Each of them represents an echo sounding made by some ship passing through the area. Lacking the time to make a careful survey and fearful of running aground, the captain simply reports the reading to a hydrographic office, where it is duly recorded. More likely than not the sounding is spurious—a reflection not from the true bottom but from a "phantom bottom" now known to exist throughout most of the seas. A ship later passing one of the supposed shoals may find blue water to depths of two or three miles. But hydrographers are naturally reluctant to erase any possible hazard to navigation, so the charts remain cluttered with fictitious banks.

The existence of a phantom sound-reflecting layer was not recognized until 1942. At that time physicists were experimenting off San Diego with underwater sound for detecting submarines. Beyond the continental shelf, over water several thousand feet deep, their transmitted sound pulses, or pings, regularly and annoyingly returned an echo from about 900 feet. Unlike the sharp echo from a submarine, it was a diffuse, soft reverberation. On pen tracings of echoes from various depths it appeared as a layer of heavy shadowing. The zone confounded experiments only during the day; at sundown it would rise nearly to the surface and diffuse. With the first light it would re-form and descend to its normal depth. It never reflected all the sound energy striking it; the echo from the ocean bottom could always be detected through it, although sometimes only faintly. The source of the unexplained reverberation was named the deep scattering layer, or DSL (soon amended to DSL's because three and sometimes as many as five layers are often found).

At first it was supposed that the DSL's, like the D layer of the ionosphere, which fades away at night, had a physical cause, such as a temperature discontinuity. No one, however, could suggest a physical effect that would account for the diurnal migration. Martin W. Johnson, a zoologist at the Scripps Institution of Oceanography, surmised almost at once that the echo must come from marine animals that rise to the surface at night and return to the depths in the morning.

This interpretation is now universally accepted. What animals they might be, how they survived the enormous changes of pressure and temperature during their migrations, what physiological mechanisms were involved in the process—all these were mysteries. Even now the answers are far from complete. Nevertheless, this nuisance to the physicist has presented the biologist with a powerful new ecological tool for understanding the mass distribution of life in the sea.

Among the first animals suggested as a deep sound-scatterer was the squid, which lives throughout the oceans. Some investigators have wishfully supposed that the DSL's consist of vast schools of large, commercially valuable fish. It has been found, however, that such large fishes, which are sometimes present in the scattering bands, return hard echoes that stand out against the soft reverberations of the DSL's. Those of us who study echograms call the animals "tent fish" or "blob fish," depending on whether their echo is traced as an inverted V or an irregular bulge [*see illustration on pages 36 and 37*]. Almost always these larger fishes rise to the surface at night. Sometimes they can be "seen" on the echogram working their way toward the surface in the evening along with the DSL's. Curiously they rarely show up in the descending scattering layers of early morning. Instead they suddenly appear at full depth.

Today it is clear that the scattering layers consist of small, nocturnal marine organisms. They cannot be too small; at the frequencies used by echo

DEEP SCATTERING LAYERS are well developed in this echogram, made in the deep

sounders nothing shorter than about an inch will scatter sound effectively. Among habitats only the sky offers as little refuge as the sunlit upper regions of the sea. An animal sighted by a predator in either place is almost as helpless as grass in a meadow. In the open sea many small organisms conceal themselves by assuming the same transparency and refractive properties as the water itself. Others contain poisons or have hard shells or sharp spines. The animals of the deep scattering layers hide in the dark. In the daytime they seek the deep water, where sunlight hardly penetrates; they rise only at night to browse in the plankton-rich surface waters.

During the day DSL's lie at depths roughly between 700 and 2,400 feet. At night they rise almost to the surface and diffuse, or they may merge into a broad band extending down to 500 feet. The nature and the complexity of the bands vary with time and place. Off California they generally lie at about 950 feet, 1,400 feet and 1,700 feet. Most places in the ocean usually have three layers, the deepest at an average of 1,900 feet.

In studying echograms covering about 200,000 miles of ship tracks, the deepest one I found was at 2,350 feet south of the Aleutian Islands. This must mark the approximate boundary of the twilight zone even in the clearest waters. From bathyscaph dives we know that the last glimmer of light fades into blackness at about 2,000 feet. The twilight organisms have more sensitive eyes than ours and their threshold of vision extends somewhat deeper. Even if their eyes can respond to single photons of light, as has been suggested, the greatest depth at which they can possibly see anything is no more than 3,000 feet.

Occasionally discrete echoes have been observed from depths below 3,000 feet. They may be caused by animals that sometimes descend this deep, but no one really knows. Scattering bands also appear in the waters above the continental shelves, which are no more than 600 feet deep. None of these compare in importance or stability with the "true" DSL's, which may be defined as the layers between 700 and 2,400 feet, most of which rise to the surface at night.

Sometimes the echograms show very diffuse layers that stay at the same depth day and night. They may in part represent stay-at-homes among normally migrating species—immature forms or individuals that are resting or breeding. Some of the layers must consist of nonmigrating organisms that always stay in the twilight zone, probably salps, ctenophores, medusae, pteropods, pelagic worms and others. In any case they too are less important than the migrating DSL's.

Although we are far from knowing all the animals that make up DSL's in all parts of the world, it is clear that two groups predominate: (1) myctophids, or lantern fish, so named because of the luminous spots on their bodies, and similar small bathypelagic fishes; (2) shrimp-like crustaceans called euphausiids and sergestids. Nets towed at night near the surface commonly pick up from one to five of these individuals per cubic yard; exceptionally rich bands bring in as many as 20 per cubic yard.

Whether the fish or the crustaceans

Pacific off the coast of Peru. The three layers can be seen descending to depth (*starting at right*) with the coming of dawn. Echogram covers about one hour. This echogram and the ones that follow were made by vessels of the Scripps Institution of Oceanography.

play the most important role in sound-scattering is an open question. Edward Brinton of the Scripps Institution of Oceanography has found the euphausiids to be 10 times more abundant than fish off San Diego. In their effect on sound pulses the lantern fish may offset the abundance of the euphausiids by their larger size—several inches compared with one inch for the crustaceans. Furthermore, many lantern fish have swim bladders, each of which contains a minute bubble of air that can resonate with the sound waves to make a highly effective scatterer.

The striking changes in animal population with depth in the sea were known long before the discovery of the scattering layers. Most biologists have tried to explain the phenomenon in terms of temperature. The late Danish oceanographer Anton F. Bruun divided the oceans into two temperature levels separated by the 10-degree-centigrade isotherm. The upper, warmer zone he termed the thermosphere; the lower, colder region, the psychrosphere. But the DSL's suggest that light exerts a considerably more important control than temperature does. From the evidence of scattering layers the sea can be divided into three light zones: the sunlit zone, inhabited by both plants and animals, extending down to about 500 feet; the twilight zone, populated only by animals, from 500 feet down to 2,400 feet; and the black abyssal region with its few and highly specialized animals.

The lower boundary of Bruun's thermosphere rises with distance from the Equator, eventually reaching the surface in the higher latitudes. The scattering layers show no such warping. Of course toward higher latitudes tropical species fade out and boreal species become predominant, but all the animals move in response to light, not to water temperature.

Both the crustaceans and the lantern fish of the DSL's display acute sensitivity to light. On the darkest nights dip nets pick them up at the very surface, but even moonlight sends them down many feet. Echograms occasionally show scattering layers descending to moderate depth as the full moon rises. With the first glimmer of daylight, often an hour before actual sunrise, the various organisms begin their descent. Off San Diego the myctophids, the most sensitive to light, generally start down first, because they must dive to the greatest and darkest depth. A short time later the second layer, probably consisting of euphausiids, takes form and settles, and it is soon followed by the third layer, probably composed of sergestid shrimps. (The order cannot be exactly specified because it depends on the species present in a given population at a given time.) The layers never cross one another; each appears to be precisely adjusted to some particular level of twilight. Diving at speeds of as much as 25 feet a minute, the scatterers are at least halfway to their ultimate levels at sunrise. Within one hour after sunrise they attain their

preferred depth. As the sun approaches the zenith, however, they sink a bit lower to their maximum depth to avoid the penetrating light rays. When passing clouds darken the sky, the scatterers react by rising to somewhat higher levels than normal for the daytime.

While aboard the U.S.S. *Cacapon* in 1947 I traced the DSL's all the way from California to the Antarctic. By day the ever present layers, sometimes dense and at other times thin, hung like decks of stratus clouds between 900 and 2,100 feet. Each night the scatterers rose toward the surface and diffused. As the days grew longer the diurnal migrations remained precisely synchronized to sunrise and sunset. But near Antarctica, where the nights were reduced to a mere four hours, the migrations seemed to break up in confusion. The organisms of the shadows apparently could not cope with a 20-hour day.

The DSL's are almost, but not quite, universal in the deep ocean. In the nearly lifeless central South Pacific the bands become extremely faint and may even disappear completely. They do not exist in the Arctic Ocean, where the permanent cover of pack ice cuts off so much sunlight that diatoms cannot flourish. The Arctic Ocean is the most sterile of all the seas.

Most attempts to photograph the scattering layers on film or by television have failed. Deep trawling with nets has also met with indifferent success in catch-

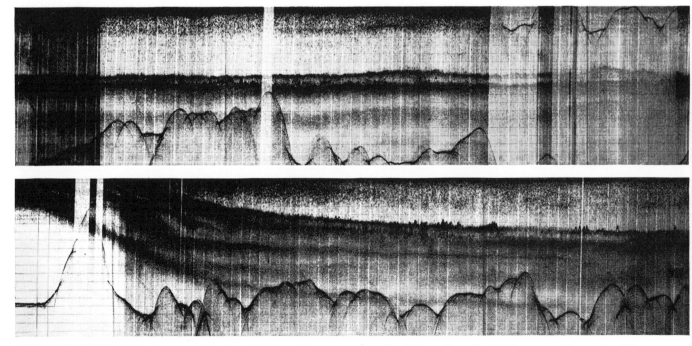

FULL DAY'S CYCLE of the deep scattering layers begins at far upper right with descent from surface at daylight. The organisms remain at depth throughout the day, in separate layers. At night they rise again to feed (*far lower left*). "Tent fish" and "blob fish"

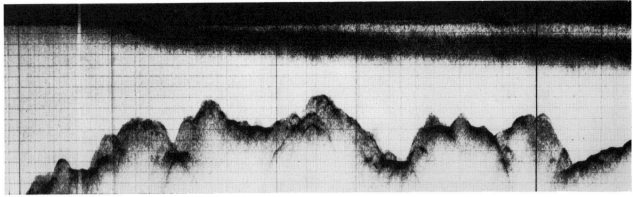

EFFECT OF BRIGHT MOON on deep scattering organisms is to hold them at shallow depth. Deep scattering layer can be seen rising to surface (*toward left*) after moon sets. This echogram was recorded between 11:45 p.m. and 1:15 a.m. in the eastern Pacific.

ing and identifying the DSL organisms. They are too small and too widely dispersed. Bathyscaph divers looking through portholes have fared somewhat better, at least when they did not expect to see great schools of fish. Crustaceans and lantern fish have appeared occasionally at the depth of the scattering layers but seldom in real abundance. On a bathyscaph dive to 3,600 feet in the Mediterranean I saw no organisms big enough to be effective sound-scatterers. Down to 2,100 feet there was an increasing abundance of minute suspended detritus from living plants and animals called sea snow; it scattered our underwater light beam as dust motes scatter a shaft of sunlight. Just below 2,100 feet the water abruptly became crystal clear; apparently this is the boundary between the ocean's twilight zone —the realm of the scattering layers and their light-sensitive organisms—and the almost lifeless, eternally dark waters of the abyss.

In bathyscaph dives off San Diego, Eric G. Barham of the U.S. Navy Electronics Laboratory has had better luck than I have had. He has repeatedly seen concentrations of four-to-six-inch fish between 650 and 1,000 feet; they were probably young hake. Hard echoes (tent fish) often come from these depths. Euphausiids have been found in the stomachs of these fish, so it seems likely that the hake feed on the scattering layer organisms.

In a recent bathyscaph dive into the San Diego Trough, Barham reported that he saw much sea snow but no large organisms between 850 and 1,200 feet. Then he entered a zone inhabited by deep-sea prawns. From 1,200 to 1,500 feet he saw so many of these sergestids that he could not count them. In the next 200 feet he encountered a large number of lantern fish. Below 1,700 feet the bathyscaph entered a region relatively free of large organisms. Then from 2,150 to 2,300 feet it sank through a zone containing the greatest concentration of fish he had seen on any dive. Again these appeared to be lantern fish, as many as eight in view at a time until he could not keep count of the sightings. Most tended to avoid the lighted area, and Barham saw them only at the edge of the cone of light. Within the abyssal zone, from 2,300 feet to the bottom at 3,900 feet, he saw only an occasional red

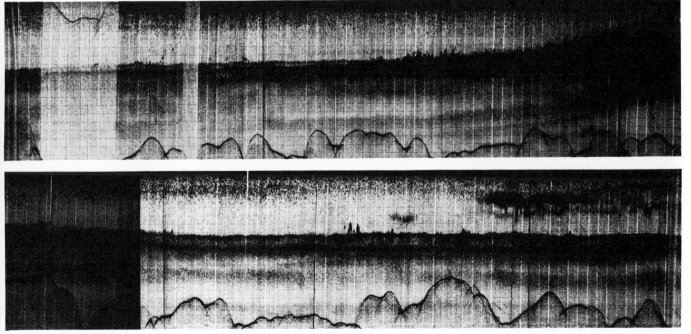

appear among the layers. The "mountains" represent echoes from the sea bottom, exaggerated 20 times in the vertical dimension. The echo-recording apparatus repeats its cycle for every 2,400 feet of depth; the bottom echo is actually on the fifth cycle.

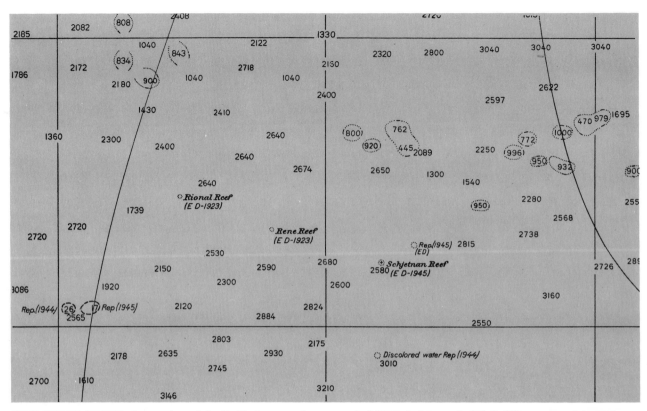

DEEP-OCEAN CHART of a portion of the Pacific between the Hawaiian Islands and the Marshall Islands displays four "reefs" marked "ED" (existence doubtful). Those reported in 1945 were found by echo sounders and could represent scattering phenomena.

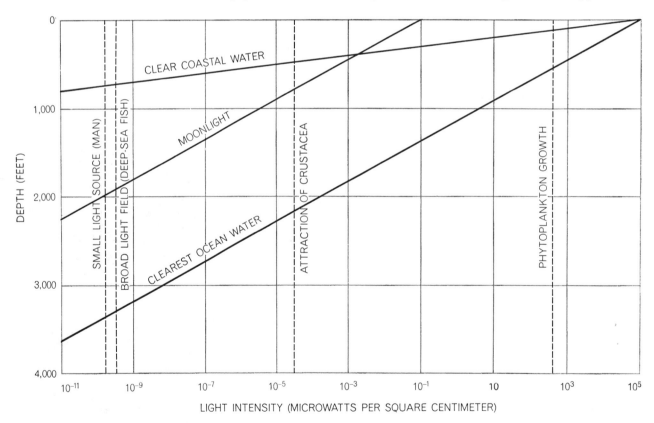

INTENSITY OF LIGHT decreases with depth. The lines "Clear coastal water" and "Clearest ocean water" refer to sunlight. Coastal water may appear clear to the eye but is always turbid. Black vertical lines denote amount of light necessary for each function. "Small light source (man)" is the faintest light source a man can detect. The broad light field for deep-sea fishes indicates minimum quantity of general daylight these fishes can see. More light is needed to attract crustaceans, and phytoplankton require light many orders of magnitude more intense in order to grow. The diagram is adapted from a paper by G. L. Clarke of Harvard University and the Woods Hole Oceanographic Institution and by Eric J. Denton of the Marine Biological Association laboratory in Plymouth, England.

prawn, some shrimplike mysids, medusae and two types of worm.

In spite of their distribution, the DSL's can never be a sea-food cornucopia for man. The echo effects indicate a concentration no greater than a few animals per cubic yard. Nevertheless, the animals of the scattering layer must constitute an important source of food for commercially important fish; surely they are an important link in the food chain of the oceans.

The DSL forms themselves feed on tiny surface organisms: euphausiids and sergestids are herbivores, grazing on diatoms; lantern fish eat other crustaceans that are nourished by diatoms. The DSL population in its turn falls prey not only to larger deep-sea fishes but also to surface swimmers. One reason oceanic banks and shelf margins make good fishing grounds may be that they are often richly populated by animals from the DSL's. During the night, when they are near the surface, many of the little creatures must drift over the banks and shores. In the morning they are trapped in the shallow lighted waters and are quickly devoured by predatory fishes. Above the submerged tops of sea mounts the DSL's sometimes practically disappear, presumably as a result of such devastating grazing.

As human beings we sometimes tend to forget that we are not the only end of nature's branching food chain. Until quite recently the source of food of the fur seal was a mystery. Every year during the breeding season some three and a half million of these voracious mammals migrate from their deep-water habitat far out in the northern Pacific and come ashore on the Pribilof Islands. Experience with fur seals in zoos indicates that it must take three and a half billion pounds of fish a year to feed the enormous wild population. Yet the fur seals never seem to compete with commercial fishing. Clearly they are tapping some vast reservoir of noncommercial sea food. Recently we have learned that in the deep ocean they feed largely on lantern fish, probably catching them at night when they rise to the surface.

Another eventual consumer of the twilight animals is the cachalot, or sperm whale. Unlike the seal, it goes down to hunt for its food. Although it is a mammal, forever tied to the surface, it sometimes takes a single gulp of air and dives as much as two-thirds of a mile to forage. The cachalot does not feed on the scattering-layer organisms but eats the squids and larger fishes—the echogram

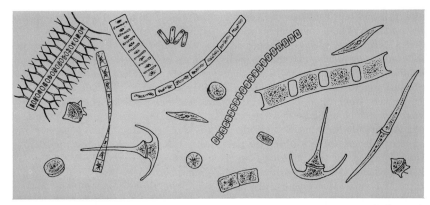

PHYTOPLANKTON, tiny drifting plants, are the "grass" of the sea: they form the basis of the food chain. These are diatoms and dinoflagellates, enlarged approximately 200 diameters.

ZOOPLANKTON are tiny animals that feed on phytoplankton and in turn furnish food for larger animals. These copepods, enlarged about 10 diameters, drift near the sea's surface.

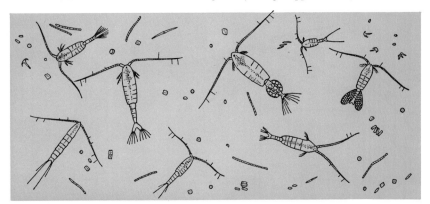

"DEEP SCATTERERS" rise from depths at night to feed in photic zone. The myctophids (lantern fish) eat copepods; the euphausiids (crustaceans) consume the phytoplankton.

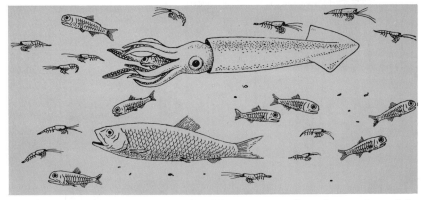

LARGER ANIMALS, such as this squid and herring, may well eat the organisms of the deep scattering layers at night near the surface. Fur seals also feed extensively on myctophids.

blob fish and tent fish—that do feed on the scatterers. These constitute its whole diet. Many rare species of deep-sea fishes are known only from the stomach contents of the sperm whales that abound around the Azores and are still harpooned from open boats as they were by the New Bedford whalers of old. Apparently the animals of the deep scattering layers are safe during the day from foraging by the plankton-sieving baleen whales. According to the whale expert Raymond M. Gilmore of the Museum of Natural History in San Diego, Calif., these whales limit their dives to the upper 200 to 300 feet of the sea, or considerably above the DSL's.

A basic tenet of oceanography is that parcels of water throughout the ocean, although they look exactly the same, in reality differ from one another. We do not need precise chemical analyses to prove the point: the organisms of the sea are remarkably sensitive indicators. Nothing illustrates variation in the sea more dramatically than echograms and their DSL's. For example, off Peru the echograms are heavily blackened by the teeming life of the Humboldt Current, but a little farther out the life dwindles to nothingness. The scattering is an index of organic productivity and in turn of those chemical factors that control the distribution of living things.

It is unfortunate that the world's expanding population cannot look to the deep scattering layers as a direct source of food. Nevertheless, the organisms of the DSL's are well up in the pyramid that requires 1,000 pounds of diatom fodder to support the growth of a pound of commercial fish. The deep scattering layers play a major role in the biological economy of the seas.

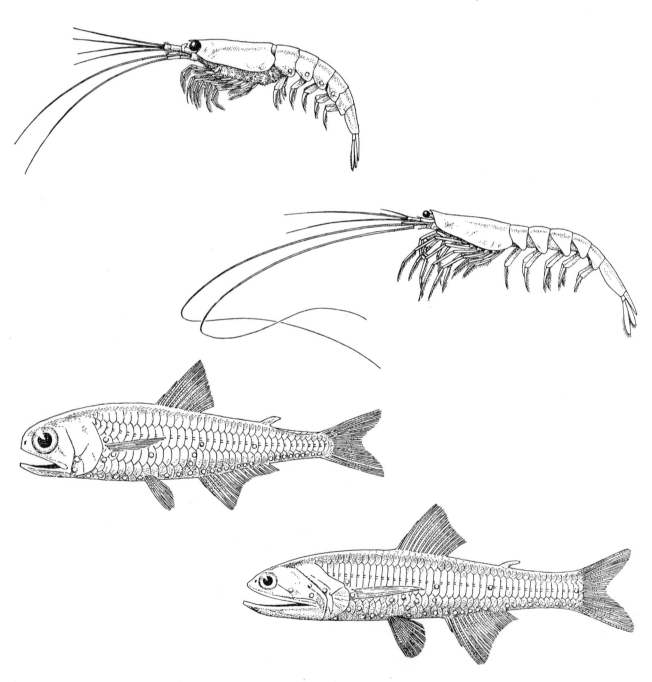

SCATTERING-LAYER ORGANISMS include crustaceans and fishes. Among them are (*top to bottom*) a euphausiid, a sergestid and two forms of myctophid, or lantern fish. The animals range in size from about an inch for the euphausiid up to three inches for the lantern fish. The four spots on the euphausiid and the similar spots on the myctophids are photophores, or light-producing organs.

Active Animals of the Deep-Sea Floor

4

by John D. Isaacs and Richard A. Schwartzlose
October 1975

*Baited automatic cameras dropped to the bottom
of the ocean reveal a surprising population of large
fishes and other scavengers that find and consume
dead animals that fall from the waters far above*

Photographs made by automatic cameras that have been dropped to the deep-ocean floor have confirmed a finding that was first suggested by evidence gathered from baited traps some years ago. A deep-sea population of large, active animals thrives in what was generally assumed to be a province inhabited mainly by small, feeble creatures such as worms, snails and sponges. The thousands of pictures make it clear that much of the deep-sea floor teems with numerous species of scavengers: vigorous invertebrates and fishes, including some gigantic sharks, that are supported by a marine food web whose extent and complexity is only beginning to be perceived.

The celebrated expedition of H.M.S. *Challenger* in the 1870's laid to rest the old idea that the deep waters of the open ocean were a lifeless desert. The *Challenger*'s trawls and dredges brought to the surface and to the attention of biologists a vast collection of deep-midwater and bottom-dwelling creatures from even the deepest ocean trenches [see "The Voyage of the 'Challenger,'" by Herbert S. Bailey, Jr., SCIENTIFIC AMERICAN, May, 1953]. The inventory included some of the most grotesque forms into which higher animals have ever been molded by adaptation to extreme conditions, and several animals that had been thought to be long extinct.

In this century much has been added to that inventory and to our understanding of deep-living animals. The investigations were conducted, however, by means of trawls and dredges, by direct observation from deep-diving submersible vehicles and by inspection of photographs made by cameras lowered on cables from ships, and it is characteristic of these methods that they ordinarily sample only the stationary or slow-moving animals of the deep-sea communities; the deeper the investigation is, the more difficult and the less effective it is. That gave rise to the prevalent assumption of a few years ago that the ocean bottom was sparsely inhabited by weak creatures specially adapted to live on the only food material then thought to be available at great depths: a terminal food web supported by the thin but constant rain of detritus that sifts down from the surface layers and that is metabolized by primitive filter feeders or bacteria and deposit feeders in the bottom ooze. A proper reconnaissance of the active deep-sea creatures whose presence was revealed by the baited traps required new tools.

Over the past seven years Meredith Sessions, Richard Shutts and the authors, working at the Scripps Institution of Oceanography, have designed and constructed robot motion-picture and still cameras and other instruments with which to explore the bottom and to study the nature, distribution and behavior of its inhabitants. We find, first of all, that the population of sea-floor invertebrates is by no means sparse and that many of its species are far from weak. And they are not alone. Their domain is shared by a population of scavenging fishes and crustaceans adapted for the prompt discovery and consumption of larger falls of food: the bodies of dead animals descending from above, mid-water creatures that happen to approach the bottom and juveniles of their own species that return to the deeps from the shallower water where they undergo their early development.

The cameras we have devised for the ocean-floor study are free-fall devices connected to a recovery buoy, a floodlight and a bait holder. Released at the surface, the camera falls to the bottom and remains there, anchored by the bait holder for between 12 and 48 hours, making still photographs or short bursts of motion pictures at five- to 15-minute intervals. The bait is in the foreground of each frame and the camera's view is either vertical or oblique. At the end of the mission the bait ballast is released and the camera and its buoy rise to the surface, where a radio transmitter broadcasts a signal that aids in recovery. The free-fall technique allows a research ship to conduct various other missions while a number of cameras it has distributed keep functioning on the sea floor. And the work can be done from rather simple, inexpensive vessels, provided only that they have reasonable navigation and sonic-sounding equipment.

In a typical drop the camera is released in water that is between 400 and 7,000 meters deep. Greater depths call for a special camera housing and flotation gear because of the immense pressures. The camera operates during the descent, which may take as long as several hours, but ordinarily not much is seen in photographs made on the way down except for the "snow" of small particles, which are ubiquitous in the midwaters of the sea, and a few fleeting small crustaceans.

Sometimes the very first photographs

FIFTEEN-FOOT SHARK attacks crushed bait can in the color photograph on the following page, obtained by the authors from a depth of some 750 meters in the eastern Mediterranean Sea. Such large fish typically arrive at the bait from three to eight hours after the camera system reaches bottom, frightening off most other feeding creatures. The line (*left*) leads to the camera.

on the bottom, made a few minutes after the camera reaches it, show great activity, with fishes and invertebrates already tearing at the bait. More commonly, however, the first scenes show only brittle stars, large shrimps, amphipods (small crustaceans) and perhaps a fish or two. On more than half of the missions at least one fish has been photographed within 30 minutes of the camera's arrival on the bottom, even at the greater depths. (Curiously the first photographs sometimes record a fish of a species that is not seen again in the sequence, as though the camera had invaded the territory of a creature that was not a part of the population of active scavengers and wanted nothing to do with the subsequent activity.)

Usually the number of fish gathered around the bait increases slowly, reaching a maximum after a few hours. Often the scene develops into one of furious activity, with several species of fish competing for the bait, thrashing and tearing at it and sometimes attacking one another. Shrimps, brittle stars, amphipods and other invertebrates encroach on the melee. In almost half of the sequences from drops down to 2,000 meters the party ends abruptly after three to eight hours, when some creature, usually a large shark, moves in, frightens off the other fish and consumes the bulk of the bait. In any case the time comes when most of the bait has been eaten. The fish depart, and slowly crabs, sea urchins, snails and other such creatures arrive to complete the task of sanitizing the sea floor.

Sometimes the direction of the near-bottom currents has been determined and it can be seen that these latecomers to the banquet plod upstream toward the bait—an indication that they are probably following a scent. The fishes also probably depend on scent for close-in detection of the bait; on some occasions we could relate the number of fish gathered around the bait to the strength of the current, which suggests that an increased current had carried the scent more widely. Like wide-ranging scavengers on land, however, fish that are farther away are probably led to the area of the bait by other cues. They may sense the successive collapse of loosely established territories as the scavengers that held them move closer to the bait, each invading the area once held by an absent neighbor. The western prairie wolf, vultures and other terrestrial scavengers respond to just such a territorial collapse to converge on large kills.

Surprisingly, we find particularly abundant bottom populations in regions of the Pacific where the surface populations are the least abundant, and vice versa. For example, under the least productive surface waters of the North Pacific Gyre, on an austere bottom paved with manganese nodules nearly 6,000 meters down, more than 40 large fish and shrimps were attracted by the bait within a few hours; at least four species were represented. Moderately productive areas of the Indian Ocean, the California Current system and biologically poor areas around the Hawaiian Islands have also shown remarkably large numbers of deep-sea creatures. On the other hand, only 4,000 meters down under the most productive oceanic waters of the world, in the Antarctic over a bottom of soft organic ooze, the bait was visited by only a few eelpouts, brotulids and rattail fish (grenadiers), along with some small crustaceans, the louselike isopods. The brotulids (fishes distantly related to cods) remained almost motionless for hours, nibbling gently at the bait. Camera drops along the underwater ridges of the Line Islands, a highly productive equatorial area of the Pacific, revealed only a few eels. Photographs from the bottom below the rich Peru Current showed very few fish but large numbers of invertebrates: furious masses of amphipods that stripped the bait in a few hours.

The explanation for such a puzzling distribution must lie in the nature of this part of the marine food web. These abyssal roving scavengers must depend for their sustenance in substantial part on

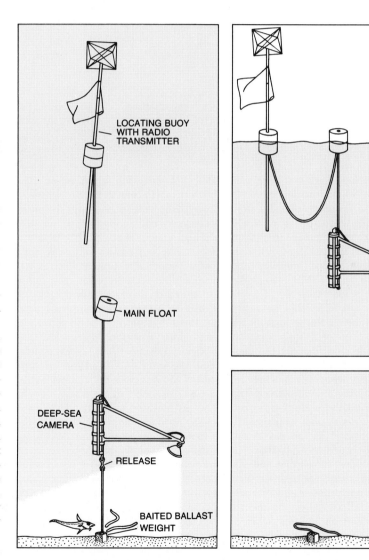

FREE-FALL CAMERA is dropped to the ocean bottom, where it is held in position by a baited ballast weight and a float made of foamed plastic (*left*). The camera can be suspended above the bait as shown here or can be positioned off to the side. At a preset time a clock mechanism cuts the camera system loose from the ballast and the system rises to the surface (*right*). An attached buoy with a flag, a radar target and a radio transmitter marks the site.

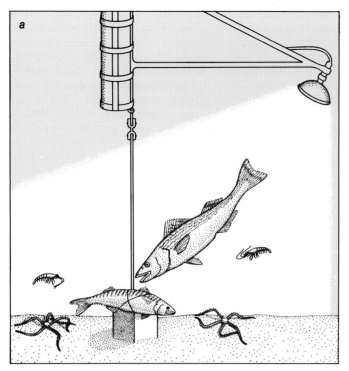

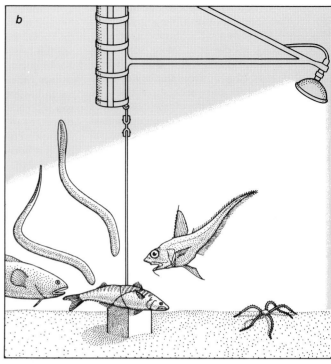

ONE LIKELY SEQUENCE OF EVENTS following the arrival of a baited camera system on the ocean floor is shown in the idealized drawings on these two pages. The first few photographs made after touchdown usually show only brittle stars, shrimps and other small crustaceans (*a*). Within 30 minutes or so the larger fishes begin to arrive on the scene (*b*). The number of fish gathered around the

falls of large food particles. Such particles include fish and marine mammals that have died, fragments of forage from surface-feeding schools of fish, vertically migrating creatures, garbage from ships and (nowadays) animals killed by underwater explosions, ships' propellers and whaling. (The North Pacific Gyre photographs were made below an area patrolled by a weather ship, and her garbage may be partly responsible for the large numbers of animals observed there.) Falling food material can support a population of deep scavengers only if the food descends to the bottom, and it may be that the mid-water population below highly productive regions is sufficiently dense and continuous to consume such food on the way down. Regions of low productivity, on the other hand, may support such sparse and discontinuous mid-water populations that a substantial proportion of dead surface creatures do fall all the way to the sea floor. And there on the floor even a meager fall can support a sizable population, whose scavenging can be much more intensive than that of fish living on the relatively brief passage of falls through the immense volume of the midwaters.

Somewhat different explanations are possible, perhaps in combination with this mid-water-population effect. The western north-central Pacific, where so many fish were photographed over a manganese-nodule bottom, is an area through which tuna and other large surface fishes and whales and other cetaceans periodically migrate. Just as terrestrial deserts are traps for aged, infirm or injured creatures, so may this marine desert be a trap. The number of large migrants that succumb in traversing its unproductive surface waters may be sufficient to support the bottom population. We know nothing of the "natural" end of the largest fishes and cetaceans; it could well be that these vast areas of low productivity inflict the final stress on aging and infirm members of the populations of great marine creatures, exhausting their ultimate reserves. Deep below areas of high productivity, on the other hand, the principal descending food may be rather small particles, sufficient to support a sizable and continuous population of mid-water creatures as well as throngs of small bottom scavengers that do not need to rove far for sustenance. Moreover, in such waters the large near-surface animals have plentiful food and are not in such a precarious situation as the schools of large migrants crossing the great marine deserts. Areas of high productivity might

therefore fail to support the larger scavengers of the deep-sea floor; few active, roving fish would be searching for large food falls such as our bait represents.

Some of the fishes now being observed for the first time at great depths at low latitudes turn out to be species that are well known in near-surface waters at high latitudes. The flatnose codling, the sablefish and the arctic sleeper shark are common inhabitants of the bottom off the coast of southern California and Lower California; the sablefish, which is found in commercially valuable numbers off the coast of Washington, British Columbia and Alaska, is particularly abundant on the deep-sea floor of the Southern California Bight thousands of miles to the south. On the basis of the incidence of this species in a random series of pictures made with unbaited cameras, we have estimated that there are 800,-000 tons of sablefish at depths of between 800 and 1,500 meters in the southern California waters; they seem to be of the same race as the commercially harvested variety far to the north. At least along the Pacific coast of North America, then, some species that are generally considered arctic fishes apparently represent mere near-surface outcroppings of populations that inhabit

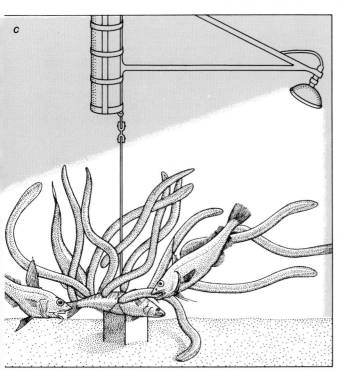

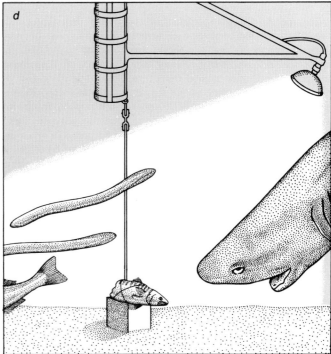

bait increases slowly, reaching a maximum after a few hours (c). After three to eight hours some much larger creature, usually a shark, moves in and consumes the bulk of the bait (d). Afterward sea urchins and snails accumulate to finish off the scraps. The latecomers to the feast appear to sense some kind of territorial collapse as successive waves of scavengers converge on large kills.

the cold, deep waters of continental borders to the south. If that is true of other coasts as well, the size and extent of a number of fish populations may be dramatically larger than has been thought. The total numbers of large shrimps in particular must be immense, since they are almost always attracted to the bait at all depths and in all regions we have sampled.

In one of our sequences a small brown shark appears in the bottom waters of the Santa Barbara Basin, which contain virtually no dissolved oxygen. We have also photographed dead mid-water fish on the bottom there. The combination suggests another source of food in some deep waters. Perhaps anoxic basins serve as traps for unsuspecting vertically migrating animals that have never encountered pockets of suffocating water. A scavenger adapted for short forays into such a basin could find a rich harvest. Perhaps the brown shark is so adapted, able to survive a brief period of oxygen deficit.

Sometimes no fish appear in our pictures; the bait is completely consumed by vast numbers of invertebrates such as shrimps and swarming amphipods. Motion pictures from the Santa Cruz Basin off southern California show the amphipods accumulating until the bait is totally covered and the surrounding water is nearly saturated by a roiling mass of these crustaceans. In that sequence one sablefish came near the bait but left immediately, gaping as if to rid itself of amphipods. Is it possible that where there are swarms of aggressive invertebrates such as amphipods, fish are not able to compete for the small amount of food that reaches the bottom? Amphipods can sustain themselves on small food particles, but they are also capable of quickly and completely devouring the much larger pieces that fall to the ocean floor. Deep-living octopuses are other invertebrates that tend to monopolize the bait. In one motion-picture sequence made at about 4,000 meters a small octopus squats on the bait, keeping the grenadiers at tentacle's length, and in an unusual sequence made with a still camera off Cedros Island in Lower California two octopuses fend off a single grenadier from the bait.

The bait is often taken over aggressively and completely by innumerable hagfish, primitive eyeless and jawless chordates that drill into dead creatures and consume them from the inside out [see "The Hagfish," by David Jensen; SCIENTIFIC AMERICAN Offprint 1035]. Hagfish thrive at a depth of from 200 meters to nearly 2,000 meters. At first we were puzzled by the reluctance of other fishes to penetrate the Gorgon's-head tangle of hagfish and feed on the bait. Closeup motion pictures gave the answer: the hagfish enclose the bait in a thick cocoon of slime that other fishes apparently find distressing. On a number of occasions fish emerge from the feeding mass making frantic efforts to clear their gills of slime. The spectacular ability of the hagfish to exude slime has long been known; the exudation of a single hagfish can convert a large container of water into a slimy gel. Their employment of this defense mechanism to sequester food, however, had not been suggested.

Some of the scavenging bottom fishes display a feeding capability that could scarcely be predicted from an examination of their jaw and tooth structure. The grenadiers are quite capable of tearing out the abdomen of a bait fish; sablefish shake large baits as terriers do and spin furiously to twist off mouthfuls of food.

The large sharks that frequent the deep-ocean floor have been photographed to depths of about 2,000 meters. Their behavior in approaching the bait appears to be mediated in part by a sense of smell: the fish execute slow, deliberate geometrical maneuvers that apparently combine to establish a complex search pattern. On some occasions, when the bait holder has been hung somewhat above a rough bottom, the sharks are quite unable to discover it. Clearly they are accustomed only to food resting on the sea floor; when they cannot find it, they nudge and bite at rocks or other sea-floor objects under the bait. Even their search for bait that is on the ocean floor usually requires more than one sortie before the bait is found. Several picture sequences end just as the shark is in a position to seize the bait, with the bait dead ahead within the width of the shark's jaws. Yet the next sequence, made five minutes later, may show the bait untouched and the shark still engaged in a slow, deliberate search. Like sharks, hagfish are unable to discover the bait when it is a meter or so above the bottom, whereas eels and grenadiers have no difficulty locating bait that is well above the bottom. Knowledge of such feeding limitations will help in the development of better techniques for some deep-water fisheries, where the depredations of sharks and hagfish greatly limit the catch.

One of the shyest fishes we have photographed, and perhaps one of the deepest-living, is the brotulid. In a number of photographs from the deepest locations brotulids lurk at the outer edge of a group of grenadiers. In one motion-picture series a brotulid hovers like a motionless blimp through sequence after sequence, facing the feeding grenadiers. Unlike most of the deep-water fishes, the brotulids have very reduced eyes, and it may be that their behavior is related to diminished visual acuity. Creatures living more than several hundred meters below the surface are maneuvering in a profound darkness that is lighted only by the faint glow or brighter brief flashes of bioluminescent organisms. Only fishes with a highly developed visual system can be expected to be guided by visual cues. Brotulids may be responding not to the presence of the bait but to the sound of the feeding fish.

Many sequences yield fresh insights into feeding behavior. A crab gingerly lifts some sea urchins off the bait, holds them away from its body like a spider ejecting a distasteful insect from its web and drops them. Sheltered by the empty bait holder, a small spiny lobster flails its antennae in a strong current, apparently grasps a small swimming crustacean between the antennae and conveys it to its mouth by some movement too quick for us to make out. A grenadier goes after small food particles in the sediment with a sudden explosive thrust into the bottom, throwing a cloud of sediment through its gills.

The uniformly large size of most of the fish photographed at great depths presents something of a puzzle. Very few small fish are ever seen. The grenadiers photographed at from 750 to 6,000 meters in the Pacific, Antarctic and Indian oceans are all large, mature fish, some measuring more than a meter in length. We have so far recovered no free grenadier eggs or juvenile fish from collections made off California anywhere between the surface and the bottom, although females with ripe ovaries have been collected near the bottom. All the arctic sleeper sharks photographed off southern California and Lower California have been very large, but most of the photographs have shown only a small portion of their total length, which we can only estimate as being between five and eight meters. They also must be quite common, since they have been photographed in nearly half of our missions off California down to 2,000 meters; on a number of occasions when more than one large shark was photographed during a mission, we could tell from distinguishing scars and other markings that the sharks were different individuals.

It is probable that the juveniles of these deep-living fishes inhabit much shallower depths than the adults. The young of the sablefish, for example, are numerous in many places at depths of 100 meters or so. It may be that the rarity of juveniles is merely the result of great adult longevity and low fecundity. On the other hand, juveniles of many species must return to the deep bottom environment; indeed, their return may constitute a meaningful importation of food from the more productive upper layers for the nourishment of the total adult population. We usually think of the relation between juveniles and adult

SWIRLING MASS of shrimps and amphipods (deep-sea relatives of sand fleas) completely covers the bait and fills the surrounding water in this oblique photograph, made at a depth of 7,000 meters in the Peru-Chile Trench. Exhibiting surprisingly aggressive behavior, the small crustaceans stripped the bait in a few hours. Curiously no fishes appeared in this sequence of pictures, filmed under one of the most highly productive fisheries in the oceans.

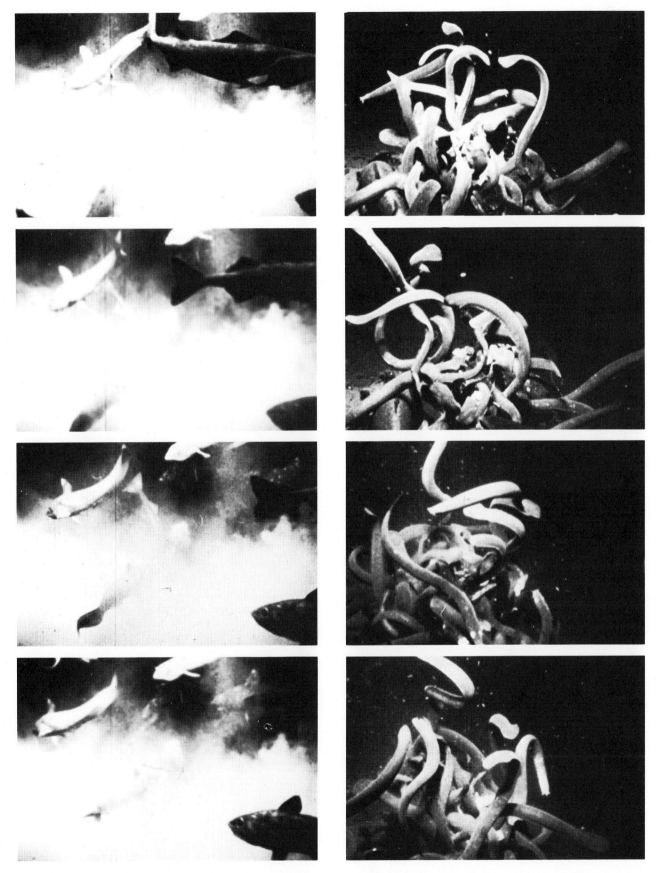

TWO MOTION-PICTURE SEQUENCES show the frenzied activity that develops around the bait. These pictures were made with a 16-millimeter camera dropped to the Pacific floor at a depth of 1,300 meters some 50 miles off the coast of southern California. In sequence at left a variety of fishes attack bait, stirring up the sediment on the bottom. In the sequence at the right the bait is entirely hidden by a mass of hagfish: primitive eyeless and jawless chordates that drill into the bait and consume it from the inside out.

stocks only in terms of replacement. Juveniles may also be important as prey; indeed, in some freshwater environments the young of a species are a principal prey of adults of the same species. To the degree that this process, which has been called Faginism, is important to the food economy of the deep-sea populations, we would of course expect to observe a paucity of juveniles in the populations: most of them are quickly consumed by their elders!

The baited camera suffers from some limitations common to many simple exploratory tools. Its sample is highly selective. Quantification and even identification can be doubtful. There are several ways to deal with these problems. One can, for example, set out fishlines and traps in order to retrieve specimens for sure identification. A particularly promising means of quantification is an unbaited drifting camera that lightens itself until it rises just a few meters above the bottom. It drifts with the current, making overlapping photographs and meanwhile recording the direction and distance of drift. Such a camera has been successfully operated in waters about 1,000 meters deep. We have plans to develop a drifting camera that will remain submerged for several months, making photographs along a drift track as long as 200 kilometers. The drifting camera should yield a meaningful census of the creatures that have been observed by stationary baited cameras and thus help us to assess the potential food harvest from the deep-ocean floor and to understand the ramifications of this remarkable branch of the marine food web.

Meanwhile vast areas await investigation with the baited camera. A series of photographic "sections" across the continental shelves along all the major land masses and down the slopes to the abyss would fill many gaps in our knowledge of bottom-dwelling animals and should reveal entirely new fishery resources. Among the environments that are of special interest to oceanographers and marine biologists are the floor of the Arctic Ocean, the deep delta of the Congo River, the Antarctic slope, the deeps of the Mediterranean, the top of seamounts and mid-ocean ridges and the slope and bottom of deep oceanic trenches. Since the free-fall cameras can be operated from inexpensive, unspecialized craft, they are particularly suitable for studies by investigators in underdeveloped countries who want to know more about the creatures and the potential deep fisheries off their coast.

5

The Buoyancy of Marine Animals

by Eric Denton
July 1960

*The ability of the creatures of the sea to float
is usually taken for granted, but to keep from sinking
they must either keep swimming or be equipped
with special flotation devices*

Many fishes, like some people, must keep swimming just to keep from sinking. Bone and muscle are denser than sea water and so tend to drag the animal to the bottom. It is obvious that the ability to float can be advantageous for animals that live in the sea. Accordingly most fishes are equipped with swim bladders which give them neutral buoyancy—that is, an average density equal to that of sea water—and save them the labor of continuous swimming. Two other animals—the cuttlefish and the cranchid squid—have developed quite different kinds of flotation organs. They anticipated man in using the working principles of the submarine and the bathyscaph, the one endowing the cuttlefish with active control of its buoyancy, the other permitting the squid to live at great depths.

The swim bladder, the most familiar of the flotation organs, has a fundamental limitation. It normally occupies about 5 per cent of the volume of the fish; the low density of the gas that inflates the bladder offsets the higher density of muscle and bone, giving the fish an average density close to that of sea water. But the volume of the bladder changes with depth. As a fish swims downward in the sea, the pressure exerted by the water increases by one atmosphere (about 15 pounds per square inch) every 33 feet. Each such change in external pressure reduces the volume of the gas inside the swim bladder in accordance with Boyle's Law; if the fish dives from the surface to 66 feet, for example, the volume of its swim bladder shrinks by a third. With the change in volume the average density of the fish, and its buoyancy, must change. In principle the fish thus resembles the well-known Cartesian diver: a perfectly immobile fish might have neutral buoyancy at one depth, but the slightest

change in equilibrium must tend to push it upward or downward with steadily decreasing or increasing buoyancy. This seemed such an impractical arrangement to early investigators that they thought the fish must be able to exert some active muscular control over the volume of its swim bladder, contracting the bladder when it seeks to descend or permitting the bladder to expand when it seeks to rise. In 1876, however, the French physiologist Armand Moreau showed that the fish has no muscular control over its swim bladder.

What the fish can do, as Moreau first discovered, is to change the mass of the gas inside the swim bladder and thus keep its volume constant. It secretes more gas into the swim bladder when it goes deeper and resorbs gas from the swim bladder when it rises. But this is typically a slow process, which occurs in response to long-lasting changes of pressure and takes days rather than minutes to reach completion. The fish thus lives in a state of unstable equilibrium and compensates for transient changes in pressure by swimming.

Most fishes can tolerate some change in the pressure of the surrounding water, provided it is not too large. But if a fish is brought up rapidly from, say, 125 feet, it meets a violent death. With the restraining external pressure reduced fivefold, the swim bladder suddenly expands, crowding and rupturing the fish's internal organs. It is for this reason that whiting, cod and hake that have been trawled almost always arrive at the surface dead. Yet under stable conditions of external pressure, the swim bladder can produce extraordinary internal gas-pressures. Some fishes have been seen from bathyscaphs swimming and hovering, with all the ease of a goldfish in its pond, at a depth of 6,500 feet, where the pressure is 3,000 pounds per square

inch. Still other fishes have been caught at depths below 15,000 feet, where the gas in their swim bladders must have exerted a pressure of more than 7,000 pounds per square inch to withstand the pressure of the sea.

The precise nature of the gas in the swim bladder was first determined by the French physicist Jean-Baptiste Biot in 1803. Biot's curiosity was aroused in the course of a survey voyage in the Mediterranean; he noticed that fishes that had been hauled up rapidly from great depths reached the surface with their swim bladders projecting from their mouths and their bodies distended with gas. He introduced a quantity of the gas, together with hydrogen, into the glass tube of a eudiometer, an instrument used for gas analysis. He fired a spark, in accord with the routine of this procedure, and was startled by an explosion that shattered the instrument. Biot realized at once that the swim bladder must have contained a high concentration of oxygen. With a new instrument Biot discovered that although the gas taken from fishes that live near the surface often contains a smaller fraction of oxygen than air does, the gas from fishes that have been brought up from appreciable depths consists largely of oxygen.

It is, of course, the fish's circulatory system that brings the oxygen to inflate the swim bladder. The remarkable countercurrent mechanism of the "wonderful net" of capillary vessels that establishes the often high gas pressures in the swim bladder has been elucidated by P. F. Scholander, now of the University of California at La Jolla, and Jonathan Wittenberg of the Albert Einstein College of Medicine in New York [see "The Wonderful Net," by P. F. Scholander; Scientific American, April, 1957]. In a sense the swim bladder has its human analogue in the air-filled lungs of a skin-diver who carries his air supply in a tank. The diver, however, can adjust the volume of gas in his lungs at will; the fish has no such immediate control of its swim bladder.

A surprising number of fishes get on successfully without a swim bladder. The turbot, which is about 5 per cent denser than sea water, simply hugs the bottom. The dogfish and the plaice can be seen to swim hard whenever they want to stay off the bottom. More surprising are the mackerel and some tuna, fishes of the open sea that can only maintain their level by steady swimming. Observation of a mackerel in an aquarium shows that whenever it slows down, it sinks. In this sense it is not better off than the prawn and the krill, which must paddle unceasingly to maintain their place in the plankton layer, or the oceanic crab *Polybius,* which can be seen from a ship's side swimming far from shore and with the ocean bed more than two dark miles below it. The lack of a swim bladder, however, gives the mackerel facility in moving up and down in the top layers of the sea. A sudden ascent of 60 feet would increase the volume of a swim bladder three-fold; quite apart from the danger of internal injury, the fish would have to exert a force equal to 10 per cent of its weight to go down again.

In some mid-ocean fishes the swim bladder is invested with fat instead of gas. Fat offers the advantage of maintaining constant volume under varying pressure. But fat has about .9 the density of water, and fish that depend upon it for buoyancy must carry appreciable amounts of it outside the swim bladder. Indeed, it appears from the work of N. B. Marshall of the British Museum that the fat of *Cyclothone,* perhaps the world's most common fish, amounts to about 15 per cent of its volume and gives the animal an over-all density close to that of sea water.

There are yet other bathypelagic, or deep-sea, fishes that have neither swim bladders nor a high fat content, and yet have a density that is within .5 per cent the density of sea water. These fishes embody another solution to the problem of buoyancy. Instead of having special organs to buoy up their heavier parts, those parts have been diminished. Their skeletons, especially in the tail and trunk, are light, and the swimming muscles that pull on the lightened bones are correspondingly attenuated. Such fishes may

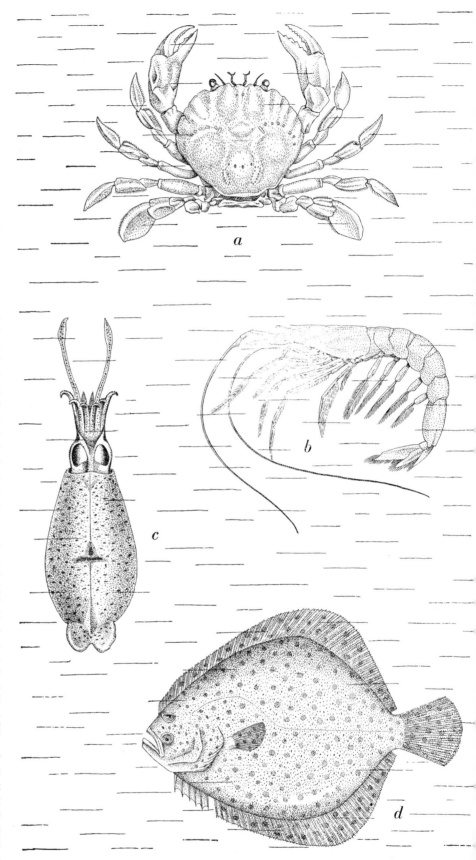

SOME BUOYANT AND NONBUOYANT SEA CREATURES are depicted in this drawing. Both the oceanic swimming crab **Polybius henslowi** (a) and the deep-sea prawn **Sergestes** (b) are denser than water, and so must swim to keep from sinking. The turbot (d) is about 5 per cent denser than sea water, but is adapted to life on the ocean floor. Gas in a chamber of its shell gives **Nautilus pompilius** (h) buoyancy. A gas- and fluid-filled

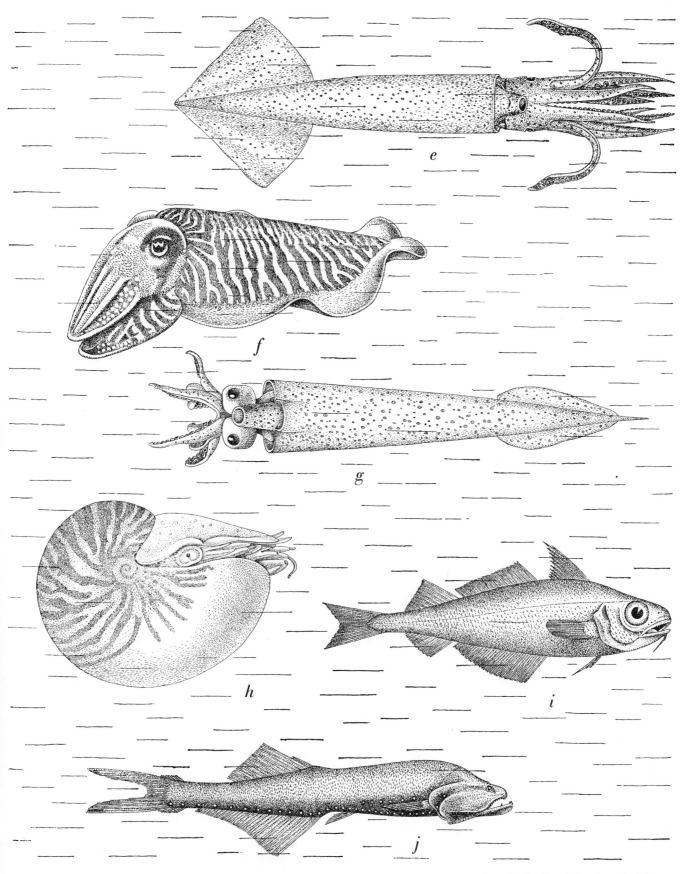

e

f

g

h

i

j

bone that occupies almost 10 per cent of its volume gives the cuttlefish *Sepia officinalis* (*f*) variable buoyancy. *Gadus minutus* (*i*) is representative of a class of fish with buoyant, gas-filled swim bladders. *Cyclothone* (*j*) also has a swim bladder, but the bladder is filled with fat instead of gas. Its fat comprises about 15 per cent of its volume. Squids of the family *Cranchidae* (*c and g*) have a fluid-filled body cavity that lowers their average density and enables them to float. The common ocean squid *Ommastrephes illecebrosa* (*e*), having no such device, must swim to stay afloat. Organisms, seen from various angles, are not drawn to scale.

be less than 5 per cent protein, compared to the 17-per-cent-protein constitution of coastal fishes. The presence of a little fat and of dilute body fluids goes far toward offsetting the net weight in water of scant bone and muscle and brings these fishes close to neutral buoyancy.

In evading or almost evading the buoyancy problem by giving up their swimming muscles, it might seem that these bathypelagic fishes have struck a poor bargain. Yet they manage to catch and eat prey as large as themselves because they are sufficiently equipped

with bone and muscle for the peak effort of their mode of predation. Though their trunk and tail are attenuated, their gill arches and main structures for seizing and swallowing prey are the most heavily ossified parts of the skeleton and are well muscled. These fishes are simply floating traps; they lie in the dark of the deep ocean and attract their prey with luminous spots and appendages around their head and mouth. The arrangement has inspired my colleague E. D. S. Corner to liken them to the practitioners of television ("telly" in the United Kingdom):

The use of its luminous lures
An adequate diet ensures
Like people on telly
It fattens its belly
By feeding on gullible viewers

The so-called lower orders of animals have developed other solutions to the problem of buoyancy that are no less elegant than the swim bladder of fishes. One of the great steps in evolution was that by which the ancestor of the modern squid evolved a gas chamber at the apex of its cap-shaped shell. We may imagine that the chamber came to function as a

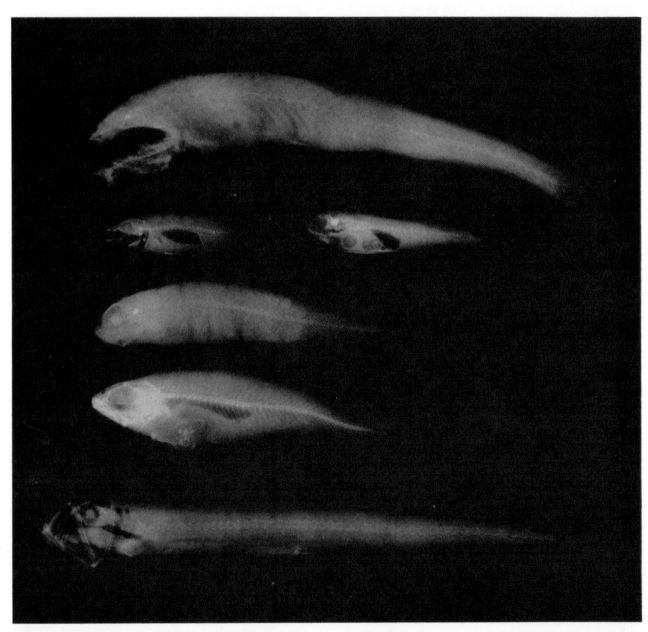

X-RAY PHOTOGRAPH OF FISHES illustrates the fact that a heavy skeleton is usually accompanied by a swim bladder. The buoyancy imparted by the bladder makes the heavier skeleton biologically feasible. Members of the families *Labridae* and *Bathylagidae* (*two fishes below fish at top*) and the family *Gadidae* (*second from bottom*) have both swim bladders and heavy skeletons. The swim bladders appear as dark areas just beneath their spines. Specimens of the families *Chauliodontidae* (*bottom*), *Gonostomatidae* (*top*) and *Alepocephalidae* (*third from bottom*) have no swim bladders, and also have relatively light skeletons.

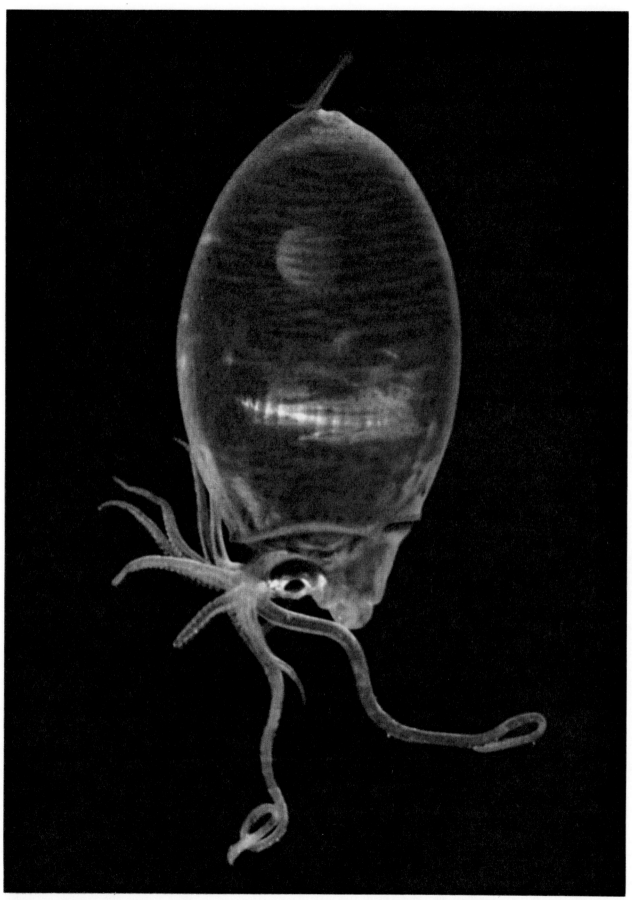

DEEP-SEA SQUID *Heliocranchia pfefferi* hovers head-down. Its buoyancy is imparted by a body fluid that is lighter than water. The fluid comprises two thirds of the squid's volume. Below edge of mantle (*horizontal fold across lower quarter of body*) is squid's head and one of its eyes (*middle*). Bulge to right of eye and below it is siphon through which squid squirts water to propel itself.

buoyancy tank, permitting the animal to take off easily from the bottom, propelled by a jet of water. Its foot, freed from the task of crawling, spread forward and surrounded the mouth, its edge becoming a fringe of tentacles. Thus, perhaps, did the distinctive cephalopod form emerge. The chambered, gas-filled shell took many and beautiful forms, and various species of this large family ranged and dominated the Paleozoic and Mesozoic seas. But except for one descendant (the rare Nautilus, the shell of which is one of the most coveted ornaments of the conchologist's cabinet) there are no survivors of these spectacular animals. There is, however, another cephalopod whose buoyancy device clearly derives from the gas-filled shells of the nautiloids. This is the cuttlefish, *Sepia officinalis,* a close relative of the octopus, that thrives off the western shores of Europe.

Beneath the skin along the back of the cuttlefish lies a large bone that serves as the animal's buoyancy tank. The cuttlebone is a soft, chalky structure, as fanciers of canaries and parakeets are well aware. (A piece of it is placed in a cage to provide a surface on which the bird can groom its beak.) The cuttlebone is built up of lamellae, thin plates that, with the pillars which hold them apart, form the walls of independent chambers. The animal lays down lamellae throughout its growth, and by the time it has matured, its cuttlebone is a beautiful structure of about a hundred delicate layers. A thick, calcified outer layer seals most of the bone and extends backward to form a curved, fin-like structure. The back part of the undersurface of the bone is formed by the turned-up ends of the lamellae, and is covered with a yellowish membrane [*see illustration at right*].

The cuttlebone constitutes about 9.3 per cent of the total volume of the cuttlefish; its density is usually around .6 that of water. In a cuttlefish weighing 1,000 grams, therefore, the bone will give an upthrust of approximately 40 grams, balancing the excess weight in sea water of the rest of the animal.

The first clue that my colleague John B. Gilpin-Brown and I had that the cuttlebone is something more than a static and unchanging organ, passively buoying up the fish, came from the behavior of cuttlefish that we kept in aquariums. They were sometimes so buoyant that they seemed to have difficulty in staying at the bottom of the tank; at other times they settled down and re-

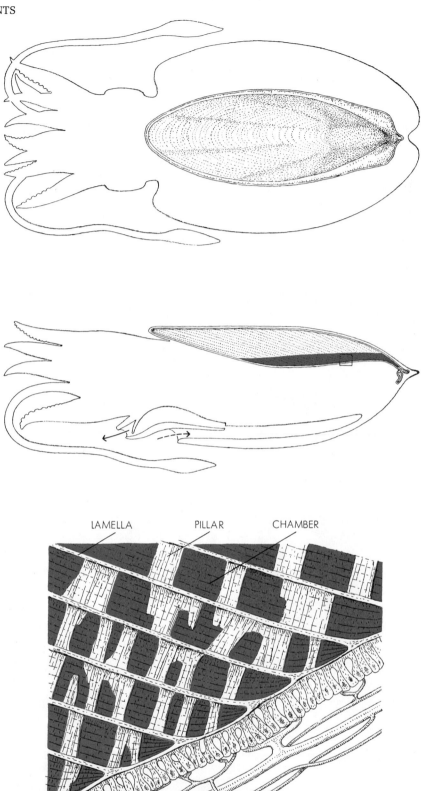

LAMELLA   PILLAR   CHAMBER

LIQUID-FILLED CUTTLEBONE buoys up the cuttlefish because it is less dense than sea water. Top drawing shows position of cuttlebone from above. Middle drawing is lateral section; color represents bone fluid. Arrows indicate path of water cuttlefish uses to propel itself. Area in small rectangle appears in bottom drawing. Membrane against underside of bone pumps ions from bone fluid (*color*) into blood. Increased osmotic-pressure difference between the two holds out water that sea pressure tends to force in.

mained at the bottom with apparent ease. Cuttlefish can be shepherded into enclosures without being too much disturbed, and we took advantage of their docility to weigh them under water in boxes filled with sea water. We found that the surface specimens were less dense than sea water, and that the bottom specimens were more dense. Upon dissecting the specimens, we found that the difference could be wholly attributed to the varying density of their cuttlebones and that the change in the density of the bones in turn was accounted for by the relative volume of gas and of liquid they contained. In a "heavy" cuttlefish the cuttlebone might have a density as high as .7 and contain about 30 per cent liquid by volume; on the other hand, the cuttlebone of a "floating" specimen might have a density of .5 and contain 10 per cent by volume of liquid.

The response of well-fed cuttlefish to light suggested to us that they can effect changes in density quite rapidly. When the light is bright, they usually bury themselves in the gravel at the bottom of their tank; after twilight they emerge and swim about until dawn. We repeated our weighing procedure with animals subjected to rapid alternations of light and dark and found that their density varied as much as 1 per cent within a few hours.

The cuttlefish thus uses its cuttlebone as a submarine commander uses the buoyancy tanks of his craft. When the submarine is to submerge, its buoyancy tanks are filled with water; when it is to surface, compressed air blows the water out of the tanks. The corresponding mechanism in the cuttlefish must meet stern specifications. In the waters off Plymouth, England, the cuttlefish is most frequently found between 100 and 250 feet and is thought to go down as deep as 600 feet. The animal is thus commonly exposed to pressures around eight atmospheres and may occasionally come under a pressure of 20 atmospheres. The pump that empties the cuttlebone must not only be capable of moving water in and out of the bone at the surface of the sea, but also must be capable of balancing the considerable pressure of deep water. Since compressed air works so well in the submarine and even in the swim bladder of fishes, this is the first mechanism that suggests itself.

We soon found that this suggestion is a false one. From cuttlefish that had just been hauled aboard ship from about 230 feet we quickly dissected the cuttlebones, placed the bones under water and punctured them with needles, confidently expecting to see bubbles stream from the punctures. We found instead that water rapidly entered the holes. Obvi-

CHAMBERED SHELL OF NAUTILUS forms a logarithmic spiral in this X-ray photograph. The animal makes a partition after each period of growth. It occupies outer chamber, which has volume equal to all the other chambers. Gas in shell helps animal float.

ously the gas in the cuttlebones was not under pressure. Far from being at the eight atmospheres of pressure we had expected, the gas was in fact closer to .8 atmosphere. Moreover, we found that the gas pressure falls when the cuttlefish pumps water out of its cuttlebone. Clearly it is not the gas that expels the water. It seems rather that the gas, which is mostly nitrogen, plays a completely passive and incidental role in the working of the cuttlebone. It presumably arrives there merely by diffusing into a space created by some other force.

We guessed that this force might be an osmotic one, since osmosis occurs very commonly in biological systems. Our next step was to secure samples of the fluid from active cuttlebones. We immersed a cuttlebone taken from a freshly trawled animal in liquid paraffin in a sealed container. When we lowered the pressure in the container, fluid seeped from the rear undersurface of the cuttlebone. Upon analysis we found that the fluid, which is chiefly a solution of sodium chloride, has a lower concentration of salts than does the animal's blood. The difference in the concentration of salts is enough to provide an osmotic force sufficient to balance the hydrostatic pressure of the sea. Since the fluid issued only from the rear undersurface of the cuttlebone and nowhere else, it was apparent that the yellowish membrane covering that region is active in the osmotic process.

Microscopic examination shows that this is indeed the case. The membrane has not only a copious blood supply but also numerous ampullae, or sacs, close to the bone. Small ducts connect the ampullae to the veins in the membrane, and these constitute a special drainage system. The arrangement apparently enables the membrane to pump salts from the liquid within the cuttlebone into the blood, and so causes fluid to flow from the bone into the bloodstream. The salt pump can thus increase or decrease the osmotic pressure between the animal's blood and the cuttlebone liquid in response to changes in the hydrostatic pressure exerted by the sea, and can keep fluid from diffusing into the bone even when the cuttlefish is exposed to heavy external pressure. It cannot protect the bone from the crushing weight of the sea. But the bone itself is well designed to withstand compression. One bone under test in a pressure tank imploded only after it was exposed to pressure greater than the maximum pressure to which the animal is exposed in life.

In its cuttlebone the cuttlefish thus possesses a buoyancy-regulating device

of a most ingenious kind. It is true that the bone is bulkier than a swim bladder, but it also serves as a skeleton. As compared to the swim bladder, quick changes of depth affect its buoyancy only slightly, and changes in hydrostatic pressure affect its volume not at all. A very small change in the salinity of the fluid contained in the bone restores the balance between the external hydrostatic pressure and the opposing osmotic pressure. But there is a limit to which this process can go. If the liquid inside the cuttlebone could be completely desalted, the osmotic pressure could withstand the hydrostatic pressure of no more than 800 feet of water.

There is no such limit on the function of the buoyancy organ of the common deep-sea squids, the *Cranchidae*. These are common denizens of the ocean and grow to large size. Perhaps because they live at great depth, the largest have been caught only second-hand, in the stomachs of sperm whales; at least one specimen has been recorded that measured more than 10 feet, not counting the two very long tentacles that are usually included in descriptions of giant squids to make the length seem more impressive. With the facilities of the Plymouth Marine Laboratory ship *Sarsia* at our disposal, my colleagues T. I. Shaw and John B. Gilpin-Brown and I were able to study the nature and function of the buoyancy organ in several varieties of squid kept in shipboard tanks.

A squid propels itself by shooting a jet of water from its siphon, and, to a lesser extent, by means of two small fins at its rear. Confined in tanks of sea water aboard ship, however, our specimens did little but remain still. They hung headdown and almost motionless without any apparent effort. Since the squid possesses neither a swim bladder nor a cuttlebone, it must have some other organ to balance the excess weight of its muscles.

When the outer mantle of the animal is cut open, there appears a large fluid-filled cavity, called the coelomic cavity. If the membrane containing the fluid is punctured and the fluid is drained away, the squid loses its buoyancy and promptly sinks. It is clearly the fluid that gives the animal its buoyancy. From measurements aboard ship and more exact measurements ashore we found that the coelomic fluid of the several varieties studied had a density between 1.01 and 1.012. Since the density of sea water is about 1.026, and since the coelomic fluid comprises fully two thirds of the weight of a squid, this is enough to offset the weight in sea water of the animal's pro-

teins and to give it neutral buoyancy.

What gives the fluid its low density? First inspection did not disclose the answer. The fluid is clear, colorless and bitter-tasting. At first we thought that it might be hypotonic, or less "salty" than sea water, and so have a lower density. A shipboard test of the freezing point of the two liquids showed, however, that this is not the case; the coelomic fluid is isotonic to sea water.

This finding and the bitter taste of the fluid pointed to the one ion that could give a solution isotonic to sea water but with a sufficiently low density, that is, the ammonium ion. Analysis showed ammonia present in the astonishing concentration of nine grams per liter.

The source of the ammonia is the squid's peculiar metabolism. Unlike mammals, the squid excretes the nitrogen from the breakdown of proteins in the form of ammonia instead of urea. It seems to employ a simple chemical stratagem to trap the ammonia in its coelomic cavity. The acidity of the coelomic fluid is high; in consequence the ammonia diffuses from the bloodstream into the cavity and there dissociates into its ions. The ammonium ion does not pass so readily through living tissue, and so remains in the cavity to reduce the density of the fluid and float the squid.

The squid's method of achieving buoyancy has the great merit of being virtually unaffected by external water-pressure. As compared to fishes, the squid need not alter the quantity of its coelomic fluid as it changes depth. As compared to the cuttlefish, the squid can dive much deeper. The great defect of the device is that it is so cumbersome, for while the swim bladder occupies only 5 per cent of the volume of the fish, and the cuttlebone only 10 per cent of the volume of the cuttlefish, the coelomic cavity has a volume equal to 200 per cent of the rest of the cranchid squid.

There is a striking resemblance between the cranchid squid and the bathyscaph designed by Auguste Piccard that has descended into the deeper parts of the ocean. In the bathyscaph a large chamber filled with gasoline plays the role of the coelomic cavity and its fluid; below hangs the observing chamber, which corresponds to the denser working parts of the squid. For this reason, we have described the cranchids as "bathyscaphoid squid." Though the etymology of the term is dubious, it may serve to remind us of an important lesson: Our pride in man's latest discoveries must be tempered by the knowledge that other animals may have been using them from time immemorial.

The Buoyancy of the Chambered Nautilus

by Peter Ward, Lewis Greenwald
and Olive E. Greenwald
October 1980

*The animal trims its weight and gains mobility by
dividing its shell into compartments and removing
their watery content. It succeeds in spite of the ocean's
pressure, which tends to drive water back in*

Before the end of the Cretaceous period some 65 million years ago life in the oceans was quite different from that of today. Among the most important large inhabitants of the seas were chambered cephalopods: mollusks that dwell in the outermost compartment of a shell they divide into compartments by secreting a sequence of septa, or walls, inside it. What gave them their dominance was their development of the ability to achieve neutral buoyancy: an overall density, or weight per unit volume, virtually equal to that of the seawater around them. In particular the chambered cephalopods evolved an organ capable of removing water from the inner compartments of their shell. With the advent of this ability (and also the evolution of a directional water jet) the chambered cephalopods were freed from the confines of their bottom-crawling ancestors and became the first of the large free-swimming carnivores in the sea.

The chambered cephalopods reached their greatest diversity during the Triassic, Jurassic and Cretaceous periods, from 225 to 65 million years ago. In the latter part of the Cretaceous their numbers began to diminish, perhaps in response to increasing numbers of a new type of mobile predator, the modern bony fishes, which maintain their neutral buoyancy in an entirely different way: they have an inflatable bladder and a gland that can fill it with gas. Whatever the reason, the extinction of the chambered cephalopods was almost complete by the end of the Cretaceous. Only the genus *Nautilus* now remains. The animals of the genus provide an opportunity to study the ancient buoyancy system. The precision of the system in the nautilus is remarkable: the difference between the weight of a mature nautilus (as much as 1,400 grams, or more than three pounds) and that of an equal volume of seawater can be as little as a gram.

The surviving species of *Nautilus* are found in the sea just outside the coral reefs that surround the islands of the tropical western Pacific. They are rarely seen near the surface, but they have been trapped at depths as great as 600 meters. Since they live below the typical range of human divers, little is known about their behavior and their ecology. Nevertheless, some observations have been made. For one thing, the stomach of a dissected nautilus usually contains pieces of bottom-dwelling crustaceans. Moreover, on the rare occasions when we ourselves have observed the nautilus in the sea the animal has been on the bottom or close to it. Finally, the nautilus is a slow swimmer; a diver can easily keep up with one. The impression is strong that the nautilus feeds by slowly foraging on the slopes of the fore reef.

The living soft parts of the nautilus consist of two principal sections, the head (more properly the head-foot) and the body. The head is covered by the hood, a tough, fleshy tissue that acts as a shield. The nautilus can retract into its shell, leaving only the hood exposed.

The most notable features of the head are its tentacles, which number more than 90, far more than there are in any other living cephalopod. Each tentacle is lodged in a sheath, into which it can be retracted and from which it can be protruded. The surface of the tentacle lacks the suckers found on the tentacles of other cephalopods. Instead it is covered with a sticky substance that aids it in holding prey.

The tentacles surround a massive pair of jaw parts that resemble the beak of a large parrot. The jaws are heavily calcified, allowing the nautilus to break up even the most heavily armored exoskeleton of a crustacean. Unlike the saliva of the octopus, a fellow cephalopod, the saliva of the nautilus contains no toxins to incapacitate a struggling prey animal. The jaws of the nautilus are its only offensive weapon. Below the tentacles and the jaws is a fold of tissue called the hyponome, which is used for locomotion: it expels a jet of water. The hyponome is actually a pair of muscular flaps that curl around each other to form a highly flexible funnel.

One other feature of the head deserves mention. Like other cephalopods the nautilus has prominent eyes, one on each side of the head, but unlike the eyes of the others, the eyes of the nautilus are poorly developed. For example, they have no lens. A tiny opening admits light, and presumably images form the way they do in a pinhole camera. Our own encounters with the nautilus suggest that its eyes may serve solely to detect changes in the intensity of light. On the other hand, the tentacles seem to bear cells that are sensitive to the presence of chemical substances. These cells may serve the animal in place of accurate vision.

The other main section of the living nautilus, the body, includes a large sac that contains the animal's organ systems. Much of the space is occupied by the systems for digestion and reproduction. The sac itself is enveloped by the mantle, a sheet of tissue that secretes the shell. The posterior mantle secretes the septa that divide the shell into compartments. The space between the mantle and the sac is open under the body, creating a large cavity that communicates with the hyponome. The cavity contains four large gills, in contrast to the single pair of gills found in all other cephalopods. The cavity also receives the exit pores for the digestive and reproductive systems.

The sexes are distinct in the nautilus. The male is slightly larger than the female because of the presence of a large organ, the spadix, which during copulation introduces a packet of sperm into the mantle cavity of the female. The females of many cephalopod species produce thousands of eggs in a year; an average female nautilus produces no more than 10 large eggs.

The shell in which the nautilus lives

is approximately a hollow cone wound tightly around itself. The shell is made of layers of aragonite, a crystalline form of calcium carbonate, in alternation with layers of a proteinaceous substance having a chemical composition resembling that of fingernail.

The internal division of the shell into chambers immediately suggests the possibility that the nautilus might change its buoyancy as a submarine does. As early as 1696 Robert Hooke proposed that "the animal has a power to fill or empty each of [the chambers] with water, as shall suffice to poise and trim the posture of his vessel, or shell, fitteth for that navigation or voyage he is to make; or if he be to rise, then he can empty these cavities of water, or fill them with air." We ourselves, however,

have never found a nautilus in shallow water to refill a chamber with water; it can only empty the chambers. Moreover, it can empty them only slowly; the maintenance of buoyancy is in fact a lifelong effort. As the animal grows in its shell it builds a septum behind itself, thereby sealing off a new chamber. Each new chamber is filled at first with a watery fluid called the cameral liquid, but the liquid is slowly removed. This provides the buoyancy the animal needs to counteract the growing weight both of its living parts and of its shell.

The removal of the liquid from the chambers is accomplished by the siphuncle, a strand of living tissue enclosed in a calcareous tube that spirals from the posterior mantle of the animal through all the chambers of the shell, including the earliest chambers. At the

center of the strand is a network of blood vessels. The surrounding seawater exerts a hydrostatic pressure on the body of a nautilus, and this pressure is transmitted to the blood that circulates in it. Hence the pressure of the blood in the siphuncle is equal to the blood pressure generated by the heart plus the pressure of the seawater, which increases by one atmosphere for every 10 meters of depth. At 400 meters, which is a typical depth for the nautilus, the inside of the siphuncle would have a pressure of more than 40 atmospheres, or nearly 600 pounds per square inch. In an emptied chamber, therefore, where the pressure is always less than one atmosphere, the calcareous tube surrounding the tissue of the siphuncle must keep the tissue from bursting.

One aspect of our research has been to

CHAMBERED NAUTILUS was photographed at night at a depth of 30 feet near the coral reef surrounding the South Pacific island of New Caledonia. The animal hovers almost effortlessly because its weight is virtually equal to that of the seawater it displaces. A few feet from where the photograph was taken the reef begins to drop off to a depth of several hundred feet. The nautilus may have migrated upward during the night (the time when it is active), or it may have hidden near the top of the reef during the day. Because living nautiluses are seldom seen, little is known about their behavior. The diver at the right is Pierre Laboute, a participant in the authors' research.

examine the mechanism by which the siphuncle removes the cameral liquid that initially fills each chamber. The liquid closely resembles nautilus blood, and also seawater, in that the concentrations of potassium, sodium and chloride ions are virtually the same in all three. It is therefore impossible to say with certainty whether the liquid is a filtrate of blood, a secretion of the siphuncle or of the mantle or perhaps seawater the animal has modified. Nevertheless, if siphuncular cells were to transport some of the ions from the liquid into the blood vessels at the center of the siphuncle, the concentration of ions in the liquid would fall below that in the blood. As a result the water in the liquid would flow by osmosis from the chamber into the blood. It could then be removed from the blood by the animal's kidneys.

The trouble with this hypothesis is that the hydrostatic pressure deep in the ocean favors the tendency for water to pass from the siphuncle back into the chamber. Indeed, even if the siphuncle could reduce the concentration of ions

in the cameral liquid to zero and thereby maximize the osmosis of water out of the chamber, a calculation shows that at a depth of just over 240 meters the hydrostatic pressure becomes great enough to drive water in the opposite direction. Since nautiluses are found with completely empty chambers at depths as great as 600 meters, we were prepared to rule out the simple osmotic mechanism.

We also wanted to rule it out experimentally. Experiments in the open ocean were not practicable. Instead we drilled a small hole through the shell of the nautilus, piercing a nearly empty chamber. We removed any liquid that was present and then added five milliliters of a solution whose concentration of ions was greater than that of nautilus blood. If the movement of water across the siphuncle were governed by simple osmotic forces, water would diffuse from the siphuncular blood into the artificial cameral liquid. Actually we observed the opposite. The siphuncle was capable of removing the liquid from

the chamber even when the concentration of ions in the liquid was almost twice that of the blood.

At this point we turned to the work of Jared M. Diamond. In the 1960's Diamond and his colleagues at Harvard University demonstrated that the gall bladder of the rabbit could transport water against both osmotic and hydrostatic gradients. The water passed from the lumen, or hollow, of the bladder into the blood vessels in the bladder wall. The investigators then showed that the cells forming the wall of the bladder abutted one another at the surface facing the lumen, but that there were spaces between the cells at the places in the wall of the bladder where the cells were closest to blood vessels.

The Harvard workers proposed that the cell membrane bounding each space might include enzymes that transport ions across the membrane. These molecular pumps could build up a high but local concentration of ions in the spaces between the cells. A mathematical anal-

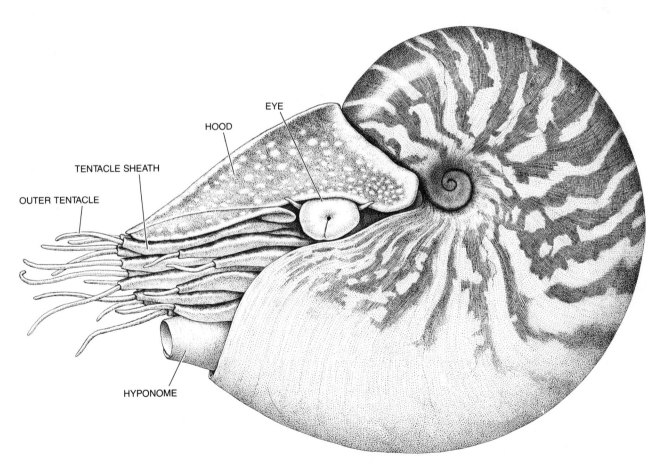

GROSS ANATOMY of *Nautilus macromphalus*, the species studied by the authors, is displayed in two views. The view above shows the features that are visible externally. The head of the animal protrudes from the shell. It has more than 90 tentacles that are arrayed in two whorls on each side of the mouth. The eyes are prominent but

the vision they provide is not acute. The head is protected by a tough fleshy hood. The view on page 60 shows the internal structure of both the shell and the living animal. The outermost living part is the mantle, a tissue whose secretions enlarge the shell. The posterior mantle secretes the succession of septa, or walls, that divide the shell

ysis showed that under these conditions water would diffuse from the lumen of the bladder into the cells forming the wall of the bladder and from there into the regions of high ion concentration in the spaces between the cells. If the ion pumps were in the cell membrane along the blind ends of the intercellular spaces, the water that entered the spaces would flow toward the open ends of the spaces against either osmotic or hydrostatic pressure. Ultimately it would enter the blood vessels.

Because this hypothesis involves osmosis into small regions in a cell layer it is called a local-osmosis model. The significance of the model with regard to the nautilus is that it could account for the ability of the siphuncle to transport cameral liquid against either osmotic gradients (such as we produced in the laboratory) or the hydrostatic gradients an animal faces deep in the ocean. The question is whether the siphuncle has a cellular geometry consistent with local osmosis.

Evidently it does. In 1966 Eric J. Denton and John B. Gilpin-Brown of the Marine Biological Association of the United Kingdom at Plymouth published drawings of the siphuncular tissue that bore a striking resemblance to Diamond's low-power electron micrographs of gall-bladder tissue. Denton and Gilpin-Brown later suggested that local osmosis might function in the siphuncle. In our own examination of siphuncular tissue we too have noted that segments of siphuncle taken from chambers that were emptying or had already emptied have prominent intercellular spaces. Our electron micrographs suggest to us, however, that the large intercellular spaces may serve merely as drainage channels for fluid from a much finer set of canals leading into the larger spaces. Our reason for proposing that the cameral liquid is transported first into the finer set of canals is that they are lined (in the cytoplasm surrounding the canals) by numerous mitochondria, the intracellular organelles supplying energy to a cell. We hypothesize that in this case the mitochondria provide energy for the transport of ions or some other solute.

Tissue taken from the part of a siphuncle that traverses a chamber whose septum is still being built lacks both large and small intercellular spaces. On the other hand, the part of the siphuncle in chambers that have been emptied maintains its network of spaces. Hence the siphuncle appears to be capable of bailing out those chambers, which might otherwise be filled by water forced inward by the substantial hydrostatic pressure at depths greater than 240 meters.

In the emptied chambers the liquid has been replaced by a gas that is essentially air from which most of the oxygen has been removed and into which some extra carbon dioxide has been secreted, presumably by the siphuncular cells. Denton and Gilpin-Brown found that when a chamber is being emptied, the gas pressure inside it can be as low as .1 atmosphere. This led them to speculate that the gas is not transported into the chamber by a process requiring en-

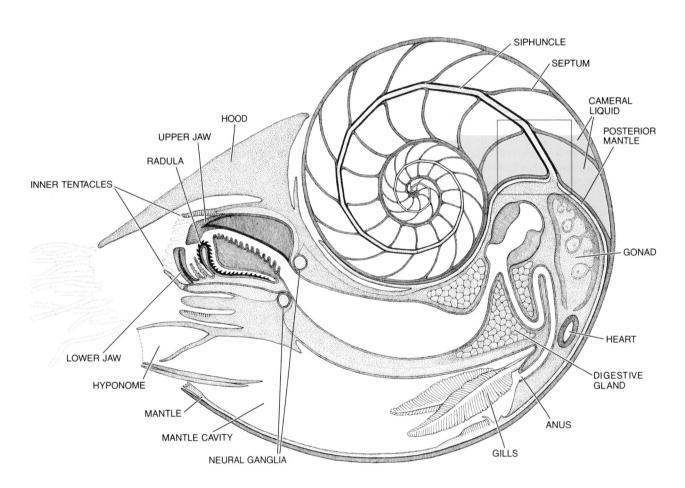

into chambers. The siphuncle, a strand of living tissue encased in a calcareous tube, extends from the posterior mantle in a spiral that takes it through every chamber. The animal's two pairs of gills are deployed in a cavity between the mantle and the viscera. The cavity connects with the hyponome, a flexible funnel through which water is expelled in a jet for locomotion. When a nautilus of the species *N. macromphalus* is fully grown, it is several inches across and weighs nearly a kilogram. At that stage the animal's shell has 30 or more chambers. Colored rectangle in the view above marks the part of the shell that appears in the illustration at the top of page 61.

ergy. Perhaps it simply leaves the siphuncular blood by diffusion.

Because the seawater in which the nautilus swims is in chemical equilibrium with the atmospheric gases at the surface of the ocean, and since nautilus blood is in chemical equilibrium with seawater, it is not surprising that the pressures of nitrogen and argon in the chambers filled with gas are the same as the pressures of those gases in sea-level air. In any case the gas does not contribute to the buoyancy of the animal. Actually it slightly decreases the buoyancy, because the gas after all has weight whereas a vacuum weighs nothing. The point of putting gas in a balloon is to establish the balloon's size and shape. Those functions are served in the nautilus by the rigid shell itself.

A nautilus weighs from one gram to five grams in seawater, regardless of its size. Hence the animal maintains its neutral buoyancy as it grows. It does so by keeping two rates in balance: the rate at which its growing shell and soft parts add weight and the rate at which it empties the shell's chambers of their liquid. To understand how these rates are related Desmond H. Collins of the Royal Ontario Museum, G. E. G. Westermann of McMaster University, Arthur W. Martin, Jr., of the University of Washington and we, in a variety of studies, have considered various aspects of shell growth in the nautilus.

X-ray images of a growing nautilus show that the growth involves a sequence of processes. First the part of the mantle investing the forward part of the body at the open end of the shell secretes new shell material. The open end thus grows both longer and wider, thereby enlarging the body chamber for the growing soft parts of the animal. When the open end of the shell has got long enough, the animal moves forward in the shell, leaving a volume of cameral liquid behind. The mantle forms a seal between the animal and the walls of the body chamber. A thin layer of a slimy substance may complete the seal.

Next the posterior mantle secretes a new calcareous septum, starting at the edges of the shell and progressing inward toward the centrally located siphuncle. At the same time the siphuncle secretes its own calcareous covering. The cellular geometry of the siphuncle in the newly formed chamber is not yet of the type we hypothesize is appropriate for the emptying of a chamber by local osmosis. At this stage, however, the newly forming septum is not yet fully thickened, and the liquid in the newly formed chamber is acting as a brace against the pressure of the surrounding seawater, which is transmitted to the septum through the body of the nautilus. If the liquid were removed be-

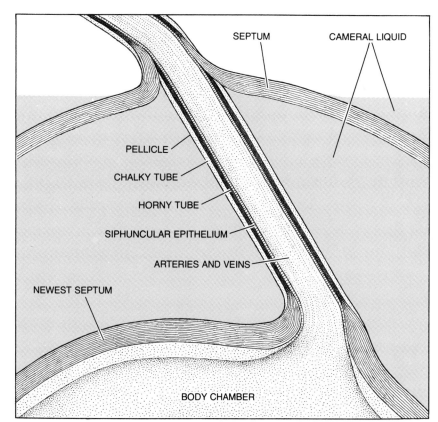

NEWEST CHAMBER in the shell is the one that lies behind the living animal and whose floor is formed by the newest septum. At first the chamber is filled with a watery fluid called the cameral liquid, but the liquid is slowly removed. Hence as the animal grows it adds buoyancy to counter the increasing weight both of its living parts and of its shell. At the time the newest chamber begins to empty, the next-newest chamber still contains a few milliliters of liquid.

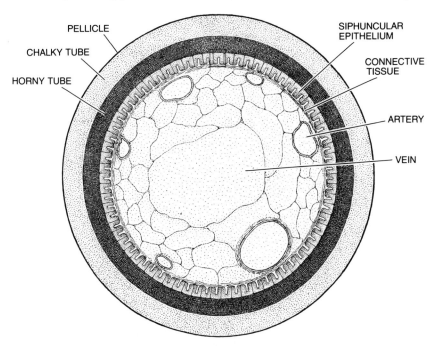

SIPHUNCLE, the organ that spirals through the chambers of the shell and removes the liquid from them, is shown in cross section. The outer layers of the organ are porous and calcareous. They encase living tissue: an epithelial layer (that is, a layer of closely packed cells) inside of which is a core consisting of arteries and veins. The pressure of the blood in the siphuncle corresponds to the pressure of the sea around the animal. In an emptied chamber, therefore, the outer layers of the siphuncle must keep the inner tissues from exploding. The rectangle in color shows the part of the siphuncle that appears in the illustration at the top of page 62.

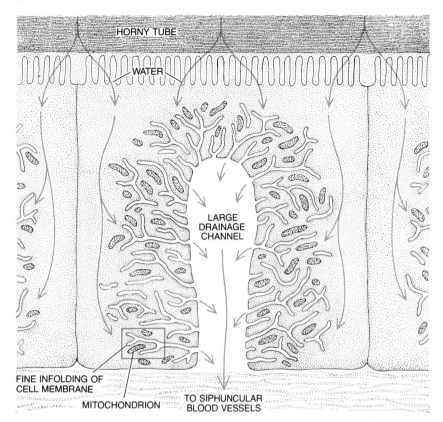

EMPTYING OF A CHAMBER requires that water molecules pass through the porous outer layers of the siphuncle and then through the cells of the siphuncular epithelium. The membrane of each such epithelial cell forms a network of fine infoldings. The infoldings communicate in turn with a larger extracellular space that may serve as a drainage channel, allowing the water to enter the blood vessels at the center of the siphuncle. The water can later be removed from the blood by the kidneys of the animal, which then excrete the water into the mantle cavity. The rectangle in color shows the part of an epithelial cell that appears in the illustration below.

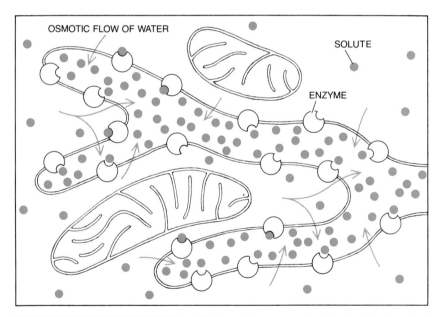

DETAILS OF THE MECHANISM by which the authors propose that the chamber is emptied appear in this schematic drawing of the network of fine infoldings in the membrane of a cell of the siphuncular epithelium. The emptying begins when substances dissolved in the cameral liquid (*colored dots*) diffuse into the siphuncle. Enzymes on the membrane of the siphuncular cells then pump the solute into the infoldings. The energy for the pumping is supplied by mitochondria, the intracellular organelles that make energy available to cells. When the concentration of the solute in the network of infoldings is sufficiently great, water enters (*arrows*) by osmosis in spite of the pressure deep in the ocean, which tends to drive water back out of the siphuncle.

fore the septum were sufficiently thick, the new chamber would collapse. Moreover, the siphuncle would explode from the pressure of the blood inside it if the cameral liquid were absent while the siphuncle lacked its calcareous sheath.

When the septum is complete, the structure of the siphuncle has become the one we think is appropriate for local osmosis, and the chamber begins to empty. When the chamber is full, it contains as much as 30 milliliters, or about an ounce, of liquid. Meanwhile the next-newest chamber, and sometimes the one or two before that, contains a few milliliters of liquid. The older chambers are empty. In chambers where the level of the liquid has fallen below the siphuncle the cameral liquid can nonetheless be removed. It can reach the siphuncle by capillary action along the inner surface of the shell.

The removal of liquid from the newest chamber and from the adjacent chambers adds buoyancy to the nautilus at a rate that counters the decrease in buoyancy due to the addition of weight to both the shell and the animal's tissues. This relation is apparent when the length of the body chamber is graphed against the volume of cameral liquid in the chamber just behind it. Immediately after the chamber is formed the body chamber is relatively short and the volume of cameral liquid is maximal. As the body chamber lengthens, the volume of cameral liquid decreases. The cycle begins whenever a new chamber is formed.

The nautilus does not, however, grow indefinitely large. Like many animals it reaches a certain final size, coincident with the attainment of sexual maturity. In the nautilus the approach of maturity is marked by a reduction in the spacing between the final two or three septa. The reduction is quite variable from one animal to another, and it may reflect an ability of the nautilus to make final adjustments to its buoyancy. In any case the crowding of septa can be seen not only in the nautilus but also in the remains of many fossil chambered cephalopods. It follows that many or even all chambered cephalopods had definite limits in size and were not ever-growing.

When the final septum is complete, the nautilus enters on the final processes that act to trim its buoyancy. The open end of the shell again is enlarged, and the body of the animal fills the space. Meanwhile the last of the cameral liquid is being removed from the newest of the chambers. Growth has now ended. The animal with its shell is several inches across. The shell has 30 or more compartments. The animal is thought to be at least three years old. The enlarging of the shell and the secretion of septa in it has led up to this point: the attainment of neutral buoyancy in the fully grown nautilus. The buoyancy will remain neutral if the siphuncle continues to bail out the shell against the hydrostatic pressure of the ocean. How long the animal lives is not known.

We have not yet learned whether there is a biological feedback mechanism in the nautilus that keeps the growth of the shell and the emptying of the chambers in phase. If there is, we will find, for example, that the animal is inhibited from starting to make a new chamber when we prevent the previous chamber from emptying. We have already noted that when the level of the cameral liquid in the newest chamber falls below the level of the siphuncle, the secretion of a new septum begins.

Moreover, we have shown that the nautilus can compensate for artificial changes in its buoyancy. In our earliest experiments we had been confronted with the problem of rates of emptying so low that they could barely be

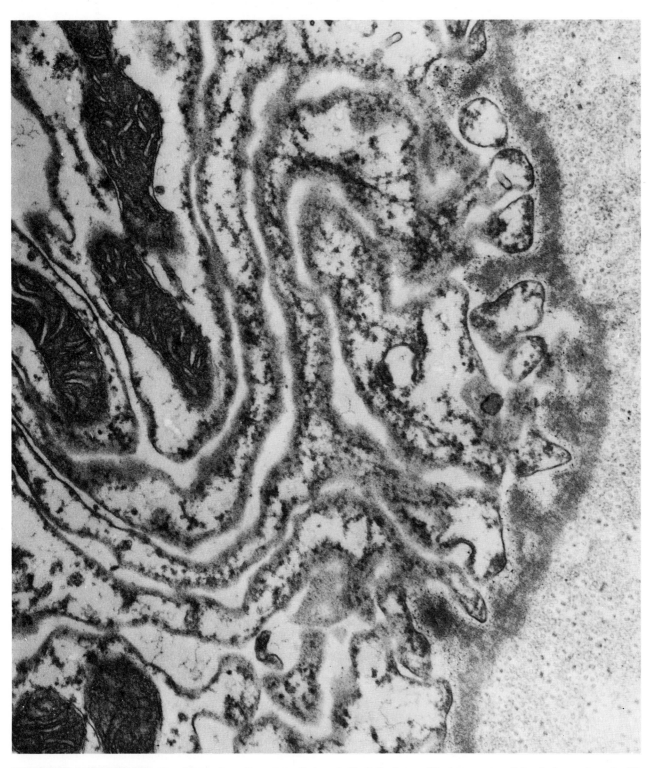

ELECTRON MICROGRAPH of a siphuncular cell confirms that mitochondria congregate near the fine infoldings. The mitochondria are the darkest objects; five appear in whole or in part in the left half of the image. The larger extracellular drainage channel with which the infoldings communicate runs from top to bottom at the far right. The magnification of the micrograph is 40,000 diameters.

a

b

c

d

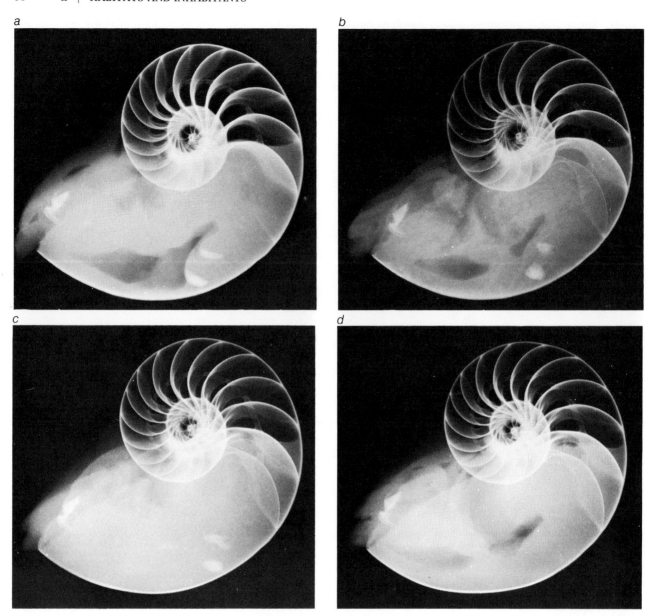

SECRETION OF A SEPTUM and the emptying of chambers in the shell can be followed in this sequence of four X-ray images of a living nautilus. In *a* the level of the cameral liquid in the newest chamber has fallen below the base of the siphuncle in that chamber. A much smaller amount of liquid remains in the chamber above it. All the other chambers are empty. In *b* (made three weeks after *a*) a new septum is faintly apparent. The chamber it defines is filled with cameral liquid. Meanwhile the level of the liquid in what had been the newest chamber has continued to fall. Although the liquid in that chamber no longer makes contact with the siphuncle directly, it can reach the siphuncle by wicking along the inner surface of the shell. In *c* (made two weeks after *b*) the newest septum is nearly complete. In *d* (made two weeks after *c* and seven weeks after *a*) the newest septum is complete. Two dark regions at the upper left of the newly formed chamber are gas bubbles. Their presence shows that the newest chamber is now being emptied. Presumably the newest septum has become strong enough to withstand the pressure of the ocean without the hydrostatic brace the cameral liquid provides. In each image the living nautilus occupies the body chamber at the bottom of the shell. Two whitish spots at the right of the body in *a* are kidney stones. They have disappeared in *d*. The white lines at the left in each image are the jaws of the animal. The dark structures are part of the mantle cavity.

measured. We therefore gave our animals an "incentive" to empty the experimental chamber (the chamber in which we made measurements) by adding liquid to a second, empty chamber, thereby increasing the animal's weight in seawater. In all the animals made heavy by flooding a chamber we observed a much more rapid emptying of the experimental chamber. When the same animals were later made abnormally buoyant by emptying the flooded chamber, we observed a depression in the emptying rate.

The most rapid emptying we observed in the laboratory was less than a milliliter per day, a value consistent with the estimate that the animal needs at least a month to empty a chamber.

Since in all our observations of the nautilus in shallow water we have never noted an animal adding liquid to a previously emptied chamber, we conclude that it has only the option of increasing or decreasing the rate of chamber emptying (and perhaps the rate of shell formation). Why then does the animal

maintain its living siphuncle back to the very first chamber? Its life would be more efficient if it had developed instead a mechanism for permanently making each chamber waterproof after it has been emptied. Perhaps the part of the siphuncle that extends to the early chambers is the vestige of an organ that could flood and empty chambers for the ancestral chambered cephalopods. Those creatures, now extinct, would then have been far more mobile than the ones into which they evolved.

7

The Head of the Sperm Whale

by Malcolm R. Clarke
January 1979

It can represent a quarter of the animal's length and a third of its total weight. The oil-filled spermaceti organ housed within it may keep the whale neutrally buoyant during dives

Among the great whales the sperm whale is most clearly recognizable in having a head that seems disproportionately large. The sperm whale's head can make up more than a third of the animal's total weight (50 tons for the average adult male) and more than a quarter of its total length (an average of 60 feet). There is good reason for this apparent disproportion. The sperm whale's skull accounts for perhaps 12 percent of the weight of the head. The other 88 percent consists mainly of a peculiar anatomical feature located in the whale's snout above the upper jaw: the spermaceti organ. The organ is a complex mass of muscle and oil-filled connective tissue. The oil is what gave the sperm whale its name; in a large male the organ may hold four tons of spermaceti oil.

Such a mighty organ clearly must have a very important function in the life of the sperm whale. The purpose of the great "case," as whalers often call it, has long been a subject of speculation. Even the structure of the spermaceti organ was scarcely known until the past decade or so. Before then guesses about its function could rest only on generalities and on what was known about similar but far smaller organs in a few other toothed whales. Today enough is known about the organ to make it possible to suggest its main purpose: to enable the whale to remain neutrally buoyant when it is submerged.

Why were the anatomical details of the organ not known much earlier? Part of the answer is that pioneers in the field gathered their data by dissecting sperm-whale fetuses. It happens, however, that both the skull and the snout of the fetus differ greatly in proportion from the same components in the head of the adult whale. Another part of the answer is that it is no small task to dissect an adult sperm whale. Without such commercial whaling facilities and tools as flensing platforms, steam winches, five-meter steam-driven saws and razor-sharp flensing knives the dissection would be quite impossible. Even with

such aids and the cooperation of the commercial whalers this is not a trivial task.

On the flensing platform half-ton masses of fat, flesh and fiber are cut, rolled and pulled off this way and that until the observer's sense of orientation is easily lost. It is no wonder that even in this decade a book has been published that presents the principal structures of the sperm whale's head upside down. Only after watching and photographing many sperm whales being cut up on the flensing platform could one hope to clarify the anatomy of the adult whale's snout, and that is the task I undertook. My work was greatly advanced by the capture of one small adult whose head the commercial whalers cut for me in a series of transverse sections 20 centimeters thick. I was then able to photograph the sections and measure them in detail.

Many functions for the spermaceti organ have been suggested. Among them are that it is a means of generating and focusing sounds (and for receiving them), a means of moving air between the whale's lungs and its nostrils when the whale is deep underwater, a means of opening and closing the whale's long nasal passages, a means of absorbing nitrogen from the bloodstream in the course of deep dives and even a means of attack and defense.

Certainly this complex organ may have more than one function. Variations in the chemical composition of the spermaceti oil in different parts of the organ suggest that one function may well be the channeling (or focusing) of sound generated by the whale. At the same time it is hard to accept the suggestion of one worker that such focusing can concentrate sound intensely enough to stun the squids that are the sperm whale's main prey. More than one of the other suggestions, however, seem reasonable, if difficult to demonstrate.

Although an understanding of the structure and proportions of the head of the adult sperm whale is a necessary preliminary to studying the function of the

spermaceti organ, one must also know something about the biology and behavior of sperm whales, in particular how these toothed whales differ from other whales that do not have a large snout. Moreover, the large quantity of oil contained in the spermaceti organ suggests that the oil itself must serve some special function; one needs to know something of the physical properties of the oil to understand what role it plays in the life of the sperm whale.

Let us begin by reviewing the biology and behavior of the sperm whale in search of clues to the function of the animal's snout. One immediately apparent fact is that sperm whales are unusual, although not unique, on the roster of toothed whales in being distributed worldwide. They have been hunted for centuries in every ocean from as far as 60 degrees north latitude to 40 degrees south. Since the development of modern whaling techniques the sperm whale has also been pursued to the high latitudes of the Antarctic. Together with its baleen-whale cousin the right whale, the sperm whale was of particular importance when the whaling industry depended on hand harpooning from open boats: unlike many other whales, these two stayed afloat after being killed.

The food of the sperm whale consists almost entirely of that Concorde of the snail family, the squid. To catch these speedy denizens of the depths the sperm whale dives deep and stays down for long periods. A large sperm whale is typically submerged for 50 minutes of an hour-long diving cycle. During a 10-minute surface interval between dives the whale will take 50 to 60 breaths of air. Dives longer than 50 minutes have been observed; the record dive is somewhat more than 80 minutes.

Sperm whales not only stay submerged for long periods but also frequently go deeper than 1,000 meters. One sperm whale, watched by sonar, was observed to go below 2,250 meters; even deeper descents can be inferred from the presence of bottom-dwelling

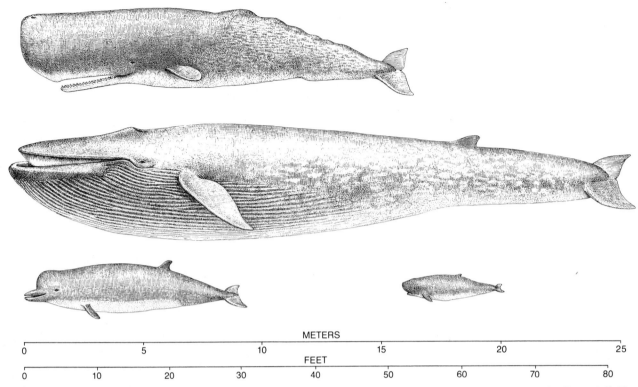

METERS

| 0 | 5 | 10 | 15 | 20 | 25 |

FEET

| 0 | 10 | 20 | 30 | 40 | 50 | 60 | 70 | 80 |

THREE TOOTHED WHALES that possess spermaceti organs are compared in this drawing with the largest of all whales. They are the sperm whale *Physeter catodon* (*top*), the pygmy sperm whale *Kogia* (*bottom right*) and the bottle-nosed whale *Hyperoodon* (*bottom left*). The largest whale is the blue, or sulfur-bottom, whale *Balaenoptera musculus*, one of the suborder Mysticeti: the toothless baleen whales.

sharks in the stomach of a sperm whale captured in an area where the bottom depth was greater than 3,000 meters.

Why should a sperm whale need to dive so deep? Squids are found at all depths in every ocean. They are a principal food of many other air-breathing vertebrates: seabirds as well as such marine mammals as seals and lesser toothed whales, including porpoises. Perhaps the reason for the sperm whale's diving behavior is that the deeper a squid-eater can descend, the farther it can outdistance its competitors and the greater the stocks of squids within its reach are. Certainly the sperm whale can catch deep-living squids that are beyond the reach of seabirds, seals or porpoises. The whale can also reach bottom in the zone where the sea floor drops steeply to the abyssal plain from the edge of the continental shelf. Here on the continental slope, at depths of between 200 and 3,000 meters, many squids lay their eggs and can be found in large numbers and dense concentrations. Few air-breathing animals other than the sperm whale can hope to reach such easy pickings.

A peculiarity of the sperm whale's diving cycle is that the animal frequently surfaces within a few hundred yards of the point where it began its dive. The reason is not that its underwater time has been occupied by a slow descent and ascent. The whales are known to descend at a speed of about four knots

(120 meters per minute) and to ascend at a speed of about five knots. Hence a round trip to a depth of 1,000 meters would not take more than 15 minutes. As we have seen, the duration of a deep dive is some three times longer. Both the duration of the dive and the fact that the sperm whale's place of emergence is close to its place of submergence suggest that when the whale comes to the bottom of its dive, it must sometimes lie almost still in the water.

Many, although by no means all, squids are fast swimmers over short distances. One may therefore wonder why a sperm whale would lie still at depth instead of actively pursuing its swift prey. The whale's lower jaw is long and narrow. It is the jaw of a snapper; even when it is open, it offers little water resistance. Perhaps the sperm whale's hunting strategy relies less on active pursuit and more on silent hovering followed by a quick pounce into a passing shoal of squids. Little or no daylight penetrates these hunting depths, but most of the squids on which the whale preys are luminescent. In its efforts to catch these speedy invertebrates a still, silent whale may well have the advantage over a swimming one.

A whale can lie still underwater only by being very nearly neutrally buoyant, that is, by having the same density as the surrounding water. In the older classes of man-made submersibles

that displaced from 1,600 to 2,000 tons buoyancy had to be controlled within 40 liters of water (between two and three hundred-thousandths of the displacement weight) to enable the vessel to lie still in the water for listening purposes. Many water-dwelling animals can also achieve buoyancy within very fine limits and so can lie still at the depth where they live. For example, some fishes counter the sinking effect of those body tissues that are denser than water by storing low-density fats; other fishes manage neutral buoyancy by means of an air-filled swim bladder. Many squids do the same by replacing the dense sodium ions in their body with less dense ammonia ions.

In its biological and behavioral aspects, then, the sperm whale exhibits some unusual features. Among them are migratory behavior that takes the males from equatorial waters to polar ones, deep dives of long duration, the ability to lie still when submerged and the property of floating when dead. No one feature is unique to the sperm whale, but only the sperm whale is known to combine them all.

Of these four features two concern buoyancy and the other two—great range both horizontally and vertically—involve changes in ambient water conditions that are accompanied by changes in buoyancy. Although the sperm whale could carry the right amount of fat or air to be neutrally buoyant at a particular geographical location and depth, if the whale depended on such a static system for the control of buoyancy, the different water densities at other depths and geographical locations would push it either down or up.

Is this the clue we are seeking? Is the spermaceti organ a device for controlling buoyancy over a wide range of conditions? If it is, the organ must be able to vary its density. How could it do so? Only one substance is present in the organ in large quantities that is also able to undergo a substantial change in density: the spermaceti oil itself. In point of fact this oil has long been known to have properties different from those of other whale oils. When the liquid oil is dipped out of a dead whale's head and exposed to ambient air temperatures, it soon loses its clarity and becomes a soft crystalline solid.

Temperature probes of freshly killed sperm whales show that when the whales are resting on the surface, the temperature of the spermaceti oil is 33 degrees Celsius (90 degrees Fahrenheit). The oil begins to crystallize, or congeal into a solid, when its temperature drops below 31 degrees C. Unlike the crystallization of water, which is almost instantaneous at the freezing point, the crystallization of spermaceti oil is a gradual process that is not completed until the temperature drops several degrees. When spermaceti oil freezes, it becomes denser and therefore occupies less volume. And occupying less volume, it displaces less of the ambient seawater and is less buoyant.

If the temperature of the spermaceti oil could be varied, then, the changes in density that accompany changes in temperature might be enough to let the sperm whale control its own buoyancy. Are such changes in temperature physiologically possible? The problem is of course one of the loss and gain of heat. With that in mind let us consider the anatomy of the spermaceti organ.

The tissues that house the spermaceti oil in the whale's snout have a dense network of capillaries supplied with blood by large arteries that enter the snout at the rear. The circulation of the arterial blood is therefore the principal means of conveying heat to the oil. The same circulation at the capillary level is also the main distributor of heat within each block of spermaceti tissue; when the tissue is cooled locally, the movement of blood through the capillaries helps to spread the cooling effect.

In addition the larger arteries and veins in the snout of the sperm whale lie side by side; this countercurrent system assists the cooling of the spermaceti tissue by the exchange of heat between the warmer incoming arterial blood and the cooler outgoing venous blood. The arteries that supply blood to the snout are

surrounded, particularly at the point where they pass through the skull, by a dense network of veins carrying cooler blood away. The heat exchange can maintain a sharp difference in temperature between the snout (where the blood is normally below 34 degrees C.) and the rest of the whale's body (where the blood is normally above 37 degrees C.).

The sperm whale can lose heat through the surface tissues of the snout, either by "passive" conduction through the blubber and skin or by "active" heat transport: circulation of the blood to the papillae, minute fingerlike structures within the skin. As we shall see, passive conduction by itself is much too slow a process for the achievement of neutral

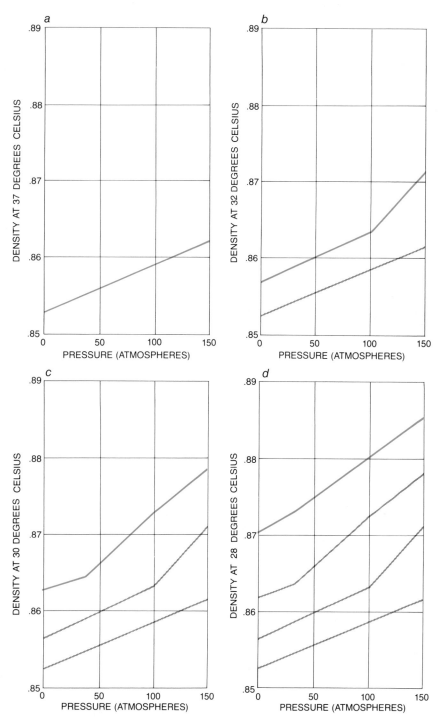

SPERMACETI OIL varies in density according to its temperature. These graphs show the density of the oil at progressively lower temperatures: at 37 degrees Celsius (a), at 32 degrees (b, with the 37-degree reading included for comparison), at 30 degrees (c) and at 28 degrees (d). Pressure also affects the density of the oil. Each 10 meters of additional depth adds one atmosphere (14.7 pounds) of pressure; thus at a depth of 500 meters the pressure is 50 atmospheres. The slopes of the four graphs trace the increasing density of the oil up to the equivalent of 1,500 meters. The findings are from a study made at the British National Physical Laboratory.

buoyancy in the time available. Active heat transport is a more promising possibility.

The anatomy of the sperm whale's head suggests still another means of heat loss: the sperm whale's nasal passages. Asymmetrical nasal passages are characteristic of toothed whales, but no other toothed whale has passages like those of the sperm whale. The simpler of its nasal passages, the left one, runs back from the cavity under the animal's single blowhole, curves to pass the left side of the spermaceti organ and enters the skull just in front of the brain case. The sperm whale breathes through this muscular tube, which can be expanded until it becomes circular in cross section.

The right nasal passage is totally different. Although it too begins in the cavity under the blowhole, it at first runs forward and is a tubular passage of small diameter. Then it widens and flattens out as it passes downward to form a broad, flat chamber, the vestibular sac, located at the front of the snout just under the blubber. A broad horizontal opening in the rear wall of the sac, often called the "monkey's mouth," gives entry to the continuation of the nasal passage, a wide, flattened tube that runs back the length of the snout to a point just in front of the brain case. There the wide passage narrows as it enters the skull and meets the left nasal passage in a common cavity. Just forward of this narrowing the right nasal passage opens upward to connect with a second sac, the nasofrontal sac, located above the central part of the skull crest.

The right nasal passage is thus often more than a meter wide; its roundabout course through the snout is five meters or more long. Not only does it pass through the core of the spermaceti organ but also its two sacs cover the front and rear ends of the organ.

The interior of the right nasal passage is lined with a delicate layer of black tissue. Under the black layer are two layers of white tissue, first an elastic layer and then a fibrous one; the three layers together form a wall that is between .6 millimeter and one millimeter thick. Spermaceti tissue lies directly in contact with the wall of the nasal passage on all sides, and capillaries from the spermaceti tissue enter the white elastic layer.

The intimate relation between the sperm whale's right nasal passage and the spermaceti organ is such that if seawater enters the nasal passage, the spermaceti oil will be markedly cooled. One can calculate the rate of oil cooling for any given temperature of seawater (on the basis of heat exchange between the blood in the capillaries and the cooler seawater) from the total area and the thickness of the nasal-passage wall. One can go on to calculate the rate of heat loss via the skin of the whale's snout, via the right nasal passage or via both heat-exchange areas combined and so determine the time required for the whale to reach neutral buoyancy over a range of selected depths in either an Antarctic environment or an equatorial one.

In either environment, with minimum values in the calculation, the right nasal passage proves to be a slower heat exchanger than the skin of the whale's snout. The calculated difference between the two surfaces, however, is probably not significant; the nasal passage is elastic, and its wall could be expanded to present a greater heat-exchange area than the minimum I have calculated.

My calculations show that a 30-ton

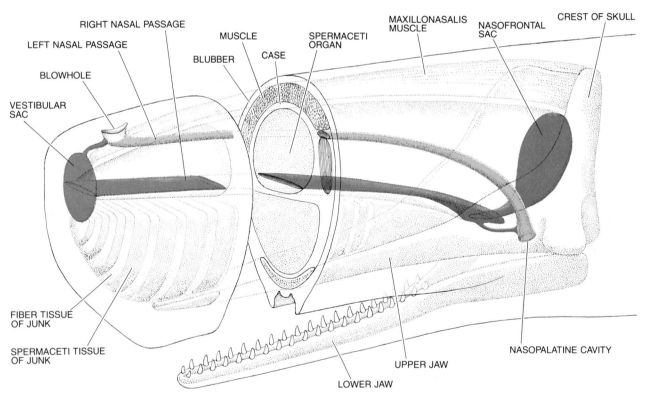

NASAL PASSAGES of the sperm whale are not only asymmetrical but also intimately associated with the spermaceti organ. The left nasal passage (*dark gray*) is the simpler of the two. Beginning in a cavity under the blowhole, the passage curves to pass along the left side of the spermaceti case and terminates in the nasopalatine cavity of the skull. The right nasal passage (*color*) also begins under the blowhole; it then runs forward and widens to form a vertically oriented sac, the vestibular sac, at the forward end of the spermaceti case. A narrow horizontal opening at the back of the sac provides for the continuation of the passage rearward through the interior of the spermaceti organ until the passage approaches the scooplike crest of the whale's skull. There the nasal passage gives rise to a second vertically oriented sac, the nasofrontal sac, located at the back of the spermaceti case, before narrowing to enter the skull cavity that is shared in common with the left nasal passage. The complex route followed by the right nasal passage is some five meters long, and the passage itself is in places more than a meter wide. Over most of its length vessels from the network of capillaries in the spermaceti-oil tissue extend into the wall of the nasal passage. A major muscle of the snout, the maxillonasalis, runs from the crest of the skull to the forward half of the spermaceti case.

sperm whale, exploiting both heat-exchange areas simultaneously, could adjust to neutral buoyancy in less time than it normally takes to swim to a depth of 500 meters. In dives from 200 to 1,000 meters deep, exchanging heat via the snout skin alone, the whale would reach neutral buoyancy within five minutes of attaining the desired depth. If both heat-exchange areas come into play, the interval would be shortened to three minutes.

In the Antarctic, because lower water temperatures mean greater buoyancy at and near the surface, the calculated time needed to reach neutral buoyancy at a depth of 100 meters, with either heat exchanger in play alone, comes to about 20 percent of the total submersion time of even a prolonged dive. At first this may appear to be too great an investment of time to make the attainment of neutral buoyancy worthwhile. There is, however, a counterbalancing factor: swimming in the cold surface water of Antarctic latitudes, the whale may maintain the temperature of its spermaceti oil at a level lower than the 33 degrees C. characteristic of equatorial waters. If it does, cooling to neutral buoyancy would be quicker. In addition, when the whale was submerged during a dive, its resorting repeatedly to the nasal heat exchanger could hasten the achievement of neutral buoyancy. My calculations show that below 200 meters a filling of the nasal passage with seawater twice rather than once would be more than enough to exchange the required quantity of heat.

Just how can seawater be drawn into the whale's right nasal passage? Surrounding the outer wall enclosing the spermaceti tissues are large muscles. They run from the front half of the "case" to attachments on the crest of the skull. Their contraction would suffice to raise the front end of the case, thereby lifting the upper half of the nasal passage. The same contraction would also open the front end of the nasal passage and draw water in from the cavity under the blowhole. The relaxation of the muscles would expel the water.

How far might the seawater travel along the right nasal passage? There are additional muscle fibers in the floor of the passage and still other fibers that run forward within the spermaceti tissue from the front wall of the nasofrontal sac. The contraction of these fibers would hold down the bottom half of the right nasal passage, ensuring that the water would travel at least as far as the nasofrontal sac.

One cannot exclude the possibility that seawater also reaches the spermaceti organ by a different route: the left nasal passage. Once drawn into this shorter tube from the cavity under the blowhole, the water could be pumped to the cavity where both nasal passages meet;

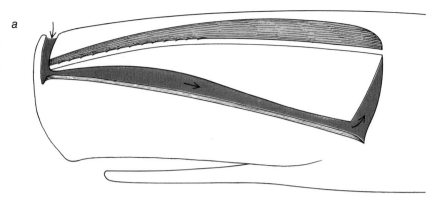

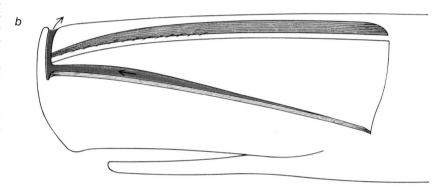

WATER INTAKE through the right nasal passage (*a*) may be accomplished by contraction of the major snout muscle (*dark gray*). Muscle action would widen the passage (*color*) so that water could enter; small muscles within the spermaceti tissue would aid the process. Relaxation of these muscles would allow the passage to narrow again (*b*), thereby expelling the water.

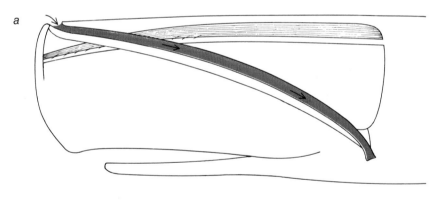

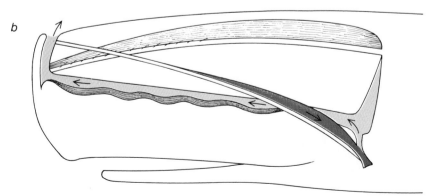

ALTERNATE INTAKE ROUTE is the left nasal passage (*color*). Contraction of a minor muscle, the nasal-plug muscle, would widen the left passage and draw water into the nasopalatine cavity (*a*). Thereafter (*b*) the action of muscle fibers in the floor of the right nasal passage (*gray*) could pump water forward through the spermaceti organ and out through the blowhole.

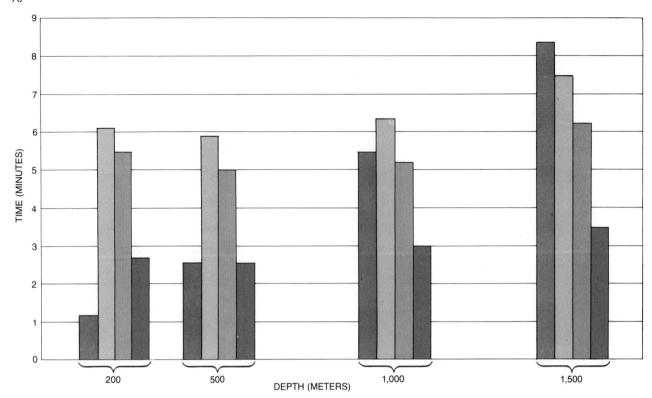

THREE ALTERNATIVES for cooling spermaceti oil when a whale is diving at the Equator take different lengths of time, as is apparent in this graph. The calculations are for a 30-ton whale diving with full lungs. The first bar (*gray*) in each of the four sets shows the time needed by the whale to swim to the indicated depth at a speed of five knots. When the dive is to 200 meters, even maximum heat exchange via the right nasal passage and the skin of the snout combined (*color*) must continue for more than an additional minute before neu- tral buoyancy is achieved. Heat exchange via the right nasal passage (*light gray*) or via the snout skin (*light color*) would have to continue even longer. With deeper dives the trend favors achievement of neu- tral buoyancy during the time the whale is swimming to depth. In a dive to 500 meters the time required to achieve neutral buoyancy through maximum heat exchange is equal to the time of descent. In a dive to 1,500 meters even the least efficient form of heat exchange, via the nasal passage alone, achieves neutral buoyancy during descent.

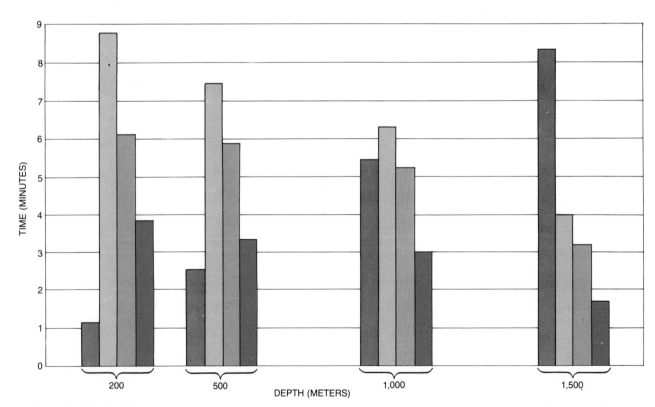

SAME ALTERNATIVES are calculated for a dive in polar waters; again the whale weighs 30 tons and is swimming down with full lungs at a speed of five knots. Maximum exchange of heat (*color*) almost achieves neutral buoyancy during a descent to 500 meters. Heat ex- change via the snout skin alone (*light color*) achieves neutral buoyan- cy during a descent to 1,000 meters. Even the least efficient form of heat exchange, via the nasal passage only (*light gray*), achieves neu- tral buoyancy four minutes before the whale reaches 1,500 meters.

the water could then enter the rear of the right nasal passage and be moved forward by the contraction of the muscle fibers in the passage floor. If that is the case, one wonders why a relatively major role has been assigned to these few fibers and a relatively minor one to the very large muscles attached to the case. Nevertheless, on the two occasions when dissection revealed the presence of seawater in a sperm whale's right nasal passage, I also found water in the left passage, and so this alternate route cannot be ruled out.

To find out what changes in the density of spermaceti oil accompany changes in temperature and pressure I asked members of the staff of the British National Physical Laboratory to conduct density measurements over a range of temperatures and pressures. They found that whereas changes in density are mainly a function of temperature, pressure also has a marked effect, particularly near the onset of freezing.

At this point in my study I faced two related questions. First, what changes in buoyancy would a sperm whale encounter over its range of diving depths in both a polar and an equatorial environment, supposing it had no means of controlling buoyancy? Second, is the quantity of spermaceti oil in the head of an adult sperm whale large enough to achieve neutral buoyancy by means of density changes over the same range of depths and environments?

Turning to the first of these questions, the factors that influence sperm-whale buoyancy include the density of the seawater, the effect of pressure on both the liquid and the solid components of the whale's tissues and the effect of pressure on the volume of the whale's lungs. Variations in these factors all follow established physical laws, and therefore their effect on a whale of any particular size can be calculated. The calculations, of course, depend on reasonably accurate estimates both of the proportions of liquid, solid and gas in the diving whale and of the buoyancy of the whale when it is floating on the surface. With such a large animal it is not easy to make precise estimates of this kind. Those used in our study have therefore included minimum and maximum values as well as mean values. The mean values are used in this discussion, but the conclusions remain valid even when the extreme values are used instead.

In moving on the surface between equatorial and polar latitudes a sperm whale experiences a change in surface-water temperature of as much as 26 degrees C. Moreover, if the whale dives to a depth of 1,000 meters at the Equator, the change in temperature may be as much as 23 degrees C. These changes in temperature are accompanied by changes in the density of the seawater. Density is mainly dependent on temperature,

but salinity and pressure also enter into the calculations. For example, if a whale that is neutrally buoyant on the surface at the Equator were moved to an Antarctic latitude such as 55 degrees south, the increased density of the surface water would give the whale an increased buoyancy amounting to a little more than .3 percent of its body weight. The increase in water density the whale would encounter during a deep dive, whether at the Equator or at 55 degrees south latitude, would also increase its buoyancy. Because of the greater change in water density with increasing depth at the Equator, however, the increase in the whale's buoyancy there would be about five times more than it would be in the Antarctic: at a depth of 2,000 meters it would amount to almost .5 percent of the animal's body weight. So much for water density. What about water pressure? Its effect on the water content of the whale's tissues proves to be very slight, as is its effect on the oil content. As for the solid components of the whale's tissues, they are almost incompressible, and so their mass gives the whale added lift as the pressure increases. At a depth of 1,000 meters the lift would equal about .1 percent of body weight. Lumping the effects of water pressure on all three tissue components, the sperm whale at a depth of 1,000 meters would have extra lift equivalent to .07 percent of its body weight.

Water pressure also acts on the gas in the whale's lungs during a dive and so affects the whale's buoyancy, particularly at depths of up to 200 meters. When a freshly killed sperm whale is floating on the surface in the Tropics, the volume of its body above the water roughly equals the volume of air in its lungs. By the same token, when a sperm whale exhales on the surface, it comes close to attaining neutral buoyancy. Sperm whales may sometimes exhale before diving; under these circumstances the reduction in buoyancy should facilitate the dive. If instead the whale dives with its lungs full, by the time it reaches 200 meters the effect of the water pressure on the gas in the lungs will have brought about the same reduction in buoyancy as an exhalation at the surface would have. Below 200 meters there is no great difference between diving with full lungs and diving with empty ones.

If one considers the combined effects of seawater density and pressure on the whale, one can calculate the lift or the downthrust the sperm whale will experience at various depths and in various environments. For example, the whale is very positively buoyant on the surface at the Equator. If the animal dives with full lungs, the rapid decrease in lung volume causes a downthrust equal to about .1 percent of its body weight at 100 meters. By the time the whale reaches 200 meters the change in the water density

will more than counterbalance this first effect, and the animal will have positive lift. The lift will increase as the whale descends until at 2,000 meters it will equal about .2 percent of the body weight. In Antarctic latitudes, since the cold surface water is denser than the warm surface water at the Equator, the sperm whale begins its dive with greater buoyancy and so does not experience much downthrust in the early part of its descent.

Calculations indicate that throughout its geographical range the sperm whale must adjust its overall density by an amount equivalent to .2 percent of its body weight in order to achieve neutral buoyancy below 200 meters. I have suggested that such an adjustment could be made by a "freezing" of the whale's spermaceti oil and have also described the heat exchangers available to cool the stored oil to a solid. But is the freezing effect adequate to the task? Among the known facts that are useful in answering this question are the findings of the National Physical Laboratory with respect to the density of spermaceti oil at various temperatures and pressures. It is also known that a 30-ton whale will have nearly 2.5 tons of spermaceti oil in its head and that at the surface in equatorial latitudes the oil will be at a temperature of 33 degrees C.

Starting with these facts, we can go on to calculate the spermaceti-oil temperature required to counterbalance the whale's natural lift at successive depths. The calculation shows that below 200 meters the temperature of the oil need be lowered by only a few degrees (and never below 29 degrees C.) to attain the required densities. As an example, if a sperm whale in Antarctic latitudes exhales before diving, the temperature of the spermaceti oil need not fall below 30 degrees in order to counterbalance the whale's natural lift. When the temperature and the density are plotted together, the slopes of the temperature lines and the slopes of the lines indicating the required densities are strikingly similar.

With both the anatomical and the physical data in hand one can consider how the dissipation and the regaining of heat might be timed during the sperm whale's diving cycle. To begin with an example in the equatorial environment, one finds that the heat generated during a diving cycle probably cannot be dissipated during the 10 minutes of each cycle that the whale spends at or near the surface. The reason is that at any depth shallower than 100 meters the difference between the water temperature and the subcutaneous temperature of the whale is less than 2.6 degrees C. Even if the active loss of heat at the surface by vasodilation (that is, an expansion of the blood vessels in the skin) is combined with passive heat loss at depth by conduction, only part of the

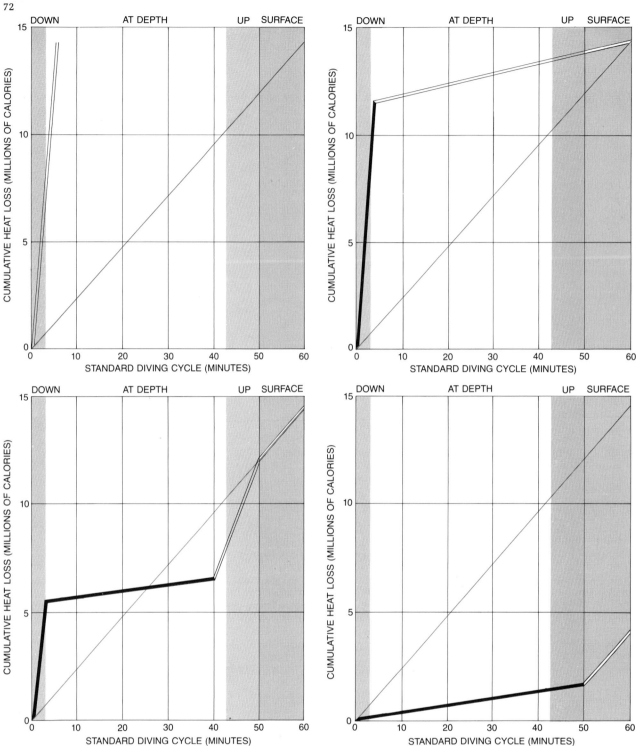

HEAT BUDGET OF A DIVING CYCLE calls for a loss of nearly 15 million calories over a 60-minute period that is divided into some three minutes for descent to a depth of 500 meters (*colored band at left*), some 40 minutes of stalking prey at that depth, some seven minutes for ascent to the surface (*second colored band*) and 10 minutes of rest at the surface (*gray band*). The loss of heat is required to balance the body-heat gain resulting from exertion during the diving cycle. Each of the four graphs plots a steady rate of heat gain (*colored diagonal line*) and the rate of heat loss that is achieved by one of four different strategies of heat exchange. Calculations are for a 30-ton whale diving to 500 meters with full lungs at the Equator. The open line in graph *A* represents the result of a dive with continuous heat exchange via body skin and snout skin until the spermaceti oil is cool enough to bring about neutral buoyancy. The heat budget would remain badly out of balance during most of the diving cycle. The line in graph *B* represents the result of a dive with heat exchange via the right nasal passage, body skin and snout skin (*solid portion of line*) until neutral buoyancy is achieved. Heat loss is then reduced by vaso- constriction. Such a strategy would produce a budget that starts badly out of balance but improves steadily thereafter. The line in graph *C* represents the result of a dive with heat exchange via the right nasal passage only during the descent; after that, heat loss is reduced by vasoconstriction until some 40 minutes have passed (*open portion cf line*). Heat is then transferred from the body to the spermaceti oil until the oil temperature rises to 33 degrees C. Thereafter vasodilation allows further heat exchange via skin surfaces. This strategy would produce alternating net heat loss and net heat gain until the whale returned to the surface. Strategies involving loss of heat via the snout skin, alone or in combination with the right nasal passage, would give rise to similar curves. One of these three strategies is the likeliest to be adopted by the whale. The line in graph *D* represents a dive with heat exchange minimized by vasoconstriction until the whale returns to the surface. The slight heat loss following vasodilation at the surface in equatorial waters (*open portion of line*), together with limited heat exchange during the dive, would produce an out-of-balance budget; the whale could not dive again within 10 minutes.

heat the whale generates during the active 50 minutes of its diving cycle will be removed. Heat cannot be retained by means of vasoconstriction during the submerged period of the dive because some of the sperm whale's time in deep cold water must be spent in losing heat and achieving neutral buoyancy.

Looking further, we find that the greater part of the heat the whale generates during the active part of the diving cycle may either be lost slowly, by a controlled partial vasodilation that keeps pace with heat production, or be lost rapidly by more extensive vasodilation during one or more short periods in the course of the diving cycle. Brief but extensive vasodilation would not call for prolonged control and would at the same time allow the whale to make use of variations in spermaceti-oil density for buoyancy control. For example, as the whale descends, the oil would be cooled by heat exchange not only via the skin but also probably via the right nasal passage. When the whale has attained neutral buoyancy, soon after reaching the desired depth, any further active cooling of its oil would be halted by means of vasoconstriction. Minor adjustments in buoyancy thereafter could be made by admitting small amounts of seawater into the right nasal passage; that would eliminate any dependence on skin vasodilation for heat loss.

During most of the 50 minutes of the diving cycle spent at depth the body of the sperm whale, and its muscular mass in particular, rises in temperature. The whale's muscles need not increase in temperature more than two degrees C. in order to store all the heat needed to bring the solidified spermaceti oil back to its normal fluid temperature. Just before the whale begins its ascent to the surface the supply of blood to its snout would be increased, carrying heat from its warmer body to the spermaceti organ. Hence at the same time that the whale's body temperature is falling to its normal level the thawing spermaceti oil is rapidly rising to its normal 33 degrees C. Heating the oil of course decreases its density and increases its volume, and so the whale's buoyancy shifts from neutral to positive; this would help to lift the whale toward the surface even if the animal were exhausted. Any muscle-generated heat in excess of that needed to reheat the oil could now be lost by vasodilation.

Using as an example a 30-ton sperm whale diving to a depth of 500 meters in equatorial waters, one can now consider five alternatives with respect to the most efficient means of heat loss. As the first alternative let it be assumed that vasoconstriction does not take place over the entire skin surface until the whale has descended to the selected depth and achieved neutral buoyancy. During the descent heat will be lost over the entire skin surface; the total heat loss will ap-

proach the total heat gain as a result of muscular activity during the diving cycle. When the passive loss of heat through the whale's blubber is added to this active heat loss, the total loss possibly exceeds the total gain. This first alternative must therefore be eliminated.

Let it be assumed as the second alternative that vasoconstriction takes place over the whale's entire skin surface as soon as the descent begins, so that almost all the cooling of the spermaceti oil is a result of heat exchange in the right nasal passage alone. Under those circumstances the time required to reach neutral buoyancy will be greater than if heat exchange were simultaneously in progress through the skin. Nevertheless, neutral buoyancy is achieved in about six minutes (12 percent of the 50-minute dive time). The dilation of the blood vessels of the spermaceti tissue before the whale begins to ascend would then shunt heat from its body muscles to its snout until the oil temperature returns to its normal 33 degrees C. Any excess body heat generated during the dive could then be lost by vasodilation.

In the third alternative the assumption is zonal vasoconstriction. The blood vessels of the whale's body skin constrict immediately after submergence but the blood vessels of the snout skin remain dilated until the spermaceti oil is cooled to the point of neutral buoyancy at depth. Such a sequence would lead to neutral buoyancy even more rapidly than the sequence outlined in the second alternative. It requires the further assumption, however, that the sperm whale can dilate and constrict the blood vessels of its snout skin and body skin independently.

In the fourth alternative heat exchange through the right nasal passage and heat exchange through the whale's entire skin proceed simultaneously. This sequence leads to neutral buoyancy faster than either the second or the third alternative. Moreover, the amount of the body heat generated during the dive that remains to be lost to the seawater is smaller.

The fifth and most efficient alternative involves heat exchange via the right nasal passage and the snout skin only.

All except the first of these five alternatives are plausible. The fourth alternative has an advantage over the third and fifth in not postulating a capacity for zonal vasoconstriction and vasodilation. Before the third and fifth alternatives are dismissed on those grounds, however, it should be noted that such zonal dilation is similar to the human facial blush. It seems at least conceivable that sperm whales too can blush.

As evidence for heat exchange via the right nasal passage I have found seawater in this passage on two occasions. Therefore the second, fourth and fifth alternatives are all likely ones, and the fifth alternative, involving both the right

nasal passage and the snout skin, is the most likely of all. In this connection, in several recently killed sperm whales the spermaceti tissue closest to the surface of the snout was found to be slightly warmer than more central tissue. This suggests that an efficient cooling system, namely heat exchange via the right nasal passage, had been operating in the center of the snout.

Changes in the density of the spermaceti oil would of course affect the sperm whale's center of gravity. As the oil increases in density during the whale's descent the center of gravity will shift forward; as the density decreases just before or during the ascent the center of gravity will shift backward. These changes would actually be advantageous during both descent and ascent because the whale often moves almost vertically down and up. The shift means, however, that the whale's fore-and-aft trim, to borrow a nautical expression, will not be the same when the whale is submerged as it is when the whale is on the surface. The main factor in the whale's trim while it is on the surface is not the spermaceti ballast but rather the considerable buoyancy provided by its air-filled lungs, so that under these circumstances the role of the spermaceti is trivial. Perhaps, however, the forward shift of the center of gravity at depth actually gives the whale its best trim during the time in the diving cycle when the submerged animal is neutrally buoyant and either lying still or swimming.

Sperm whales less than about 11 meters long have less spermaceti oil in relation to their size than larger ones. The smaller whales are also comparatively shallow divers and do not migrate to such high polar latitudes as the larger whales. As a result they encounter smaller variations in water density, and even though their supply of the oil is smaller, it would suffice to compensate for the smaller range of water densities.

For an air-breathing, warm-blooded, long-diving animal control of buoyancy has numerous advantages. At the surface extra buoyancy allows the whale to relax. At depth the animal's hunting success, which is dependent on silence and good hearing, is not jeopardized by the need to keep swimming in order to remain at a particular level. Moreover, if it were necessary for the sperm whale to keep swimming in order to remain at the same level during its 50-minute submergence, it would have to invest a larger amount of energy in its diving cycle.

These are not the only advantages derived from the control of buoyancy. Imagine for a moment a hypothetical sperm whale that was less buoyant on the surface than is actually the case. Over a part of its range of diving depths and in certain geographical locations this hypothetical whale would actually

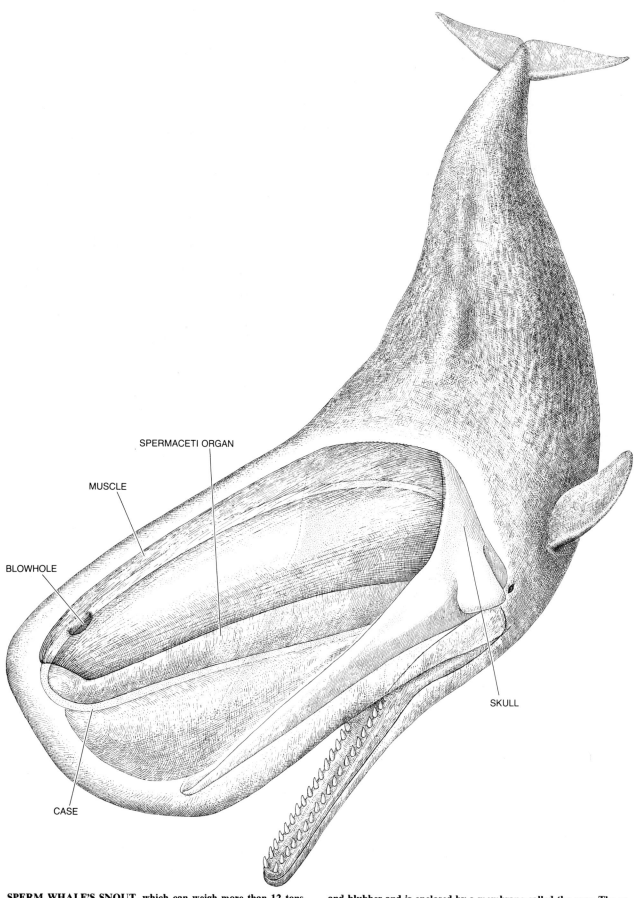

SPERMACETI ORGAN

MUSCLE

BLOWHOLE

SKULL

CASE

SPERM WHALE'S SNOUT, which can weigh more than 12 tons, is largely occupied by the spermaceti organ, a complex oblong mass of oil-filled connective tissue that is surrounded by layers of muscle and blubber and is enclosed by a membrane called the case. The organ is cradled in a long conical depression largely formed from the upper jawbones of the sperm whale's elongated skull (*light color*).

SPERM WHALE'S HEAD is peeled of its blubber on the flensing platform of a factory ship in the Antarctic; teeth of the lower jaw are clearly visible. Upper-jaw teeth are usually absent.

depths made possible by oil-density control could hardly be matched by a system that depends on an exhaustible reservoir.

Loss of heat from the spermaceti oil as a means of buoyancy control imparts still another distinct advantage. In the Tropics the water temperature at the surface may be less than three degrees C. below the subcutaneous temperature of the whale. At depths shallower than 100 meters the body heat generated by a deep dive could not be lost during the usual 10-minute resting time at the surface between dives. Both the heat exchange that is involved in freezing the spermaceti oil during the descent phase of the diving cycle and the body-heat shunt involved in thawing the oil before and during the ascent phase help to overcome this equatorial handicap.

Few whales are known to dive as deep or for as long a time as the sperm whale. Two other toothed whales—the bottle-nosed whale *Hyperoodon* and the pygmy sperm whale *Kogia*—have spermaceti organs. Both are long-duration deep divers. In view of the complicated structure of their snout it seems probable that both also use the spermaceti organ for buoyancy control. This conjecture cannot be proved, however, until much more is known about the dimensions, anatomy and diving habits of these whales.

Some other whales, for example the fin whale *Balaenoptera physalus,* can make rapid excursions to considerable depths. The dives of the fin whale, perhaps to depths of some 300 meters, seldom last longer than 10 minutes; the whale swims continuously and surfaces some distance from its point of submergence. These baleen whales usually swim faster than the three deep-diving toothed whales. It is clear that they do not submerge for long periods of time or in order to feed as the three deep divers do; their prey is the shrimplike krill found in swarms at or near the surface.

As I have noted, many suggestions have been made regarding possible functions of the spermaceti organ other than buoyancy control. None, however, accounts as adequately for the size and structure of the organ. Such a complex anatomical structure, however, could certainly serve more than one function. Several of the other proposed functions are quite compatible with the buoyancy-control hypothesis I have proposed. Nevertheless, so many peculiarities of the sperm whale's snout and its cargo of oil can be explained by the buoyancy-control hypothesis that it is hard not to accept this as the organ's main function. Final proof of the hypothesis, however, must await measurement of the temperature or the density of the oil within the spermaceti organ in the course of a deep dive. This is a difficult and costly task but not an impossible one.

experience downthrust during the diving cycle. Downthrust, of course, is a hazard to any air-breathing animal that makes deep dives of long duration. By not only having extra buoyancy at all depths but also having control over its buoyancy the sperm whale has a fail-safe diving system. Even if the animal became exhausted during a deep dive, it could still pop up to the surface for air. For example, a sperm whale that has exhausted itself, perhaps by short bursts of high-speed swimming after squids, can shunt the heat that is a by-product of its muscular activity into the spermaceti organ and again become positively buoyant. Such an exhausted whale is just as likely to reach the surface again as a fresh whale that can easily swim up.

What about alternative means of buoyancy control? For example, either collapsing the lungs or emitting air while submerged would affect buoyancy. On close consideration neither stratagem proves to be as efficient as control

by means of oil density. The degree of lung collapse depends entirely on water pressure, whereas buoyancy also depends on seawater density, which in turn is dependent on temperature. As diving depths increase, these factors work in opposite directions; only at two points in a deep dive do the two factors exactly match to yield neutral buoyancy.

As for the emission of air, that would be an ineffective means of controlling buoyancy over much of the whale's diving range. For example, below 600 meters even a small emission of air would greatly reduce the volume of air remaining in the whale's lungs for the return to the surface; in some geographical areas a whale that had emitted air at depth would experience considerable downthrust as it swam through the thermocline, the abrupt transition in water temperature that is encountered between 200 and 100 meters. Furthermore, once air is emitted it is lost. The many minor adjustments of buoyancy at different

8

The Role of Wax
in Oceanic Food Chains

by Andrew A. Benson and Richard F. Lee
March 1975

*Copepods, ubiquitous marine crustaceans, store energy
in the form of wax. It now seems that half
of the organic matter synthesized by the sea's
primary producers is converted for a time into waxes*

Waxes are known to have important functions in a wide variety of plants and animals, but until recently they were not regarded as playing a major role in the energy economy of the animal world. It was well known that waxes on the surface of leaves and fruits protect them against damage and drying. Honeybees secrete a wax for the building of the honeycomb. The human skin produces sebum, a mixture of waxes, oils and fragments of dead cells that serves to lubricate the skin and make it supple. The head cavities of the sperm whale are filled with a mixture of liquid waxes that is believed to mediate the conduction and focusing of the high-frequency sound the animal uses for communication and echo location.

In the 1960's marine biologists began to find waxes in a wide range of marine animals. The most populous group of fishes, the bristlemouths or cyclothones, are rich in waxes. Appreciable amounts of wax are also found in cod, mullets, croakers and sharks. Mollusks, shrimps, decapods, squids, deep-sea worms, sea anemones, corals and even marine birds accumulate wax. All of this indicated that wax was stored by these animals in order to fill some need, but what it was remained a mystery. Then in 1967 Judd C. Nevenzel of the University of California at Los Angeles isolated a wax from the black deep-sea copepod *Gaussia princeps*. Copepods are small marine crustaceans that are enormously abundant; indeed, in most oceanic areas they are the largest single component of the zooplankton (animal plankton). Further investigation quickly established that many species of copepods synthesize and store waxes for use as a reserve metabolic fuel when food is not available.

Copepods graze on the phytoplankton, the tiny plants that are the primary producers in the food chains of the ocean. These plants, which are algae of various types, build carbohydrates, fats and proteins by photosynthesis, and the copepods convert some of the fat into wax. Many higher forms of marine life in turn feed on the copepods [*see top illustration on page 82*]. The discovery that copepods are the main producers of the waxes in the marine food chain led to the realization that wax is a major medium of energy storage in animals of the sea. It is estimated that on a worldwide basis at least half of all the organic substance synthesized by the phytoplankton is converted for a time into wax.

The order Copepoda includes more than 2,000 species. Most copepods are herbivorous and feed only on the phytoplankton of the surface waters; some are omnivorous and feed on both phytoplankton and zooplankton; some, particularly the deep-sea species, must be carnivorous, feeding only on the zooplankton. Many species migrate vertically as much as 500 meters every day, descending to the depths in daylight and ascending to the surface at night, where they can feed on the phytoplankton that have flourished during their time in the sun.

Copepods store both fat and wax as metabolic fuel. The fats are generally distributed throughout the body and provide the animal with energy for short-term purposes. The waxes are found in more specific locations. The calanoid copepods of temperate and polar waters, for example *Calanus plumchrus*, have a large sac in which wax is stored [*see top illustration on next page*].

Both fats and waxes are lipids; they also belong to the class of compounds known as esters. Fats are called triglycerides because fat molecules consist of three long chains of fatty acid linked to a molecule of glycerol. Wax molecules have a simpler structure, consisting of a long-chain fatty acid linked to a long-chain fatty alcohol [*see bottom illustration on next page*]. A shorthand notation is employed to indicate the length of the chain and the number of double bonds in it. For example, "oleic acid (16 : 1)" refers to a chain that is 16 carbon atoms long and has one double bond between two of the carbons. When the carbon chain in a fatty acid or a fatty alcohol has one double bond or more, it is called unsaturated. The great-

FEMALE COPEPOD carrying a cluster of bright blue eggs is enlarged some 20 diameters in this photograph by Fritz Goro. Copepods are the most abundant zooplankton (animal plankton). This copepod is *Euchaeta japonica*, a species common in the North Pacific. It is omnivorous, feeding both on other zooplankton and on phytoplankton (plant plankton). The large transparent cavity visible within the animal contains a store of both wax and fat. The wax accounts for as much as 50 percent of the dry weight of the adult female. When the female produces eggs, much of the wax and fat is transferred to the eggs to provide food for the larvae. The eggs are carried by the female to protect them from predators. Red color around copepod's mouthparts and blue color of eggs are both due to same pigment: astaxanthin, which gives rise to different colors when associated with different proteins.

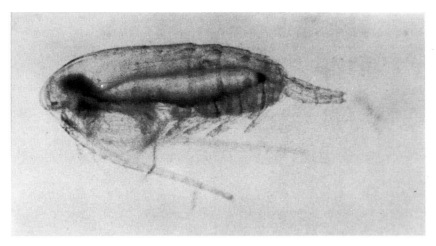

LARGE OIL SAC filled with pure liquid wax is clearly visible in this photograph of a Stage V copepodite of the species *Calanus plumchrus*, the predominant copepod in the Strait of Georgia in British Columbia. On a dry-weight basis 50 percent of the copepodite is wax.

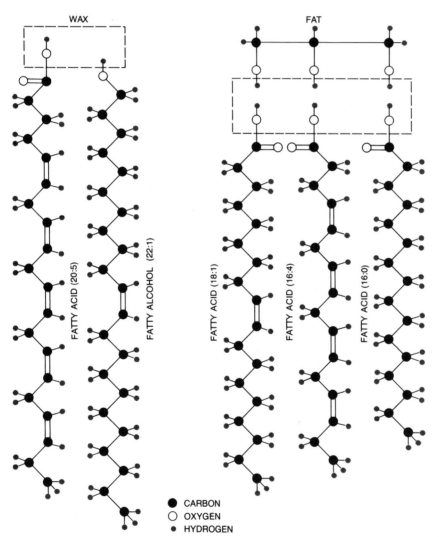

WAXES AND FATS are both esters, substances produced by the reaction of alcohols and acids. Wax is formed from a long-chain alcohol and a single fatty acid. The numbers in parentheses are a shorthand notation for identifying the carbon chain; for example, the numbers 20:5 refer to a chain that has 20 carbon atoms and five double bonds between the carbons. When a wax is formed, one molecule of water is released (*rectangle at left*). Fats are called triglycerides because they consist of three fatty acids linked to the three oxygen atoms of glycerol. Three molecules of water are liberated in the formation of triglyceride.

er the unsaturation in a fat or a wax, that is, the greater the number of double bonds, the more likely it is that the substance will be liquid at low temperatures. The fats synthesized by phytoplankton in cold waters are highly unsaturated, or polyunsaturated, and the waxes synthesized by the copepods that feed in those waters are similarly unsaturated. These waxes remain liquid even when the temperature of the water is close to freezing. In general, organisms adapted to low temperatures store more polyunsaturated lipids than organisms that live in warmer regions. Unsaturation therefore is a biological adaptation to life at low temperatures: it allows a plant or an animal to function well in a cold environment.

The fats and waxes in copepods are much alike in their physical properties. Both have nearly the same density, a similar caloric value per unit volume and similar compressibility. They do, however, differ in their coefficient of thermal expansion. Compared with a droplet of fat, a droplet of wax in a copepod expands more and becomes more buoyant as it is warmed. This change in buoyancy may assist copepods in their daily vertical migration from the cooler water of the depths to the warmer water of the surface.

Many organisms require for their metabolism fatty acids that they cannot synthesize for themselves. Such fatty acids must be obtained from food, and they are termed essential fatty acids. For man the essential fatty acids are linoleic acid (18:2) and linolenic acid (18:3). For copepods these fatty acids are also essential, particularly for the synthesis of docosahexaenoic acid (22:6), a crucial component of the phospholipids in the cell membranes of copepods, fishes and apparently all other marine animals. Herbivorous copepods synthesize docosahexaenoic acid by the addition of two-carbon units to unsaturated fatty acids obtained from phytoplankton. Such highly unsaturated fatty acids oxidize rapidly when they are exposed to air, giving rise to the characteristic odor of rancid fish. The polyunsaturated fatty acids are carefully conserved by the animal. For example, when the copepod *Gaussia princeps* is starved, its fat store is depleted in about two weeks, but there is very little change in its content of wax. Although the fat disappears, its valuable polyunsaturated chains are transferred to phospholipids in the cell membranes, where they cannot be depleted.

In most land animals fat is the only type of fuel that is stored for long-term

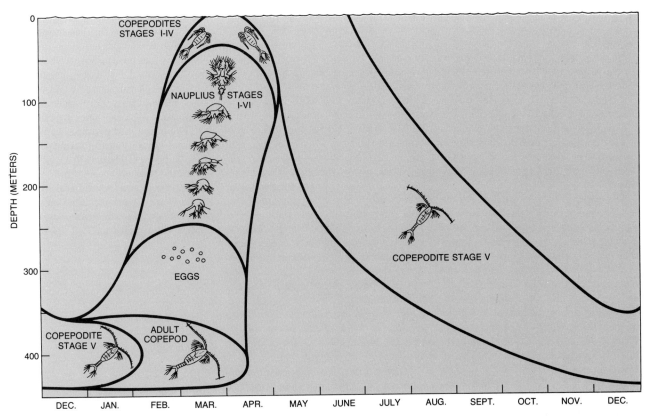

LIFE CYCLE of *Calanus plumchrus* copepods, based on the stud-ies of John Fulton and Robin LeBrasseur of the Fisheries Research Board of Canada, is depicted. In late winter the adult female begins to lay eggs. The eggs rise slowly and hatch in about two and a half days. The newly hatched larvae, called nauplii, live off the wax stored in their eggs. They go through six molting stages and then become copepodites. The copepodites feed on the phytoplank-ton near the surface and go through five molting stages. By the time they have reached the fifth stage copepodites have large stores of wax. They migrate down to the deeper waters and during the winter become adult copepods fully equipped for mating. The adult has no feeding mouthparts and must live off its stored wax.

use. Although copepods also store fat, in many species the major reserve fuel is wax; indeed, as much as 70 percent of the animal's dry weight can be wax. This alternative food-storage system provides the copepod with a food reserve that can be controlled separately from its day-to-day metabolism. The control appears to rely on the relative activity of two en-zymes: triglyceride lipase and wax li-pase. The triglyceride lipase, which cat-alyzes the metabolism of fat, is normally active in the animal at all times. For ex-ample, in freshly caught copepods tri-glyceride-lipase activity can readily be detected, but the wax-lipase activity is virtually zero. It appears that while oth-er food is available the enzyme that catalyzes the metabolism of wax is inac-tive, and that the enzyme is activated only under the stress of starvation. This provides the control that prevents the early depletion of the wax reserve. Thus copepods that accumulate wax do so be-cause they are unable to use the wax as fuel until all their fat reserve is gone.

The utilization of waxes is strongly dependent on the nature of the cope-pod's environment. Working with Jed Hirota and Arthur M. Barnett, we cap-tured 85 species of copepods from vari-ous depths in the subtropical and tem-perate regions of the Pacific and com-pared them with specimens collected under an ice island in the Arctic Ocean. In general larger amounts of wax are found in copepods as the temperature of the water in which the animals live de-creases or as the depth of the habitat in-creases. Copepods in arctic or antarctic waters, where phytoplankton bloom only during two or three months of summer sunlight, store the greatest amount of wax. In warm tropical waters, where the supply of phytoplankton is sparse but relatively constant, copepods in the sur-face water accumulate very little wax. All copepods (and most other animals) from depths below 1,000 meters contain a considerable amount of wax.

In the subtropical waters off San Diego we found that for most copepod spe-cies the total lipid accounted for only 18 percent of the dry weight, and that there was usually only a small amount of wax in the lipid. In copepods taken at depths below 600 meters the total lipid ac-counted for 40 percent of the dry weight, and well over half of the lipid was wax.

We also found that there was a transi-tion zone where about half of the cope-pods we caught had stores of wax and half had stores of fat. The explanation is that the daily vertical migration of sur-face species of copepods can result in mixtures of animals.

John Fulton and Robin LeBrasseur of the Fisheries Research Board of Canada have studied the life patterns of cope-pods in the Strait of Georgia on the Ca-nadian West Coast. One important spe-cies studied was *Calanus plumchrus*, which is a major source of food for young salmon arriving from inland streams and rivers. During the winter the adult copepods remain at a depth of about 400 meters. In late winter and early spring the female begins to lay eggs, which contain droplets of oil and are therefore buoyant. They rise slowly from the depths where they were re-leased, hatching in about two and a half days and swimming to the surface. The newly hatched larva, called the nauplius, is barely able to swim and is not yet equipped to capture phytoplankton. Fortunately it has a supply of wax that provides much of the energy it needs. After the nauplius has grown and molted, it must find phytoplankton of a size it can eat if it is to survive through its six molting stages. It then becomes a copepodite, which also molts as it grows. At about this time nutrient-rich water from the spring flood of the Fraser River gives rise to a great bloom of phytoplankton in the Strait of Georgia. The copepodites feed on the phytoplankton as they go through their molting stages. By the end of the bloom the copepodites have accumulated large stores of pure wax.

The copepodites migrate down to the deeper waters during the late summer months. At that time they feed very little, if at all. During the midwinter months they become adult copepods. The adults of *Calanus plumchrus* cannot feed and must rely on the wax formed the preceding spring for all their energy requirements during the last seven months of their life cycle. Most of the wax reserve of the female goes into the production of the eggs she discharges. The oil stored in the eggs surprisingly consists mostly of fat; it contains very little wax. Before the female breeds nearly 50 percent of her dry weight is in the form of lipid, and 90 percent of the lipid is wax. After the female has discharged between 300 and 800 eggs her lipid content has dropped to 4 percent, with only traces of wax remaining.

Arctic copepods, for example *Calanus hyperboreus*, go through an almost identical life cycle. The wax is stored during the months of July, August and September, when the phytoplankton bloom in this area. For the other nine months the copepods must survive on their reserve of lipid. The females shed eggs in January and February, and presumably they do not live much longer.

The surface-feeding copepod *Calanus helgolandicus* produces lipid amounting to 20 percent of its dry weight before its final molt into the adult stage. Although the lipid of the female of this species is 50 percent wax, the eggs she discharges are 60 percent fat and slightly denser than water. In two days nauplii having neither a mouth nor an anus emerge and valiantly struggle toward the surface, molting twice in two days before they are at last capable of feeding on small phytoplankton.

The eggs of the omnivorous deep-water copepod *Euchaeta japonica* are 58 percent wax. The female carries the eggs, which have a characteristic bright blue color [*see illustration on page 76*], to protect them from predation by other zooplankton. The early naupliar stages do not feed and must rely on the stored

wax in the egg for their energy requirements as they grow and molt. In some instances there is enough wax in the egg to carry *E. japonica* through the entire six naupliar stages to the first copepodite stage.

Copepods are a colorful group of creatures: they can be orange, red, black, blue or white. White copepods, however, always have streaks or spots of orange pigment. Whatever the color, the pigment is always the same. It is astaxanthin, which copepods produce from the yellow xanthophylls of phytoplankton. This carotenoid pigment is also seen in shrimps, lobsters and salmon. When the molecule of the pigment is associated with certain protein molecules, it assumes different configurations and gives rise to different colors. Astaxanthin can be extracted from the blue eggs of copepods or from the animal's body with an organic solvent such as chloroform, whereupon the color of the solution is always a reddish orange.

Much has been learned about the metabolism and the nutritional requirements of copepods by raising them in the laboratory. G.-A. Paffenhöfer and M. M. Mullin of the Scripps Institution of Oceanography succeeded in raising *Calanus helgolandicus* from eggs to adulthood. It was the culture of the newly hatched nauplii that required the greatest care. Paffenhöfer kept them alive (with the help of an alarm clock) by feeding them every few hours with algae. He then succeeded in bringing the nauplii through their six molting stages and the copepodites through their five molts to the adult stage.

The composition of copepod wax is altered not only by starvation but also by changes in the concentration of the food available and by the species of phytoplankton that are ingested. When laboratory-reared copepods were fed the alga *Skeletonema costatum*, which is rich in fats but contains negligible amounts of wax, the fatty acids of the wax synthesized by the copepods resembled the fatty acids found in the fat of the alga. The copepod fatty alcohols, however, did not resemble the algal fatty acids.

Since algae do not contain any long-chain alcohols, these compounds must be synthesized by the copepod and then combined with fatty acids to make wax. The wax of copepods that live in surface waters is rich in 20- and 22-carbon alcohols, whereas the wax in copepods that live in the depths is characterized by 16-carbon alcohols, which remain liquid at low temperatures.

The synthesis of wax, mediated by enzymes isolated from copepods, has been

accomplished by John R. Sargent of the Institute of Marine Biochemistry in Aberdeen in Scotland. He found that fatty acids that had been radioactively labeled were converted into fatty alcohols and waxes by cell extracts and cell suspensions of copepods supplemented with coenzyme A, adenosine triphosphate (ATP) and reduced pyridine nucleotides. In our laboratory at the Scripps Institution, R. Barry Holtz and E. D. Marquez separated the organelles from various copepod cells and found that the biosynthesis of wax is accomplished on plasma membranes, possibly membranes associated with the oil sac.

CANDACIA AETHIOPICA	COPEPOD
EUPHAUSIA SUPERBA	KRILL
HYPERIIDAE SP.	HYPERIID AMPHIPOD
GENNADAS SP.	GENNADAS
STOMIAS ATRIVENTER	DRAGON FISH
GIGANTOCYPRIS AGAZZI	OSTRACOD
ARGYROPELECUS SP.	HATCHET FISH
OEGOPSIDAE SP.	SQUID
HIRONDELLA SP.	GAMMARID AMPHIPOD
CYCLOTHONE SP.	BRISTLEMOUTH FISH
EUPHAUSIA CRYSTALLOROPHIAS	KRILL
PHYSETER MACROCEPHALUS	SPERM WHALE
LATIMERIA CHALUMNAE	COELACANTH
EUKROHNIA	ARROWWORM
GNATHOPHAUSIA INGENS	MYSID SHRIMP
GONIASTREA RETIFORMIS	BRAIN CORAL
CHONDYLACTIS GIGANTEA	SEA ANEMONE
ALCIOPIDAE SP.	POLYCHETE WORM
LAMPANYCTUS SP.	LANTERN FISH
CALANUS PLUMCHRUS	COPEPOD

AMOUNT OF WAX in marine animals varies considerably. All animals that live at

Man-made oil slicks at sea have become commonplace in modern times, but it is not generally realized that there are substantial biological oil slicks as well. On April 7, 1971, crewmen aboard the U.S. Coast Guard cutter *Minnetonka* saw a red oil slick surrounding the ship, which was on station between Hawaii and Japan. The slick extended in all directions to the limits of visibility (six nautical miles), and it was seen repeatedly over the next two weeks as the ship proceeded to other stations in a grid 10 miles by 60 miles. A sample of the slick was collected and immediately frozen. It contained the remnants of copepods,

identified as a species of *Calanus*. Chromatographic analysis of the lipid fraction revealed that it consisted of 82 percent wax, 6 percent triglyceride, 4 percent cholesterol, 2 percent phospholipid, 2 percent astaxanthin and 3 percent hydrocarbons. The principal fatty alcohols of the oil slick were found to be similar to those found in the wax of *Calanus plumchrus* and other *Calanus* species. The presence of highly unsaturated fatty acids, which oxidize rapidly, suggested that the massive copepod kill had been fairly recent. It is believed the copepods had died in the Japan Current to the west, and that the remains of the

animals, including their wax, had been borne to the location of the Coast Guard cutter by the easterly drift of the North Pacific Current.

W ax has also been observed on the surface in Bute Inlet, a deep fjord 120 miles northwest of Vancouver. In cold winters large quantities of wax accumulate on the shores of the inlet. The source of the wax was a complete mystery, and it was hypothesized that it came from a submarine deposit of "fossil wax." The predominant copepod in the inlet is *C. plumchrus*. Comparison of the length of the carbon chains in the inlet

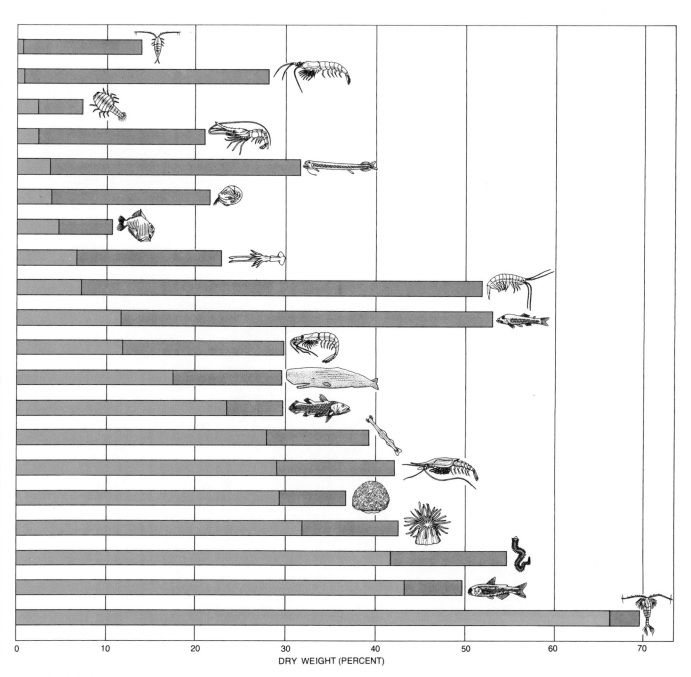

depths below 1,000 meters contain some wax. **Each bar represents the total lipid content (wax and fat) of the animal as a percent** of its dry weight. The colored portion of the bar represents the average amount of lipid in each species that is in the form of wax.

DRY WEIGHT (PERCENT)

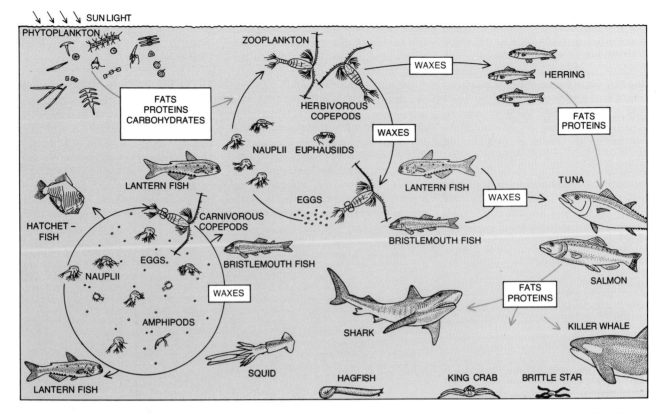

ROLE OF WAX IN THE FOOD CHAIN in Temperate Zone oceans is depicted. The food cycle begins near the surface, where the phytoplankton utilize sunlight to manufacture carbohydrates, proteins and fats. Copepods feed on the phytoplankton and store large amounts of wax. The copepods in turn are the primary food of fishes such as anchovies, sardines and herrings. These fishes have an enzyme that enables them to metabolize wax, but they do not store it. Larger fishes such as tuna prey on the smaller, copepod-eating ones. The deep-water food chain differs in that the copepods are carnivorous: they feed only on other zooplankton. Considerable amounts of wax are stored by these copepods and by deep-water fishes such as bristlemouth fish, lantern fish and hatchetfish.

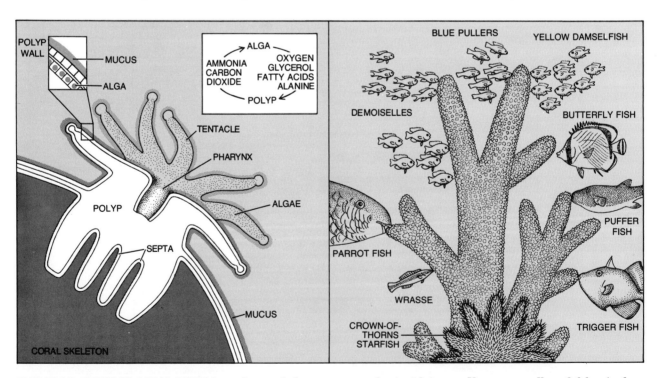

WAX FOOD CHAIN IN CORAL REEFS has only recently been discovered. Corals are part animal (polyps) and part plant (algae). The polyp and the alga have a symbiotic relationship (*illustration at left*). The alga utilizes the carbon dioxide and ammonia waste products of the polyp, and the polyp obtains oxygen, glycerol, glucose, amino acids and fatty acids from the alga. The coral exudes a mucus that is rich in wax. Numerous small coral fishes (such as damselfish, blue pullers, puffer fish, butterfly fish and small wrasses) feed on the mucus. Larger fishes such as trigger fish and parrot fish break down and devour the coral as well as the mucus. The digestive system of the crown-of-thorns starfish, *Acanthaster planci*, has been found to be adapted to digest the wax in the coral.

wax and in the copepod wax suggests that the copepods were probably the source. The fatty acids and fatty alcohols of the inlet wax, however, are much more saturated than those of the copepod wax. That is probably the result of the degradation of the more unsaturated components by bacteria. The evidence points to some catastrophic destruction of a large population of copepods at some time in the past and the consequent sedimentation of their wax.

In the deep waters of the ocean wax is stored by many marine animals other than copepods. Indeed, most of the animals that live at depths below 1,000 meters contain wax. The coelacanth and the castor-oil fish are examples of fishes with a large wax content. The hagfish and some sharks are unique in having wax in the lipoproteins of their blood serum. (In man and most other vertebrates the same purpose is served by cholesterol esters.) Squids, chaetognaths, amphipods, ostracods and decapods also store wax in varying amounts [*see illustration on pages 80 and 81*].

POLYPS ON BRAIN CORAL retreat into their holes when they are touched. The surface of the coral is covered with mucus that is secreted by the polyps. This photograph was made by Fritz Goro on the Great Barrier Reef on the west coast of Australia.

TWO CROWN-OF-THORNS STARFISH feeding on brain coral in a bay on the island of Guam are shown in this photograph made by Thomas F. Goreau. The heavy-spined starfish, which can grow to a size of two feet or more, everts its stomach over the living coral and digests it. When the starfish has finished feeding, only the limestone skeleton of the coral remains. The wax stored in the coral animal is an important part of the diet of the starfish. In the late 1960's the crown-of-thorns starfish began to multiply rapidly throughout the South Pacific, threatening the complete destruction of the coral reefs that surround and protect many of the islands in the region.

Copepods are the principal food at some stage in the life of many fish species that are economically important to man. Anchovies, sardines and herrings prey on copepods. These fishes appear to produce a bile-activated wax lipase, the enzyme that catalyzes the metabolism of wax. The site of wax-lipase production is in the diffuse pancreas of the pyloric cecum, a mass of tubular appendages leading from the junction of the stomach and the small intestine. Neither waxes nor fatty alcohols have been found in the blood of these fishes. The fry of the chum salmon devour so many copepods that they become engorged, and their fecal pellets are so rich in oil that they float. The fry do not completely metabolize the wax, but even so no wax passes into their circulating blood.

As we have noted, copepods that live in the warm surface waters of the Tropics apparently do not store wax. We have recently discovered, however, that waxes play an important role in the food chain of tropical coral reefs.

Corals are half animal and half plant. On the surface of their intricate limestone skeletons reside colonies of polyps (the animal), and within the polyps live dinoflagellate algae (the plant). The algae, with their green chlorophyll c and orange carotenoid pigments, give corals their olive brown color.

When the algae of coral are separated from the polyps and suspended in seawater, they behave like other algae, conserving the sugar and other substances they manufacture by photosynthesis. (They are related to the algae of "red tides" in temperate and polar waters.) When the alga is in contact with the proteins of the polyp, however, its cell wall becomes leaky. Glycerol, glucose and the amino acid alanine leak out of the alga and into the polyp. The polyp also "steals" fatty acids from the membranes of its algae.

Leonard Muscatine of the University of California at Los Angeles has shown that the algae avidly scavenge the waste products of the polyps, carbon dioxide and ammonia, which they utilize for the synthesis of amino acids. The algae in turn evolve oxygen for the polyps as a by-product of photosynthesis. By keeping these products within a closed, symbiotic system, both the alga and the polyp have evolved a solution to the problems of survival in the nutrient-poor tropical seas.

Corals are not generally subject to direct predation by fishes, although some tropical fishes such as the parrot fishes and puffers gnaw at the surface of boulderlike corals and bite off the tips of the more delicate species. Corals do, however, exude a slimy mucus that is ingested by many small fishes. When we brushed mucus from corals into the water, hordes of tiny, colorful coral fishes avidly gathered to eat it. These fish can also be found sucking or nibbling at branched corals. Other fishes, such as the butterfly fishes with their elongated snouts, eat mucus out of the grooves of brain coral.

It had been thought that the fishes were only feeding on tiny zooplankton enmeshed in the coral mucus. The number of zooplankton in the mucus is small, however, and the mucus would be a poor food if it were not for another ingredient in it. We found that coral mucus contains a wax that is similar to the wax of copepods. It should not have been surprising to find that coral mucus contains wax, since the main metabolic energy store of corals is the saturated wax cetyl palmitate. It seems likely that the fishes ingest the mucus for its energy-rich wax content, but the real extent of this food transfer from coral to fishes has not been measured. The fact remains that the transfer occurs, and we may yet find it is a crucial link in the food chain of tropical reefs.

The much publicized predator of Australia's Great Barrier Reef, the crown-of-thorns starfish *Acanthaster planci*, thrives on corals and has reduced miles of productive reef to lifeless rubble. Many biologists are concerned about the reefs' ability to recover. This huge starfish (specimens more than two feet across have been collected) everts its stomach over the surface of the coral and exudes strong digestive enzymes; when the starfish moves away, the coral is completely bare. The digestive enzymes of *Acanthaster planci* are highly efficient in their action on the wax components of coral. No other animal we have examined has so thoroughly adapted its digestive system to wax nutrition. Other starfishes lack this ability; they have enzymes only for digesting proteins and carbohydrates. It can be deduced that the crown-of-thorns starfish gets much of the energy it needs for growth and reproduction by metabolizing wax from the coral. This being the case, it might even be possible to develop wax-based baits impregnated with the appropriate hormones for controlling the reproduction of *Acanthaster planci*. (Poisons would harm the reef fishes that feed on waxes.) In this way one might be able to avoid the excessive growth of starfish populations and excessive damage to valuable coral reefs.

Appendicularians

by Alice Alldredge
July 1976

These small marine animals build themselves a
gossamer house out of mucus. The house incorporates
an elaborate apparatus for filtering food particles
out of the seawater

The waters of the open ocean are, among other things, a world of beautiful and delicate organisms that are adapted to an existence of floating or feeble swimming. They are the plankton, both plant and animal. They have evolved clever mechanisms for meeting the problems of life in a totally fluid and surfaceless environment. Many of the single-celled phytoplankton (the plant plankton) are equipped with long spines or globules of oil that keep them from sinking. Some of the zooplankton (the animal plankton) float with the aid of tiny gas bubbles. The zooplankton are particularly well adapted for avoiding capture by predators that search for food visually: many of them are as transparent as the watery medium in which they live.

Since the phytoplankton are minute in size and often scarce, the herbivorous zooplankton have developed a marvelous array of methods for capturing them. Many of the copepods, ubiquitous marine crustaceans, filter the single-celled phytoplankton out of the water by passing the water through tiny bristles on their appendages. The pteropods, wing-footed marine snails, collect phytoplankton on floating webs of sticky mucus. The most remarkable and most specialized of all the feeding adaptations of the zooplankton belong to the tiny tadpolelike members of the class Appendicularia. The appendicularians build and live inside a small balloon of mucus that is equipped with miniature filters that concentrate their food. Their small house is remarkably efficient at capturing particles; it can even trap bacteria in large numbers. It also protects the animal and helps to keep it afloat.

Most appendicularians filter out and ingest phytoplankton cells that have a diameter of less than five micrometers (five thousandths of a millimeter). The phytoplankton they consume belong to many groups, including the coccolithophorids, the naked flagellates, the small diatoms and the dinoflagellates. These small organisms are known collectively as the nanoplankton (from the Latin *nanus,* dwarf). Although the nanoplankton are abundant, few of the zooplankton, including the ever present copepods, are able to capture them. The appendicularians manage the task easily and

efficiently with their bubble of mucus. Indeed, many of the nanoplankton were first described by the 19th-century German biologist Hans Lohmann after he had found them in the filters of appendicularians' houses.

Appendicularians belong to the phylum Chordata, which includes the vertebrates. They form a class in the subphylum Urochordata, or Tunicata. Appendicularians and other tunicates are primitive compared with the vertebrates (members of the subphylum Vertebrata), but they exhibit three features that are common to all chordates at some stage in their life: gill slits, a tubular nerve cord and a rodlike notochord, or primitive spinal column. Some tunicates, such as the ascidians (sea squirts) are planktonic only during their larval stage; they dwell on the bottom after they reach adulthood. The appendicularians are free-swimming and lead a planktonic existence for their entire life.

Although appendicularians have not attracted much attention from biologists, they are among the commonest members of the zooplankton community: in many bodies of water they are the second or third most abundant group. There are three families of appendicularians, which are further subdivided into 13 genera and some 70 species. The species are thus comparatively few, but many are cosmopolitan; they can be found in all the oceans of the world, including the Arctic Ocean. In fact, species that are primarily warm-water forms are sometimes collected in polar waters.

Numerous though they are in the open ocean, appendicularians are most abundant in coastal waters and over the continental shelves. It is there that the phytoplankton

are most readily available. Moreover, appendicularians are generally found in the top 100 meters of the water. That illuminated surface layer, the euphotic zone, is where the photosynthetic activity and the density of the phytoplankton are greatest. Many of the zooplankton migrate through a considerable vertical distance with the daily cycle of daylight and darkness, but this is apparently less true of the appendicularians; they tend to remain in the surface waters.

Most appendicularians are shaped like a bent tadpole. The animal has two parts: an egg-shaped trunk and a flexible and muscular tail. The tail is thin and flat, resembling the blade of a knife; it is attached to the lower surface of the trunk between half and two-thirds of the way back from the front end of the animal. It consists primarily of muscles and a stiff notochord that runs longitudinally down its center.

The trunk contains all the major organs, including the reproductive and digestive systems. The mouth, at the front end of the trunk, is connected externally to the mucus house and internally to a large pharynx. The pharynx leads into a *U*-shaped digestive tract that consists of a stomach, an intestine and an anus, which empties near the base of the tail. An elongated gland, the endostyle, is located near the mouth and manufactures mucus as the animal feeds. The gill slits are connected to the pharynx by spiracles lined with cilia, which draw water laden with food particles through the mouth and into the pharynx. The animal's thin blood circulates through a simple system of sinuses, pumped by a single muscular heart and the movement of the tail.

All appendicularians, with the exception

APPENDICULARIAN'S HOUSE is the filmy pink structure in the photograph on the following page, made by James King of the Australian Institute of Marine Science. It is about the size of a walnut. The head, or trunk, of the animal itself can be seen as the white egg-shaped object at the center of the house; curving downward from the trunk is the animal's translucent tail. This appendicularian is a member of the species *Stegosoma magnum*. The internal structure of the house, which is normally transparent, has been made visible by red carmine dye. Water is drawn into the house through two filters to the left and right of the animal's trunk. Particles too large for the appendicularian to ingest are thus excluded. Smaller particles pass into the house and are trapped on a highly complex feeding filter in the interior of the house. The aninmal draws the food from the feeding filter through the strawlike buccal tube leading to its mouth.

of the species *Oikopleura dioica,* are hermaphroditic: each individual has the reproductive organs of both sexes. Some species are even protanderous, that is, they have the remarkable ability to produce both eggs and sperm from the same gonad at different times. From the human point of view appendicularian reproduction is heroic. Sperm are liberated through tiny ducts to the exterior of the trunk. Eggs, however, are released only when the ovary and the rear of the trunk split open, resulting in the death of the animal. The eggs are fertilized externally in the water, probably when sexually mature appendicularians spawn together at the same time. The embryo develops very rapidly. Within 24 to 48 hours after fertilization a miniature appendicularian has developed, complete with the ability to build itself a mucus house.

Appendicularians are generally transparent. Only the movement of their beating tail makes them visible in the water. A few species do have color; they are shades of bright yellow, red or blue-violet. The animal proper is quite small: its trunk may be only a few millimeters long. Its mucus house, however, may be as large as a walnut. And although the house is so transparent that it is virtually invisible in itself, it can usually be identified in the water because its surface and filters have trapped phytoplankton and other particles.

The three families of the Appendicularia are quite distinct in their anatomical structure. The family Oikopleurida is by far the most studied and the best-known. In the Oikopleuridae the trunk is short and compact and the tail is long and narrow. The spiracles and gill slits are located near the anus; the stomach wall is made up of many small cells. The mucus house encloses the entire animal and is structurally more complex than the houses of the other families.

In the family Fritillarida the trunk is slenderer and flatter and the tail is shorter and broader. The spiracles are at the front end of the pharynx; the stomach consists of a few large cells. In most of the Fritillaridae the house is limited to a small gelatinous bubble that is deployed in front of the mouth.

The third family, the Kowalevskiidae, is the smallest, containing only one known genus. The members of the family lack an endostyle, a heart and spiracles. The trunk is quite short and the tail is long and leaflike. The house resembles a small umbrella. Although it completely surrounds the animal, it lacks the complex filtering apparatus found in the houses of the Oikopleuridae.

On the time scale of animal evolution the appendicularians are extremely old. Members of the species *Oesia disjuncta,* a fossil appendicularian resembling the modern Oikopleuridae, have been found dating back to the Cambrian period, more than 450 million years ago. Very few fossils of any animal group predate the Cambrian. The evolution of species before that time cannot be substantiated by evidence in the fossil record. Hence most speculations on

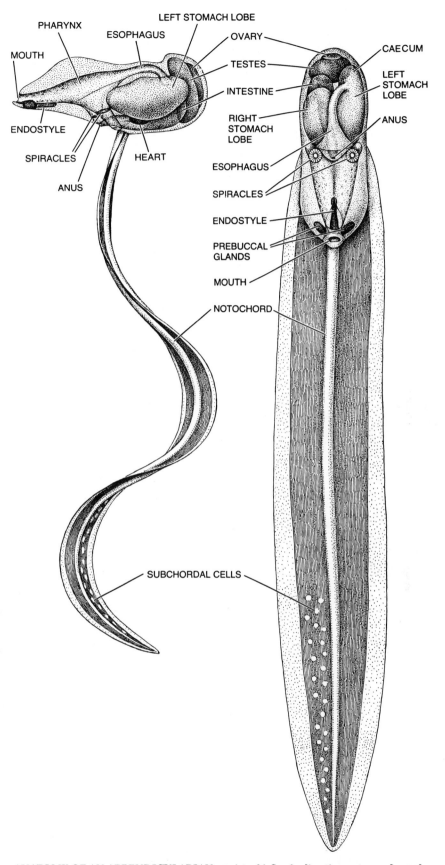

ANATOMY OF AN APPENDICULARIAN consists chiefly of a digestive system and gonads. The animal shown is the typical species *Oikopleura albicans.* The two views are slightly different to display the animal's organs. The lateral view (*left*) shows the trunk pointing forward and the tail curved; the dorsal view (*right*) shows the trunk pointing downward and the tail flat. As do all members of the phylum Chordata (including vertebrates at some phase), appendicularians have a notochord (or primitive spinal cord), a dorsal nerve cord (*not shown*) and spiracles (modified gill slits). Most appendicularians have both testes and ovaries in a single animal.

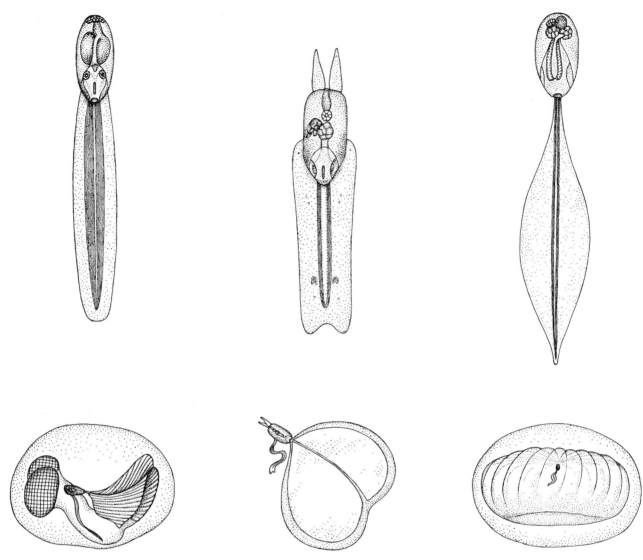

THE THREE APPENDICULARIAN FAMILIES are compared to show the differences between the anatomy of the animals and the structure of their houses. The house of the Oikopleuridae (*left*) and of the Kowalevskiidae (*right*) completely surround the animal like a balloon, whereas the Fritillaridae (*middle*) carry their house like a large bubble deployed in front of their mouth. Although the Oikopleuridae have been studied, almost nothing is known of the biology of the other families. Houses of all families are shown on pages 86, 91, and 92.

the evolutionary development of the appendicularians and the other tunicates is based on the morphology and embryology of the modern forms.

The evolution of the entire phylum Chordata has been the subject of considerable controversy. The view currently held by most vertebrate and invertebrate evolutionists is that both the vertebrates and the tunicates, including the appendicularians, arose from a sedentary filter-feeding ancestor that was attached by a stalk to the seafloor. This simple pretunicate later evolved into the sedentary tunicates of today: the ascidians. The ascidians spend their entire adult life attached to one spot on the bottom, as their pretunicate ancestors presumably did. They have a short larval stage, however, in which the animal differs markedly from the adult in appearance and behavior. The larva resembles a tadpole and is planktonic; it swims actively for a few hours or days and then settles on the bottom. Its tail, notochord and nervous system are assimilated

into its other tissues, and the animal metamorphoses into the sedentary adult.

It is hypothesized that at some time in the evolutionary past the larva developed gonads and could become sexually mature before entering the adult stage. At that point it no longer metamorphosed into the sedentary adult but began a new mode of life as a completely planktonic organism. In this way a larval pretunicate may have given rise to both appendicularians and primitive vertebrates.

Some evolutionists believe the ancestor of the tunicates and the vertebrates was never sedentary but was a free-swimming animal with a tail. That free-swimming protochordate developed along two lines, one leading to the vertebrates and the other to the appendicularians and to the class Thaliacea, which embraces the barrel-like planktonic tunicates including the salps and the doliolids. In this view the sedentary ascidians developed much later from an ancestral thaliacean that settled on the seafloor.

Although these two views have given rise to much speculation and many variations, it is widely accepted that the tunicates departed from the chordate line of evolution early. Although the appendicularians may share a common ancestor with the vertebrates, it is doubtful that primitive vertebrates developed from early appendicularians, regardless of similarities between them.

The appendicularians' method of feeding is unique in the animal world. The plankton-capturing mucus house they secrete, with its intricate filters and miniature nets, is one of the most complex external structures built by any organism other than man. The houses of Oikopleuridae are the only ones that have yet been described in detail. A house in the genus *Oikopleura*, one of the more common genera of appendicularians, is a hollow sphere of mucus interlaced with a variety of passages and filters. The animal occupies the center of the sphere with its mouth facing toward the rear, as if it were sitting backward in an automobile. The outer membrane of the house is fragile and transparent. At the front end of the house are two passages through which the water enters; these incurrent passages are covered with fibers that crisscross at right angles to form a screen. Some phytoplankton cells, particularly large diatoms and dinoflagellates, are too large to pass through the filters and never enter the house. Hence the incurrent filters serve as a sorting mechanism, excluding particles that are too large for the appendicularian to ingest. The mesh size of

the filters is a characteristic of each species. For example, in *Oikopleura fusiformis*, a particularly small species, the openings in the mesh average 13 micrometers across, whereas in *Megalocercus huxleyi*, a particularly large species, they may be as much as 54 micrometers across.

Two cylindrical passages lead from the incurrent filters into the house and join near the base of the animal's tail, forming a single passage. The animal's tail actually occupies this passage, and its sinusoidal beating draws water into the house through the filters. The passage diverges again at the tip of

the tail and joins each side of an internal feeding filter, where tiny food particles are concentrated. This feeding filter is a complex three-dimensional structure in the form of two backward-curved wings connected along one edge. The wings are a sandwich of at least two and possibly three convoluted membrane sheets, between which the water flows. The wings of the feeding filter join at a median channel. At one end the channel is connected to the appendicularian's mouth by a hollow straw-like tube; at the other end it opens into an exit passage leading back to the exterior.

Water, pumped by the appendicularian's muscular tail, enters the house through the incurrent filters, where the large particles are removed; it travels down through the tail chamber and out to the base of each wing of the feeding filter. It then flows simultaneously up both edges of the arched wing to its apex and down between the membranes to the median channel. When the water reaches the median channel, it flows out of the house through the exit passage. Sometimes the force of the outflowing water propels the house forward. All the water must go through the feeding filter before passing out of the house.

The feeding filter is a highly efficient plankton trap, but exactly how the particles are trapped is not known. They may be caught when they adhere to the sticky internal walls of the filter. Alternatively one of the membranes in the sandwich that forms the filter may let water through it but not particles. The water flowing in one direction through such a membrane would leave all the particles trapped on one side.

Once the particles are trapped in the feeding filters the appendicularian ingests them in a truly remarkable way. Once every few seconds the animal sucks the particles off the feeding filter into the median channel and up into its mouth by the action of the cilia in the spiracles on its trunk. The sucking action is analogous to drawing milk up through a straw. The particles that are carried into the animal's mouth are then collected by a feeding mechanism that is common to all the tunicates, including the sedentary ascidians: the endostyle near the mouth forms a thin, porous mucus funnel that lines the pharynx and extends into the stomach. The water entering the mouth passes through the funnel and out through the spiracles and gill slits into the tail passage of the house. The food particles are trapped because they adhere to the mucus, which is then wound into a thin string and digested in the stomach. Appendicularians can process large amounts of food in a short time. Waste is extruded as a multitude of tiny fecal pellets that can rapidly foul the house and clog the filters. When that happens, the appendicularian abandons the house.

The mucus that forms the walls and filters of the house consists of a gelatinous material that is secreted on the surface of the trunk by a layer of specialized glandular cells, the oikoplast epithelium. The pattern

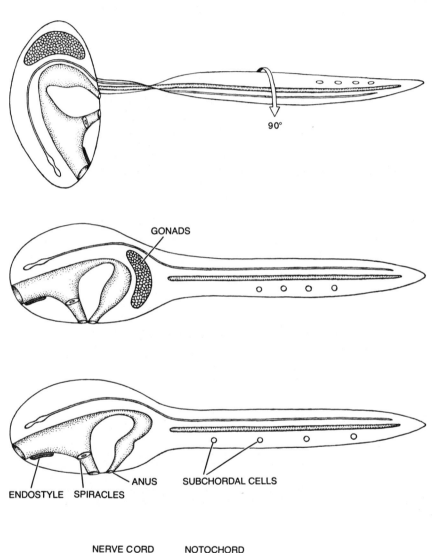

90°

GONADS

ANUS   SUBCHORDAL CELLS

ENDOSTYLE   SPIRACLES

NERVE CORD   NOTOCHORD

ANUS

MOUTH   GILL SLIT   GUT

APPENDICULARIANS MAY HAVE EVOLVED from the planktonic larva of a primitive deep-sea chordate (*bottom*), according to a hypothesis put forward by Robert Fenaux. In his theoretical evolutionary sequence the flat tail of the primitive chordate became partially differentiated from the trunk (*second from bottom*). Primitive chordate attained sexual maturity while it was in larval form (*second from top*). Eventually tail completely differentiated from trunk and rotated 90 degrees so that it was flat from front to back, as in modern animal (*top*).

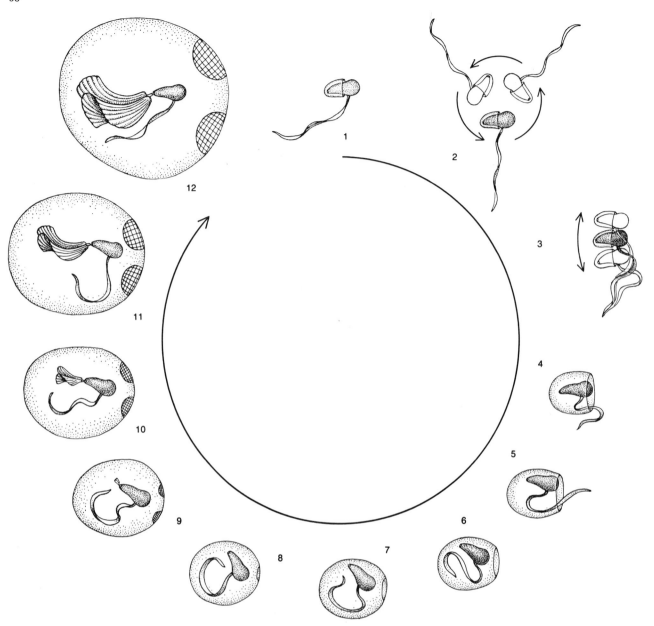

APPENDICULARIAN BUILDS A NEW HOUSE after it abandons the old one. It may build a house several times a day, in a process that takes only a few minutes. While the animal is still in its old house it has already secreted a new one complete with filters and carries the rudiment of it in collapsed form against its trunk. After abandoning the old house the animal intermittently floats and swims for 30 to 120 seconds (*1*). Soon it begins a series of violent cartwheeling (*2*) and nodding (*3*) motions that enlarge the rudiment of the house. The tail is then pulled into the rudiment (*4–8*) in less than a second. The animal enlarges the house even further with a series of sinusoidal vibrations traveling down its tail from the base to the tip. As the house enlarges, opening through which animal entered closes. Meanwhile the feeding filter expands (*9–11*). When both the house and the feeding filter reach their normal size, the animal can begin feeding (*12*).

of this epithelium, the outer membrane of the animal, is characteristic for each species and determines the structure and shape of the house. Different groups of cells can manufacture mucus of different density and elasticity, so that on expansion the mucus takes a variety of shapes that contribute to the complexity of the resulting structure. In the Oikopleuridae the oikoplast epithelium covers most of the trunk; in the Fritillaridae it is reduced to an area on the upper front portion of the trunk and around the mouth.

The oikoplast epithelium can be divided into regions based on the types of cells present and the parts of the house that are made there. One region of specialized cells, known as Eisen's oikoplast, is responsible for making the mesh covering over the incurrent openings of the house. Another region, Fol's oikoplast, produces the feeding filter. The species *Oikopleura longicauda* is unique in the family Oikopleurida in that it lacks Eisen's oikoplast; thus its house has no mesh over the water inlet.

The construction of a new house by an appendicularian begins with the secretion of gelatinous material by the oikoplast cells. The mucus accumulates in a layer at the top of each cell just under the cell membrane. A new membrane is then fabricated directly under the mucus layer, moving the mucus to the outside of the cell. There the mucus coalesces into the rudiment of a house. Special structural fibers are manufactured by both Eisen's and Fol's oikoplasts. Although the secretion of a house may take as long as four hours, some species, such as *Oikopleura cornutogastra*, can secrete one in five minutes.

After secretion the animal's trunk is covered with a thin gelatinous film, which is carried on the trunk in collapsed form until a new house is needed. The shape and final form of the house have been determined at the time of its secretion. The rudiment is now ready to be expanded.

The animal expands its new house only after it has abandoned or discarded the old one because the filters have become clogged with phytoplankton and detritus. It may also discard the house when it is threatened by a predator such as a fish larva. Appendicularians build houses with amazing frequency. Most of the species that have been studied secrete and expand a new house every four hours. When *Oikopleura rufescens* is disturbed in a collecting jar, it may build as many as three houses in 30 minutes. The house of *Oikopleura albicans* is reported to have a special hatch through which the animal can escape. Most species escape, however, simply by moving their tail vigorously and physically forcing the walls of the house apart.

The secretion of a house must require considerable energy. Members of the family Oikopleurida have developed a mechanism to unclog the incurrent filters, thus postponing the need to abandon the old house: they momentarily suspend the beating of their tail and reverse the flow of water with their ciliated spiracles. The water flowing out through the incurrent openings unclogs the filters by sweeping away the large particles trapped there.

Once the old house is discarded, however, the free-swimming animal begins immediately to expand the new one. It swims intermittently with a jerky motion for several minutes. During this period the collapsed rudiment of the new house expands slightly into a capsule that fits like a glove around the front end of the trunk. The animal then begins a series of violent movements to wriggle partially out of the house rudiment. Rapid sinusoidal motions of its tail often make it cartwheel or nod its head back and forth. The nodding pulls the trunk away from the encompassing house rudiment. When the rudiment is sufficiently ex-

panded, the animal pulls its tail into the house, base first and tip last. It is now completely surrounded and cramped in the tiny house, its tail in the shape of a question mark with the tip touching the animal's "nose." Small sinusoidal motions of the tail, beginning at the base and traveling to the tip, draw water in through the incurrent filter and slowly expand the house. The house grows larger around the animal as water is pumped into it. The expansion may take several minutes, during which the open rear end of the house through which the animal entered gradually closes. When the house attains its full size, water can pass completely through it, and the animal begins its normal feeding.

Since appendicularians are among the few multicellular organisms that are capable of feeding on nanoplankton, they hold an unusual and important position in the food web of the open ocean. As many as 50,000 phytoplankton cells may be trapped on a single appendicularian house at any one time. The entire structure and function of the house are adapted to maximize its filtering efficiency. G. A. Paffenhofer of the University of Georgia has recently raised *Oikopleura dioica* through 19 generations in his laboratory. He found that the larger members of that common and relatively small species could remove all the phytoplankton from 300 milliliters of seawater per day—more than 250,000 cells! My own observations suggest that the larger species such as *Megalocercus huxleyi* and *Stegosoma magna* filter several hundred milliliters of seawater per hour.

In areas where appendicularians are abundant they may drastically reduce the quantity of phytoplankton available to other planktonic herbivores. Even if the appen-

dicularians do not consume all the particles they filter, they substantially alter the form of the food. Normally phytoplankton are highly dispersed. In the course of collecting the phytoplankton the appendicularians concentrate them in mucus packages, thus making them unavailable to other filter feeders.

Appendicularians are an important link between the nanoplankton and the larger animals of the oceanic food web, both planktonic and neritic (large, actively swimming organisms such as fishes). Their major predators include the larvae of herring, sardines and flatfish. The larva of the plaice, a flatfish, may consume 25 or 30 small larval appendicularians per day. The sergeant major, a common yellow and black striped reef fish, often feeds on appendicularians. Since the particle-covered houses are frequently much more obvious than their transparent inhabitants, animals that seek their prey visually may actually see the house rather than the animal within and consume both together.

Jellyfishes, chaetognaths (arrowworms) and siphonophores (smaller relatives of the surface-dwelling Portuguese man-of-war) also feed on appendicularians. These planktonic predators are a great hazard to an appendicularian that is between houses and is hence not protected by its shield of mucus. At times, however, a disturbed appendicularian will desert its house and rapidly swim away, leaving the house as a decoy for a predator. Species that build particularly elaborate houses, such as *Oikopleura rufescens,* have invested large amounts of energy in the house and do not readily abandon it. The houses of these species may be poked and bounced around considerably before their occupant will desert them.

Although the appendicularian proper is

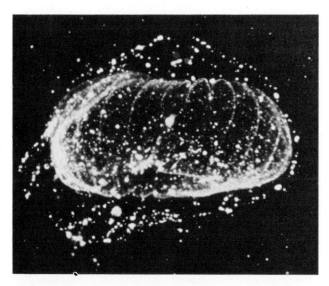

HOUSE OF THE FAMILY KOWALEVSKIIDA is particularly fragile and rare. Until recently only one such house had ever been described (by the zoologist Hermann Fol in 1872). This photograph by King is the first ever made of one. The house is shaped like an umbrella; the animal itself is the small white spot in the center. The move-

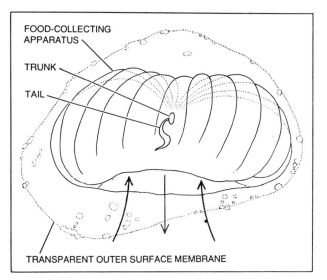

ment of the animal's tail draws water into the house along the ribbed inner walls, where particles of food are collected. The water passes out of the house in the center of the opening through which it entered. The path of the water is indicated by gray arrows, the path of food by colored arrows. The house shown is about the size of a five-cent piece.

an important link in the oceanic food web, the discarded houses may have an even greater impact on plankton ecology. A few appendicularians, each manufacturing perhaps six houses per day, can quickly saturate the surrounding seawater with their cast-off mucus dwellings. In areas where the animals are abundant discarded houses often reach a density of several hundred per cubic meter. Scuba diving among them is like swimming in a snowstorm.

Each discarded house is a concentrated package of phytoplankton, mucus and detritus, a rich food source in an environment where normally food is highly dispersed. Many kinds of zooplankton, including some copepods, euphausiid (krill) larvae, flatworms and polychaete-worm larvae, have been observed resting on the surfaces of discarded houses and grazing on the mucus and the particles trapped in it. The mouthparts and feeding appendages of these organisms are adapted to scraping rather than to filter feeding. The organisms that feed on appendicularian houses there-

fore exploit a source of food (the nanoplankton) they are not adapted to gathering by themselves. Fish feeding on occupied houses may also utilize the nanoplankton. It is in this way that the appendicularians' unusual feeding structure serves as a link between the nanoplankton and many organisms to which such phytoplankton are not normally available.

The discarded houses alter not only the form and concentration of phytoplankton but also the structure of the planktonic environment. The word plankton normally conjures up an environment where tiny plants and animals float without hindrance in a totally fluid medium. Although the planktonic organisms may bump into one another, only at the bottom do they encounter a solid physical substrate. Observations in the open ocean by divers and from submersible vehicles, however, have revealed an abundance of macroscopic organic aggregates, sometimes referred to as marine snow because of their resemblance to drift-

ing snowflakes. These aggregates may reach a diameter of several centimeters. Some are formed from material that has been deposited on small particles through a variety of processes, including bacterial action. Many, however, are the discarded and decomposing mucus by-products of various zooplankton, particularly the pteropods, the salps, the planktonic tunicates closely related to the appendicularians, and of course the appendicularians themselves.

These large, amorphous bits of mucus and detritus serve as surfaces on which many planktonic organisms can rest or feed. Bacteria and protozoans use them as a permanent habitat. Appendicularian houses and other particles of marine snow provide tiny solid substrates and introduce heterogeneity into an environment that is generally considered to be relatively homogeneous physically. Although these miniature habitats have not been completely explored, it is certain that they have influenced the adaptations and feeding strategies of many planktonic organisms.

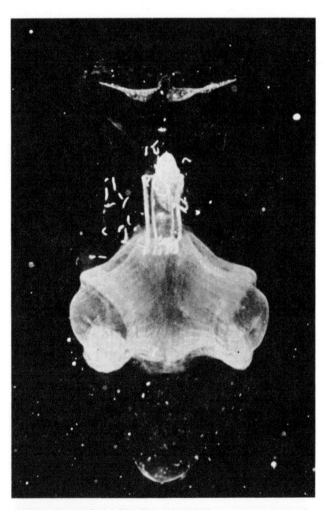

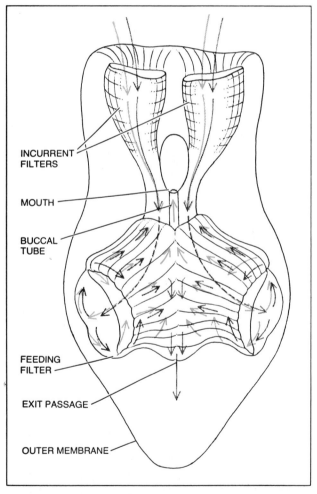

INCURRENT FILTERS

MOUTH

BUCCAL TUBE

FEEDING FILTER

EXIT PASSAGE

OUTER MEMBRANE

HOUSE OF MEGALOCERCUS HUXLEYI, a species in the family Oikopleurida, has an unusual torpedo shape. Water pumped by the animal's tail flows into the house through two cone-shaped filters. It then travels down a tail passage and diverges into two cylindrical passages leading to the feeding filter. The water, laden with food particles, enters the feeding filter at two points at the base of each wing of the filter. Particles collected on the membranes of the feeding filter are carried into the animal's mouth by the buccal tube. Filtered seawater leaves the house at the rear with enough force to propel the house forward with an erratic corkscrew motion. The flimsy outer membrane of the house is visible only at the rear end in the photograph. The house of this species may reach several centimeters in length.

Stomatopods

by Roy L. Caldwell and Hugh Dingle
January 1976

*These shrimplike creatures are formidable predators. Two
of their forward appendages are adapted either for
spearing or smashing. One blow from a smasher
can break the shell of a good-sized clam*

Stomatopods, commonly known as mantis shrimps, are predatory crustaceans that live in the shallow waters of tropical and subtropical seas. They dwell in burrows and cavities and feed on other marine animals such as annelid worms, snails, clams, shrimps, crabs and fishes. Although stomatopods can be so numerous that in some waters they are a major predator, they are seldom seen because of their retiring habits. The most an exploring diver is likely to see of a stomatopod is a pair of eyestalks peering out from a mud burrow or a cavity in rock or coral. Retiring though stomatopods are, they are also extremely aggressive, both in capturing prey and in defending their territory; as a group they are among the most pugnacious animals known. The most notable feature of a stomatopod is the two large forelimbs it uses to attack its prey. These raptorial appendages unfold and shoot forward much like the forelimbs of the praying mantis. They are adapted either for spearing or for smashing, and in the evolution of stomatopods the structure and function of these two different types of appendage have interacted closely with other aspects of the animals' morphology and with their patterns of behavior.

Stomatopods are not closely related to shrimps or other marine crustaceans. They appear to have split off from the leptostracan stock some 400 million years ago. Originally they were filter feeders, using their thoracic appendages as food strainers. Fossil evidence indicates that some 200 million years ago one forward pair of appendages, the second maxillipeds, were evolving into large folded limbs. ("Stomatopod" is from the Latin for "mouth-foot.") These animals were the first true stomatopods. During the Jurassic period, from 190 million to 135 million years ago, the stomatopods ra-

diated into the four modern families: the Squillidae, the Lysiosquillidae, the Gonodactylidae and the Bathysquillidae. The Bathysquillidae are found in deep water and have rarely been taken alive, so that nothing is known about their behavior. The remaining three families are found in shallow waters, and we have collected data on more than 40 species distributed around the world.

There are currently about 350 known species of stomatopods, and additional species are being discovered every year. The species range in length from 15 to 335 millimeters (half an inch to 13 inches), and many of them are brightly colored. They can be divided into two functional groups: the spearers and the smashers. The spearers consist of all the Squillidae, Lysiosquillidae, Bathysquillidae and several genera of the Gonodactylidae. The smashers consist of only a few genera of the Gonodactylidae.

Among the spearers the last joint of the forelimb is equipped with sharp spines, from three to 17 of them. This spined dactyl is used to spear soft-bodied prey such as fishes and shrimps. The strike is made with the dactyl open and is one of the fastest animal movements known. It is completed in from four to eight milliseconds, and the velocity of the movement is more than 1,000 centimeters per second even under water. Once the prey is stabbed it is pulled toward the stomatopod's mouthparts and ripped to pieces by its sharp, serrated mandibles and by its third, fourth and fifth maxillipeds, which are equipped with hooks. The spines of the dactyl are barbed at the tip, so that the stabbed prey is unlikely to slip off. In many species the second joint of the raptorial appendage also has spines, which help to secure the prey when the dactyl is folded.

Most spearers dig their burrows in mud or sand and wait in ambush at the entrance. For example, *Squilla empusa,* a stomatopod that is common off the coast of the southeastern U.S., digs a hole or trench and often lies in wait covered with silt so that only its eyestalks are exposed. The species is abundant in commercial shrimp beds and is believed to be a serious predator of shrimps.

Many of the larger spearers that feed on fishes have greatly elongated raptorial appendages, which give them a reach of half their body length. *Harpiosquilla harpax,* a 250-millimeter (10-inch) stomatopod of the Indian Ocean, can strike 130 millimeters. We have seen this species catch a 110-millimeter fish and devour it in four minutes.

The smashing stomatopods have a dactyl with few spines or none, but in most of them the heel of the dactyl is greatly enlarged. During the strike the dactyl remains folded, and the prey is struck by the blunt heel. These stomatopods normally feed on armored animals such as snails, hermit crabs, clams and crabs, which they batter to pieces. The strike of a large species of smasher such as *Hemisquilla ensigera* (250 millimeters long) has a force approaching that of a small-caliber bullet. A large specimen of this species from southern California broke an aquarium wall consisting of a double layer of safety glass. Even small species 80 millimeters long can break the wall of an ordinary glass aquarium.

Most smashers live in cavities in rock or coral. They visually select their prey and stalk it. When a stomatopod attacks a crab from the rear, its first blow stuns the crab; subsequent blows knock off the crab's legs and claws and smash its carapace. When the stomatopod attacks a crab from the front, its first blows are usually directed at the crab's claws. When the claws are broken, further blows crush the crab's carapace and break its legs. The stomatopod then drags the battered hulk to its dwelling cavity, where it breaks the crab into small pieces and picks the tissue from the exoskeleton. Snails, clams and hermit crabs living in snail shells are simply seized and taken to the cavity. The stomatopod wedges the prey against the wall with its third, fourth and fifth maxillipeds and repeatedly strikes the shell with its raptorial appendage until enough of the shell is broken away for it to pull out the soft tissue.

The feeding efficiency of an intertidal species, *Gonodactylus chiragra*, was studied in Thailand by one of us (Caldwell). The species lives in rock cavities and forages over the reef flat. After it has captured and eaten its prey, it dumps the broken shells outside its dwelling cavity. The broken shells made it easy to identify the various items in the animal's diet. These stomatopods normally ate snails, hermit crabs and clams. Some of them were captured and were offered different sizes of snail in the laboratory. Although they were quite capable of breaking the shells of snails up to 30 millimeters long, they preferred smaller snails. They also preferred snails to her-

mit crabs that lived in shells of the same size. Most of all these stomatopods preferred clams; it takes only a few strikes to break open the shell, and the amount of edible tissue obtained for the effort is much greater than it is with other prey animals.

The number of strikes the stomatopods needed to break open the shell of snails and hermit crabs of various sizes was recorded, together with the amount of tissue they got out of the shells. With this information the amount of edible tissue the stomatopod could get per strike with prey of various kinds and sizes could be calculated. The results showed that the stomatopods were

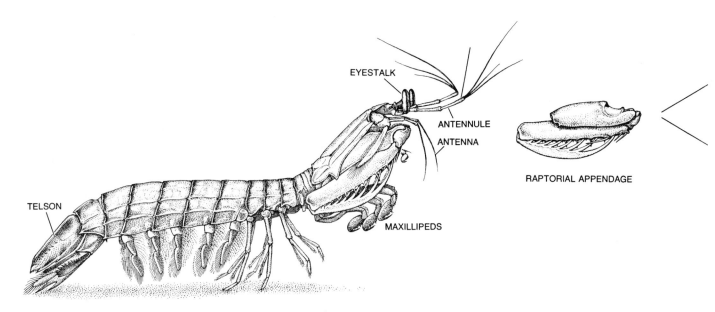

SPEARER of the species *Harpiosquilla harpax* is shown in a typical attitude. This species can grow to a length of 25 centimeters (10 inches). Like all spearers, it has two modes of striking with its raptorial appendages (*right*). In capturing prey it uses a spearing

SMASHER of the species *Gonodactylus chiragra* can grow to a length of about 10 centimeters. As do all smashers, it uses its raptorial appendages both in combat with other stomatopods and in attacking prey. If the prey is soft-bodied, the smasher will change

choosing the prey that gave them the greatest amount of tissue per strike.

Another interesting observation made in the course of the study was that individual stomatopods show considerable plasticity in learning how to open various kinds of shell. Each animal developed its own style of attacking shells and used a favorite rock as an anvil. Some would always break the apex of the shell first, others would break the lip and still others would shear off the whorls on one side. Studies of other species of smashers have shown that when they are given novel food items, they are at first inefficient in breaking open the shell but with experience their efficiency improves.

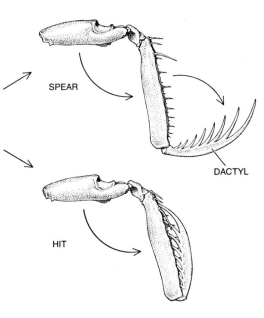

strike. In combat with another stomatopod spearers usually hit with the dactyls closed.

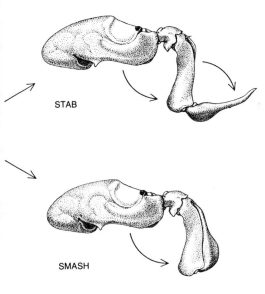

from a smashing strike to a stabbing one. It also sometimes stabs in intense combat.

The power of the smashing strike is so strong that the smasher could conceivably break off a dactyl. Some newly captured smashers are missing one raptorial appendage or both appendages, or have a dactyl with a broken heel. Such damage is not, however, the result of feeding activity but of fights with other stomatopods. When a raptorial appendage of a stomatopod is severely damaged, the animal cannot simply shed the limb, as a crab can. It literally tears the limb off with its maxillipeds, and a new appendage regenerates over the next two or three molts. Intact dactyls show surprisingly little wear. Although a smasher may make hundreds of strikes per day at hard snail shells, the heels of its dactyls show very little erosion, even over a period of several months between molts.

The tip of the smasher's dactyl, as distinct from the heel, is sharp but not barbed like the spined dactyl of spearers. On occasion a smasher will strike its prey with its dactyl tip open, usually after the prey has been struck at least once by a smashing blow. That can happen if one holds a stomatopod in one's bare hand, a point that has been painfully driven home to us. Apparently the animal gets some feedback information about the softness of the target and changes its mode of attack.

Most stomatopods are territorial and vigorously defend their dwelling space. The strike of a stomatopod is potentially lethal to other stomatopods and is usually somewhat modified when it is directed against a potential competitor of the same species or a closely related one. Spearers normally strike at another stomatopod with a closed dactyl, which is much less likely to inflict serious injury than the piercing strike of an open dactyl. Smashers also attack an intruder with a closed dactyl, but they normally direct the blow at the intruder's body armor and thus rarely inflict a fatal injury. Under certain conditions, however, many smashers will stab an intruder with an open dactyl. If blood is drawn during a fight, some smashers will try to stab and kill their opponent.

When two smashers are fighting, most of the blows are delivered to the telson, or tail shell, which is heavily armored. Spearers do not have a powerful smashing strike, and in combat they direct their blows at all parts of their opponent's body. The telson of spearers is more fragile, even though in some species it is armed with sharp spines. The telson of smashers has numerous large ridges with heavy, blunt projections instead of sharp spines. The ridges appar-

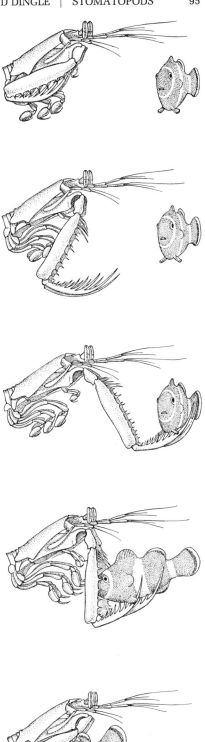

RAPTORIAL STRIKE of a spearer is depicted. When the prey, in this instance a fish, swims into range, the stomatopod strikes with both raptorial appendages. The strike is one of the fastest animal movements known, attaining a velocity of more than 1,000 centimeters per second. Spines on the second joint of the raptorial appendage secure the prey when the dactyl is folded. The captured fish is pulled to the mouthparts and ripped apart by hooks on the maxillipeds.

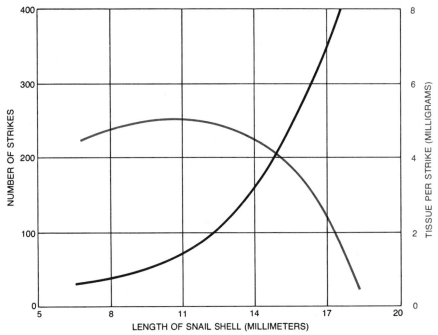

FEEDING EFFICIENCY of a five-centimeter *Gonodactylus chiragra* is shown in terms of the number of strikes it requires to completely break open snail shells of various sizes and the amount of tissue it obtains per strike. The stomatopod strikes on the average once every 30 seconds and can take three hours to break open larger shells. When offered snails of various sizes, *G. chiragra* selects the sizes that give it the highest amount of tissue per strike.

ently give the telson structural support, enabling it to withstand crushing blows. A survey of the telson morphology of various species of stomatopods shows that as the crushing power of the strike increases there is a concomitant increase in the armoring of the telson.

Smashers that have evolved heavy telson armor employ a unique defensive behavior when they are fighting. The animal being attacked lies on its back and coils its telson upward, so that blows are taken on the telson's outer surface. When there is an opening for a counterattack, the animal momentarily lowers the telson, lunges forward and strikes its opponent, which has coiled in response so that the blows fall on its own telson. After the attacking stomatopod has struck a blow, it quickly resumes its coiled posture. The armor of the telson, together with the shock-absorbing action of the coiled abdomen, minimizes physical damage.

A few species of smashers, for example *Gonodactylus chiragra*, engage in such ritualistic combat less frequently than others. Instead the animal tries to circle around its opponent and kill it by striking at the head or the thorax. The killing species tend to be found in areas where there is intense competition for suitable cavities. Their unrestrained attack behavior gives them a competitive edge over species that use the ritualized coiling form of combat, particularly since they are the largest and most powerful species in the community.

Smashers also use their armored telson to defend their domicile. They block the opening of the cavity with the telson and thereby bar the entry of competitors or predators. The species of one genus, *Echinosquilla*, have evolved a telson that mimics a small sea urchin, so that the telson serves to camouflage the opening as well as to block it.

When a smashing stomatopod is attempting to evict another stomatopod from a cavity, it will coil with its telson directed at the entrance. The telson provides protection against the strikes of the cavity's occupant. *Haptosquilla glyptocercus*, which has the heaviest armor of any species we have studied, even uses its telson to evict the occupant of a cavity. It forces its way into the cavity tail first and then stops, plugging the entrance. It holds that position for several minutes, receiving repeated blows on the telson from the occupant of the cavity. Periodically the intruding animal withdraws from the entrance for a few seconds and then reinserts its tail. This behavior may continue for several hours until the occupant flees. We do not know whether the eviction results from a reduction of the oxygen in the blocked cavity, from a fouling of the water in the cavity by aggressive defecation or from prolonged stress.

One of the largest stomatopods studied by one of us (Caldwell) is the smasher *Odontodactylus scyllarus*. The power of its blow is impressive. In the laboratory we have seen this species break open the shell of very large snails such as the fighting conch. *O. scyllarus* rarely engages in combat, however. If it did, its heavy blows could be fatal to an attacking stomatopod. Instead it presents a series of ritualized threat displays with its raptorial and posterior appendages. Interestingly, its telson is not heavily armored, and it has spines similar to those found on spearing stomatopods. It is a plausible hypothesis that this species has not evolved an armored telson and a more potent attack behavior because

COILED DEFENSIVE POSTURE of the smasher is shown at the left. The stomatopod lies on its back and presents its armored telson to the attacker, which has just uncoiled and is lunging to strike. After it has struck a blow it will recoil and the opponent will lower its telson, lunge forward and strike. This ritualized fighting minimizes physical damage.

SPEARING STOMATOPOD of the species *Squilla empusa* is found in the shallow waters off the coast of the southeastern U.S. The species is a major predator of shrimp. Specimen in this photograph is 11 centimeters (about four and a quarter inches) long.

SMASHING STOMATOPOD of the species *Odontodactylus scyllarus* is found in the Indian Ocean and the Pacific. The specimen shown here is 17 centimeters (about six and a half inches) long. This colorful species is commonly imported for display in marine aquariums. The smashing strike of this stomatopod's raptorial appendages has sufficient force to break the glass of an aquarium.

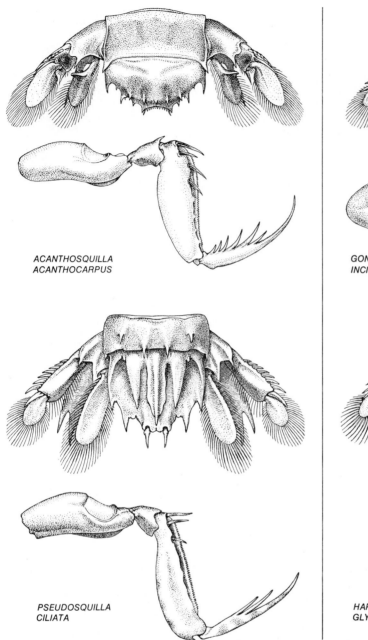

ACANTHOSQUILLA
ACANTHOCARPUS

PSEUDOSQUILLA
CILIATA

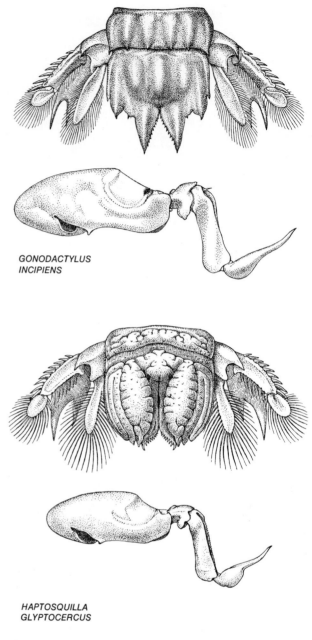

GONODACTYLUS
INCIPIENS

HAPTOSQUILLA
GLYPTOCERCUS

TELSONS AND RAPTORIAL APPENDAGES of two species of spearers (*left*) and two species of smashers (*right*) are depicted. The telson of spearers is relatively fragile. In some species it is armed with sharp spines. The raptorial appendages of spearers typically have spines, whereas the raptorial appendages of smashers have few spines or none. The telson of smashers is armored with ridges and blunt projections. Species of smashers that have powerful smashing appendages tend to have a heavily armored telson.

no amount of armor could protect it from the smashing strike of an opponent of the same species.

Most species of stomatopods have a conspicuous threat display that involves a posturing of the antennules, antennal scales and maxillipeds and the spreading of the raptorial appendages to expose a depression called the meral spot [*see bottom illustration on next page*]. The effect is to increase the animal's apparent size and to dramatically display its offensive weaponry. The meral spot of many smashers is brightly colored and edged in black. Species of stomatopods that make frequent use of a threat display typically have the most conspicuous meral spot. The meral spot of most spearers is rather dull in color. In general smashers use the meral display much more often than spearers do. The close association of frequent threat display and conspicuous coloration is another example of the coevolution of behavior and morphology in stomatopods [*see illustration on page 100*].

When several species of a stomatopod genus are found in the same area, each species usually has a distinctive meral spot. This color-coding apparently serves for species recognition in both aggressive and reproductive behavior. For example, two species common off the coast of Florida have the same body size and similar body coloration. They are often found in the same area, and the only way to readily tell them apart is by the color of the meral spot. One species, *Gonodactylus oerstedi*, has a purple spot; the other, *G. bredini*, has a white spot. We have studied four similar species of smashers in the waters off Thailand; all live on the edge of a reef and are distinguished by their meral spots, which in this instance are white, orange, yellow and purple. In the Gulf of Siam three species of spearers live side by side on mud flats. They are quite similar in appearance except for their meral spots, which are dark gray, light blue and yellow.

The threat display not only makes a stomatopod appear larger; it also signals a readiness to attack, and this in turn can inhibit attack by another animal. In one series of experiments we placed two individuals of the species *Gonodactylus zacae* in a container and observed their behavior the first time the two animals came in contact. When the animals were of equal size, not a single attack occurred when one of the animals gave a threat display during the initial contact. Attack was common when one of the animals did not give a threat display.

In many other *Gonodactylus* species smaller individuals tend to give a greater number of threat displays, both when they are matched against a stomatopod of the same size and when they are matched against larger stomatopods. Older ones are more likely to employ a "hawk" strategy, attacking first and then retreating into a defensive posture, with a threat display if necessary.

Gonodactylus smithii has the largest and most conspicuous threat display of any species we have studied. When it meets another stomatopod, its behavior is characterized by almost constant display. It is frequently able to bluff its way through initial encounters with species that are actually more aggressive.

There is evidence that threat behavior in a species is plastic, and that the frequency of threat display varies between communities of the same species. At Eniwetok atoll in the western Pacific *Haptosquilla glyptocercus* lives with two less aggressive species, and none of the three species uses threat display very often. In Thailand *H. glyptocercus* is found with two larger and more aggressive species. There it threatens more than twice as often as it does in the Eniwetok community, both against the other species and against other *H. glyptocercus* individuals.

The strike of smashing stomatopods is so potent that the survival of a species calls for the accurate communication of aggressive intent. We have found a positive correlation between the power of a stomatopod's strike and the complexity of its behavioral repertoire. Spearers typically have seven or eight different aggressive displays. These in-

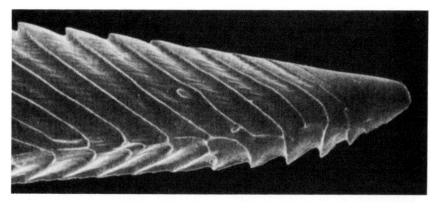

BARBED TIP of the dactyl spine of a spearer, *Squilla empusa*, is shown enlarged 125 diameters in this scanning electron micrograph. The barbs prevent prey from slipping off.

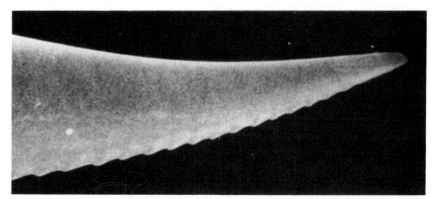

DACTYL TIP of a smasher, *Odontodactylus scyllarus*, is shown enlarged 125 diameters. Although the tip is occasionally used to spear prey, it does not have barbs. Both of the micrographs on this page were made by L. Jackson of the University of California at Berkeley.

CONSTANT THREAT DISPLAY of *Gonodactylus smithii* in the course of a fight frequently enables it to bluff its way through encounters with larger stomatopods. It has the most conspicuous meral spots and widest meral spread of any stomatopod studied by the authors.

clude acts such as approaching, meral display, striking, grasping and avoidance. Smashers, on the other hand, have complex repertoires that may consist of 30 or more different acts.

Smashers that live in areas where there are a limited number of dwelling cavities tend to be more aggressive and to have more complex patterns of aggressive behavior than smashers in areas where suitable domiciles are abundant. For example, smashers that live in rubble cavities, which are usually found only in limited numbers in any one area, are extremely aggressive, whereas those that occupy cavities in live coral, which are usually found in larger numbers in any one area, tend to be less aggressive. *Gonodactylus platysoma* often does not occupy cavities at all but takes shelter in

the large spaces under live coral. It is the least aggressive of the smashers we have studied and has the smallest repertoire of aggressive behavior.

Spearing stomatopods also show a stronger correlation between aggressive behavior and habitat. Spearers that dig long, complex burrows tend to be more aggressive than spearers that dig shallow tubes or trenches. In spearers aggressiveness appears to be correlated with the amount of energy and time put into burrow construction.

We have found that when several stomatopod species live in the same area, the more aggressive species are more abundant than the less aggressive ones. We have three clear examples. Before 1954 *Pseudosquilla ciliata*, a spearer, was the only stomatopod known to

inhabit cavities in the coral rubble around the Hawaiian Islands. Then a powerful smasher, *Gonodactylus falcatus*, was introduced (probably by Navy barges), and by 1963 the smasher had evicted the spearer from the coral rubble. The spearer is now found only on sand flats, where it constructs burrows or lives under small rocks or marine algae. When *G. falcatus* and *P. ciliata*, which are about the same size, fight in the laboratory, *G. falcatus* always wins.

On the intertidal mud flats at Ang Sila on the Gulf of Siam the spearer *Clorodopsis scorpio* is five times more abundant than another spearer, *Oratosquilla inornata*. Both are the same size, have the same diet and dig their own burrows. The burrow of *C. scorpio*, however, is more complex than that of *O. inornata*. When the two species fight, *C. scorpio* easily defeats *O. inornata*, and it is almost always able to evict *O. inornata* from its burrow.

We have mentioned that *Haptosquilla glyptocercus* is the most aggressive stomatopod in the coral-rubble habitat of Eniwetok but is dominated by other species in Thailand. We also found a relation between aggressiveness and abundance: at Eniwetok *H. glyptocercus* is the most abundant species, and in Thailand the other species are more abundant. Although the relation undoubtedly exists, it is not as simple as it might at first appear. If aggressiveness were the only factor involved, it would be difficult to explain the coexistence of several species.

It may be that the less aggressive species survive by specializing in a microhabitat that differs slightly from the microhabitats of the more aggressive ones. Other factors could also account for coexistence, among them differences in the degree to which the various species are preyed on by other animals, differences in size and differences in rate of reproduction.

The two lines of stomatopods—the spearers and the smashers—point an evolutionary moral. Our studies show an unmistakable association between changes in the form of the stomatopods' raptorial appendages and changes in their behavior, body armor, threat display, type of habitat and competitive ability. The stomatopods are an unusually clear example of the unity of evolution, reminding us that evolutionary change in one part of an organism is only one aspect of a comprehensive transformation of the entire organism and its place in nature.

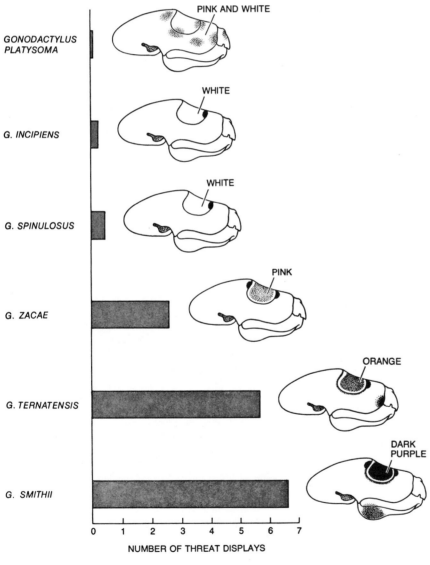

NUMBER OF THREAT DISPLAYS given by six species of *Gonodactylus* are shown. The data were obtained from observation of 40 individuals of each species during a 10-minute confrontation with another stomatopod. It was found that species with conspicuous meral spots tended to give threat displays more often than species with less prominent meral spots.

III

HOW RELATIONSHIPS WORK: SOME SYMBIOSES

III HOW RELATIONSHIPS WORK: SOME SYMBIOSES

INTRODUCTION

Competition, predation, and (in a broad sense) symbiosis are three fundamental sorts of ecological relationships. Many articles in this Reader deal with aspects or consequences of interspecific competition or of predator–prey relationships. These four essays deal with apparently mutualistic symbiotic relationships among marine organisms, and they do so in ways that make these articles instructive to read together.

"Cleaning Symbiosis," by Conrad Limbaugh, presents a behavioral relationship in which both partners modify their conduct when they meet (or in order to meet) by interacting cues and responses. The symbioses augment the cleaners' diets and free the cleaned partners of parasites and bacteria that otherwise can form fatal infections. The symbiosis depends on each partner's "right" behavior. But the behavior that marks each symbiosis manifests itself only occasionally, like the interactions of clerks and clients in a shop.

John E. McCosker's article, "Flashlight Fishes," discusses a much more intimate and continuous symbiosis, so tight that the luminescent bacterial partner is residentially committed to the species and perhaps even to the individual fish it infects. (How the infection occurs initially in either the fish's or the bacterial clone's life still remains a puzzle.) Though the symbiotic residency is permanent, the fish's use of the bacterial light is occasional and behaviorally controlled. The bacteria emit a continuous glow, but the host fish can either hide or reveal this light. To accomplish this the fish does not need the reciprocal "cooperation" of the bacteria beyond their maintenance of their own normal metabolic activity, covered or uncovered.

The symbiosis between algae and giant clams, explored by C. M. Yonge in his essay, "Giant Clams," is still more strikingly mutualistic than that of the flashlight fishes and their bacteria, although the alga–clam host specificity may be less strict. There is no occasionality on either side. The algae continuously reap the benefit of the clam's metabolites, while the clam steadily gains dissolved foodstuffs from the photosynthesizing residents in its tissues. The anatomical adjustments of host to symbiont are more extensive than, say, the pouches and lids of the flashlight fishes. If Yonge's phylogenetic reckonings are correct, giant clams have undergone extraordinary allometric transformations from their cockle-like ancestors. The whole huge host has been rearranged in its proportions to accommodate—to exploit, in fact—its partnership with zooxanthellae.

The alga–coral symbiosis described by the Goreaus in their review, "Corals and Coral Reefs," is also one in which the participants are persistently and intimately associated in nutritional interactions. While there is some evidence that the giant clams consume some of their algae, corals apparently limit their exploitation to the algal products and do not digest the algae themselves. Corals, like giant clams, raise the issue of the host's size. The tridacnid clams are individually big. Coral colonies attain mountainous dimensions, although corals are alive only in their outermost layer of small, active polyps. The rest of the reef's coral is skeleton, a geological diary of the historical interactions of organism and habitat. Among tridacnids and corals, the algal fixation of carbon and its subsequent incorporation into the hosts' skeletal calcium carbonate evidently help the hosts reach their great size (be it of one body or of accumulating clones). Of course, physiological animal–alga symbioses need not have this result, as the tiny acoel flatworm *Convoluta roscoffensis* shows. But the gigantism of tridacnid clams and of coral reefs does seem to reflect algal contributions to the hosts' particular biochemical capabilities, as well as the ease with which calcium carbonate is precipitated in warm marine habitats.

11   Cleaning Symbiosis

by Conrad Limbaugh
August 1961

The invasion of the oceans by skin-diving biologists has led to the discovery that a surprisingly large number of marine organisms either live by cleaning other marine organisms or benefit by being cleaned

While skin diving in the cool water off the coast of southern California in the spring of 1949, I observed a brief and seemingly casual meeting between a small golden kelp perch (*Brachyistius frenatus*) and a walleye surfperch (*Hyperprosopon argenteum*) twice its size. The walleye had separated itself from a milling school of its fellows several yards away and was holding itself rigid with fins extended, its body pointed at an unnatural angle to the surface of the water. The three-inch kelp perch spent several minutes picking at the silver sides of the walleye with its pointed snout. Then the kelp perch darted into the golden leaves of a nearby kelp plant, and the walleye returned to lose itself in the activity of the school. At the time I recorded this event in my notes only as an interesting incident.

Since then my studies and the observations of others have convinced me that this was not an isolated episode. On the contrary, it was an instance of a constant and vital activity that occurs throughout the marine world: cleaning symbiosis. Certain species of marine animal have come to specialize in cleaning parasites and necrotic tissue from fishes that visit them. This mutually beneficial behavior promotes the well-being of the host fishes and provides food for those that do the cleaning.

The relationship between the cleaner and the cleaned is frequently so casual as to seem accidental, as in the encounter that first caught my attention. On the other hand, one finds in the Bahamas the highly organized relationship between the Pederson shrimp (*Periclimenes pedersoni*) and its numerous clients. The transparent body of this tiny animal is striped with white and spotted with violet, and its conspicuous antennae are considerably longer than its

body. It establishes its station in quiet water where fishes congregate or frequently pass, always in association with the sea anemone *Bartholomea annulata,* usually clinging to it or occupying the same hole. When a fish approaches, the shrimp will whip its long antennae and sway its body back and forth. If the fish is interested, it will swim directly to the shrimp and stop an inch or two away. The fish usually presents its head or a gill cover for cleaning, but if it is bothered by something out of the ordinary, such as an injury near its tail, it presents itself tail first. The shrimp swims or crawls forward, climbs aboard and walks rapidly over the fish, checking irregularities, tugging at parasites with its claws and cleaning injured areas. The fish remains almost motionless during this inspection and allows the shrimp to make minor incisions in order to get at subcutaneous parasites. As the shrimp approaches the gill covers, the fish opens each one in turn and allows the shrimp to enter and forage among the gills. The shrimp is even permitted to enter and leave the fish's mouth cavity. Local fishes quickly learn the location of these shrimp. They line up or crowd around for their turn and often wait to be cleaned when the shrimp has retired into the hole beside the anemone.

Such behavior has been considered a mere curiosity for many years. The literature contains scattered reports of cleaning symbiosis, including a few examples among land animals: the crocodile and the Egyptian plover, cattle and the egret, the rhinoceros and the tickbird. As early as 1892 the German biologist Franz von Wagner had suggested that the pseudoscorpion, a tiny relative of the spider that is frequently observed stealing a ride on larger insects, is actually engaged in removing

parasitic mites from these insects. The U.S. biologist William Beebe in 1924 saw red crabs remove red ticks from sunbathing marine iguanas of the Galápagos Islands. While diving in the coral waters off Haiti four years later, Beebe also saw several small fishes of the wrasse family cleaning parrot fish. Mexican fishermen in the Gulf of California refer to a certain angelfish (*Holacanthus passer*) as *El Barbero.* They explain that this fish "grooms the other fishes" and so deserves its title as "The Barber."

Recognition of cleaning symbiosis and its implications has come only in recent years. The gear and the technique of skin diving have given marine biologists a new approach to the direct observation of undersea life. They have discovered numerous examples of cleaning behavior, enough to establish already that the behavior represents one of the primary relationships in the community of life in the sea. The known cleaners include some 26 species of fish, six species of shrimp and Beebe's crab. This number will undoubtedly increase when the many marine organisms now suspected of being cleaners have been studied more closely. It now seems that most other fishes seek out and depend on the service they render. The primary nature of the behavior is evident in the bright coloration and anatomical specialization that distinguish many cleaners. It appears that cleaning symbiosis may help to explain the range of species and the make-up of populations found in particular habitats, the patterns of local movement and migration and the natural control of disease in many fishes.

The importance of cleaning in the ecology of the waters off southern California became more and more apparent to me during the early 1950's as I accumulated observations of cleaners at work. My notes are particularly concerned with

FOUR CLEANING RELATIONSHIPS are depicted in this drawing by Rudolf Freund. In each the cleaner is in color. At top left a señorita (*Oxyjulis californica*) cleans a group of blacksmiths (*Chromis punctipinnis*). At top right are a butterfly fish (*Chaetodon nigrirostris*) and two Mexican goatfish (*Pseudupeneus dentatus*); in center, two neon gobies (*Elecatinus oceanops*) and a Nassau grouper (*Epinephelus striatus*); at bottom, a Spanish hogfish (*Bodianus rufus*) in the mouth of a barracuda (*Sphyraena barracuda*).

CALIFORNIA MORAY EEL (*Gymnothorax mordax*) has its external parasites removed by four California cleaning shrimps (*Hippolysmata californica*). At upper left is a fifth shrimp. This photograph and the one at right below were made by Ron Church.

SPANISH HOGFISH (*top*) in process of cleaning ocean surgeon (*Acanthurus bahianus*) was photographed by author in Bahamas.

LIONFISH (*Pterois volitans*) is host to a very much smaller cleaning wrasse (*purple fish in center*) of undetermined species.

the performance of the golden-brown wrasse (*Oxyjulis californica*), commonly called the señorita. This cigar-shaped fish is abundant in these waters and well known to fishermen as a bait-stealer.

Certain fishes, such as the opaleye (*Girella nigricans*), the topsmelt (*Atherinops affinis*) and the blacksmith (*Chromis punctipinnis*), crowd so densely about a señorita that it is impossible to see the cleaning activity. When I first saw these dense clouds, often with several hundred fish swarming around a single cleaner, I thought they were spawning aggregations. As the clouds dispersed at my approach, however, I repeatedly observed a señorita retreating into the cover of the rocks and seaweed nearby. Often the host fishes, unaware of my approach, would rush and stop in front of the retreating señorita, temporarily blocking its path. In less dense schools I was able to observe the señorita in the act of nibbling parasites from the flanks of a host fish. While being cleaned blacksmiths would remain motionless in the most awkward positions—on their sides, head up, head down or even upside down.

The material cleaned from fishes by the señorita and other cleaners has not been thoroughly studied. Among the organisms I have noted in the stomach contents of cleaners are copepods and isopods: minute parasitic crustaceans that attach themselves to the scales and integument of fishes. I have also found bacteria, and on several occasions I have seen señoritas in the act of nibbling away a white, fluffy growth that streamed as a milky cloud from the gills of infected fishes. Especially in the spring and summer months off California and farther south in the warmer waters off Mexico, many fishes display this infection; it ranges from an occasional dot of white to large ulcerated sores rimmed with white. Carl H. Oppenheimer, now at the University of Miami, has shown that this is a bacterial disease by infecting healthy individuals with material taken from diseased fishes.

Judging by the diversity of its clientele, the señorita is well known as a cleaner to many members of the marine community. Among the species that seek out its services I have counted pelagic (deep ocean) fishes as well as the numerous species that populate the kelp beds nearer shore. The black sea bass (*Stereolepis gigas*) and the even larger ocean sunfish (*Mola mola*) seem to come purposely to the outer edge of the kelp beds, where they attract large numbers of señoritas, which flock around them to pick off their parasites. I have also observed the señorita at work on the bat ray (*Holorhinus californicus*), showing that the symbiosis embraces the cartilaginous as well as the bony fishes.

Since first recognizing cleaning behavior in these southern California fishes, I have studied it in numerous places down the Pacific Coast of Mexico, in the Gulf of California, in the Bahamas and in the Virgin Islands. Observations such as mine have been paralleled in the literature by other skin-diving biologists and by underwater photographers. From 1952 to 1955 Vern and Harry Pederson made motion pictures in the Bahamas of cleaning behavior in a number of species of fish and in the violet-spotted shrimp that bears their name. In 1953 the German skin diver Hans Hass suggested that the pilot fish associated with manta rays ate the parasites of their hosts. Irenäus Eibl-Eibesfeldt, a German biologist, published notes in 1954 on cleaning behavior he had witnessed in fishes in Bahamian waters; he expressed the belief that it is common in the oceans of the world. In the Hawaiian and Society islands John E. Randall of the University of Miami identified as cleaners four fishes of the genus *Labroides*, two of which were new species.

A few generalizations about cleaning symbiosis may now be attempted. In the first place, the phenomenon appears to be more highly developed in clear tropical waters than in cooler regions of the seas. The tropical cleaner species are more numerous and include the young of the gray angelfish (*Pomacanthus aureus*), the butterfly fish (*Chaetodon*), gobies (*Elecatinus*) and several wrasses such as the Spanish hogfish (*Bodianus rufus*) and the members of the genus *Labroides*. Even distantly related species have analogous structures for cleaning, such as pointed snouts and tweezer-like teeth; this suggests convergent evolution toward specialization in the cleaning function. In the tropical seas the cleaning fish are generally brightly colored and patterned in sharp contrast to their backgrounds; it appears that most fishes that stand out in their environment are cleaners. Since cleaning fishes must be conspicuous, it is logical that they should have evolved toward maximum contrast with their surroundings. (The parasites on which they feed have evolved toward a maximum of protective coloration, matching the color of their hosts, and are usually invisible to the human observer of cleaning behavior.) In general these fishes are not gregarious and live solitarily or in pairs. In Temperate Zone waters, on the other hand, the cleaners are not so brightly colored or so contrastingly marked. They tend to be gregarious, to the point of living in schools, and are more numerous, though the number of species is smaller.

The cleaning behavior of the tropical forms is correspondingly more complex than that of the Temperate Zone species. Whereas the latter simply surround or follow a fish in order to clean it. the tropical cleaners put on displays not unlike those shown in courtship by some male fishes. They rush forward, turn sideways and then retreat, repeating the ritual until a fish is attracted into position to be cleaned. Frequently they sense the presence of a fish before a human observer can, and they hasten to take up their station before the fish arrives to be cleaned.

Some species clean only in their juvenile stage; none of them appears to depend exclusively on the habit for its food. Again, however, the tropical species come closer to being "full time" cleaners. One consequence of their higher degree of specialization is that they enjoy considerable immunity from predators. In an extensive investigation of the food habits of California kelp fishes I never found a señorita, a close cousin of the numerous cleaning wrasses of the tropics, in the stomach contents of other fishes. I have seen it safely enter the open mouth of the kelp bass, a fish that normally feeds on señorita-size fishes. On the other hand, the kelp perch, a more typical Temperate Zone cleaner, frequently turns up in the stomachs of fishes that it cleans. The immunity of certain cleaners is so well established that other fishes have come to mimic them in color and conformation and so share their immunity. Some mimics reverse the process and prey on the fish that mistake them for cleaners!

The same generalizations may be made in contrasting the cleaning shrimps of the Tropical and Temperate zones. Only one of the six known species occurs outside the tropics; this is the California cleaning shrimp (*Hippolysmata californica*). It is a highly gregarious and wandering animal, at the other pole of behavior from the tropical species as represented by the solitary and sedentary Pederson shrimp of the Bahamian waters. The California cleaning shrimp does not have the coloration and marking to make it stand out from its environment. So far as I have been able to determine, it does not display itself to attract fishes. These California shrimps wander abroad in troops numbering in the hundreds, feeding on the bottom at night and retiring to cover during the day. They act as cleaners when they come upon an animal, say a lobster, in need of cleaning or when a fish, perhaps a moray eel, swims into the crevice where they have found shelter. They will crawl rapidly over the entire outside surface of the animal, cleaning away everything removable, including decaying tissue. A lobster that has been worked over by a team of these shrimps comes out with a clean shell; a human diver's hand will receive the same treatment. Fishes do not seem to be bothered by these rough attentions, although the moray may occasionally jerk its head as if annoyed.

In some cases the shrimps may enter the mouth of the moray to get at parasites there, but not without risk; the stomachs of morays have yielded a considerable number of these shrimps. In

SPOTTED GOATFISH (*Pseudupeneus maculatus*) is host to the smaller Spanish hogfish. The hogfish is found in the tropical waters from Bermuda and Florida to Rio de Janeiro, in the Gulf of Mexico and around Ascension and St. Helena islands in the South Atlantic.

GARIBALDI (*Hypsypops rubicunda*) at top holds itself at an unnatural angle while being cleaned by a señorita. The latter, which is found in temperate waters from central California to central Lower California, cleans more than a dozen species of fish.

contrast, the tropical cleaning shrimps, all of them more exclusively specialized as cleaners, seem to have the same immunity from predation as the tropical cleaning fishes. With their bright colors, their fixed stations and their elaborate display behavior, they are plainly adver-

tised to the community as cleaners and attract hosts rather than predators. It is easy to visualize the evolutionary path by which the more complex cleaning symbiosis may have developed from the imperfect cleaner-host relationships such as that of the California shrimp.

In the summer of 1955, in the Gulf of California near Guaymas, I noted that cleaning behavior appeared to be concentrated at rocky points: each point was manned by two butterfly fish and one angelfish. I assumed that the concentration of other fishes arose from the

JUVENILE GRAY ANGELFISH (*Pomacanthus aureus*) at right cleans external parasites from the tail of a bar jack (*Caranx ruber*).

Below the jack is another cleaner, the Spanish hogfish. This photograph and those on the following page were made by the author.

"CLEANING STATION," consisting of a sponge (*light area with small, dark protuberances*) surrounded by turtle grass, is manned

by a juvenile gray angelfish. The station, located off New Providence Island in the Bahamas, was photographed by the author's wife.

fact that these points constitute the intersection of the communities of fishes on each side. In 1958 Randall, reporting on his studies of the cleaning wrasses in the Society Islands, observed that fishes came from comparatively long distances to the sites occupied by the cleaners, not just from the immediate community. The Pederson brothers made the same observation in the Bahamas, reporting that the cleaners congregate in regular "cleaning stations" in the coral reefs and attract host fishes from large areas.

Subsequent studies have confirmed these observations. The various species of cleaning fish and shrimp tend to cluster in particular ecological situations: at coral heads, depressions in the bottom, ship wreckage or the edge of kelp beds. Their presence in these localities accounts in great part for the large assemblages of other fishes that are so frequently seen there. Even a small cleaning station in the tropics may process a large number of fish in the course of a day. I saw up to 300 fish cleaned at one station in the Bahamas during one six-hour daylight period. Some of the fishes pass from station to station and return many times during the day; those that could be identified by visible marks, such as infection spots, returned day after day at regular time intervals. Altogether it seemed that many of the fishes spent as much time at cleaning stations as they did in feeding.

At cleaning stations inhabited by thousands of cleaning organisms, cleaning symbiosis must assume great numerical significance in determining the distribution and concentration of marine populations. In my opinion, it is the presence of the señorita and the kelp perch that brings the deep-water coastal and pelagic fishes inshore to the edge of the kelp beds on the California coast. Most concentrations of reef fishes may similarly be understood to be cleaning stations. Cleaning symbiosis would therefore account for the existence of such well-known California sport-fishing grounds as the rocky points of Santa Catalina Island, the area around the sunken ship *Valiant* off the shore of Catalina, the La Jolla kelp beds and submarine

BLACKSMITHS IN GROUP waiting to be cleaned by a single señorita (*slender fish in nearly horizontal position at right center*) assume various positions. This photograph was made by Charles H. Turner of the State of California Department of Fish and Game.

canyon and the Coronado Islands.

These generalizations of course call for further observation and perhaps experimental study. In a modest field experiment in the Bahamas I once removed all the known cleaning organisms from two small, isolated reefs where fish seemed particularly abundant. Within a few days the number of fish was drastically reduced; within two weeks almost all except the territorial fishes had disappeared.

This experiment also demonstrated the importance of cleaning symbiosis in maintaining the health of the marine population. Many of the fish remaining developed fuzzy white blotches, swelling, ulcerated sores and frayed fins. Admittedly the experiment was a gross one and not well controlled, but the observed contrast with the fish populations of the nearby coral heads was very striking. Certainly it appeared that the ailments occurred because of the absence of cleaning organisms. This impression was strengthened when a number of local fishes that had been maintained in an aquarium were found to be developing bacterial infections. I placed a cleaner shrimp in the aquarium, and it went to work at once to clean the infected fishes.

Symbiotic cleaning has some important biological implications. From the viewpoint of evolution it provides a remarkable instance of morphological and behavioral adaptation. Ecologically speaking, cleaners must be regarded as key organisms in the assembling of the species that compose the populations of various marine habitats. Cleaning raises a great many questions for students of animal behavior; it would be interesting to know what mechanism prevents ordinarily voracious fishes from devouring the little cleaners. In zoogeography the cleaning relationships may provide the limiting factor in the dispersal of various species. In parasitology the relationship between the cleaning activities on the one hand and host-parasite relations on the other needs investigation. The beneficial economic effect of cleaners on commercially important marine organisms must be considerable in some areas. The modern marine-fisheries biologist must now consider cleaners in any thorough work dealing with life history and fish population studies. From the standpoint of the philosophy of biology, the extent of cleaning behavior in the ocean emphasizes the role of co-operation in nature as opposed to the tooth-and-claw struggle for existence.

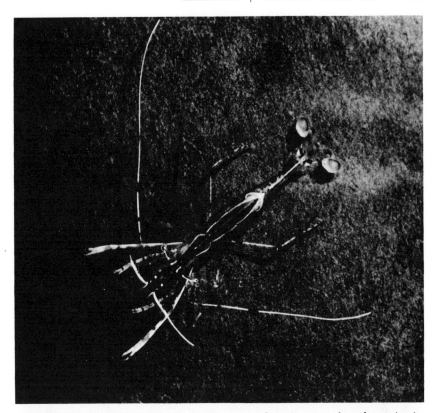

PEDERSON CLEANING SHRIMP (*Periclimenes pedersoni*) attracts hosts by waving its antennae, which are longer than its body. Shell-like objects (*upper right*) are shrimp's uropods, or "flippers." Photograph was made by F. M. Bayer of Smithsonian Institution.

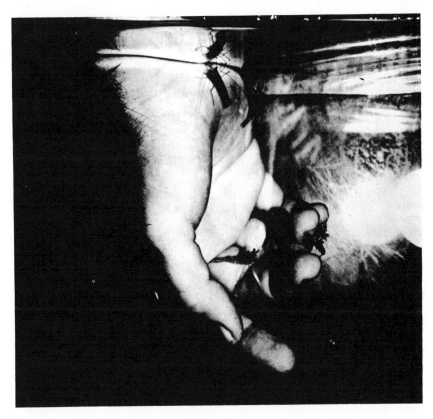

CALIFORNIA CLEANING SHRIMPS "clean" the author's hand, even to picking at his fingernails. These shrimps clean everything that is removable from the exterior of a host.

12 Flashlight Fishes

by John E. McCosker
January 1977

These marine species light up the water with a large organ under each eye containing living luminous bacteria. They use the light to see by, to communicate, to lure prey and to confuse predators

The production and utilization of light by animals, particularly by fishes, has interested biochemists, behaviorists and observers of nature for centuries. Perhaps the most spectacular example of this phenomenon of bioluminescence is found among the "flashlight" fishes of the family Anomalopidae. Whereas most bioluminescent animals normally use their light for only one purpose—be it communication, luring prey, avoiding predators or improving visibility—flashlight fishes employ their light for all these purposes.

The name Anomalopidae is derived from the Greek for "abnormal eye." The name appropriately describes the appearance of the four species in the family: *Anomalops katoptron, Photoblepharon palpebratus, Kryptophanaron alfredi* and *Kryptophanaron harveyi.* In all these species the fishes have a specialized organ below each eye that is filled with light-emitting bacteria, which collectively generate an illumination that is about as intense as the light from a weak flashlight. The fish can obscure the light in various ways, so that in effect they turn it on and off.

These luminous creatures have been shrouded in mystery since the first specimen was observed in the 18th century by the Dutch naturalist Peter Boddaert. The reason little was known about them was that few were found, and the reason few were found is that they are small and reclusive creatures that tend to be active only on dark nights and in fairly deep water. In recent years a good deal more has been learned about the flashlight fishes, and at the Steinhart Aquarium in San Francisco (a division of the California Academy of Sciences) I have been fortunate enough to be able to maintain a living collection of several of them. As a result my colleagues and I have had an opportunity to observe them and to examine their behavior experimentally. Recent advances in bacteriology, underwater observation and the transportation and maintenance of living fish have brought a synthesis of skills

that has resulted in a rapidly increasing body of information about the flashlight fishes.

The common name "flashlight fish" evolved from an article in *Science* in 1975 titled "Light for All Reasons: Versatility in the Behavioral Repertoire of the Flashlight Fish." James G. Morin of the University of California at Los Angeles and his five coauthors concluded the article: "Thus the bioluminescent behavioral repertoire of *Photoblepharon* is extensive and varied. It includes many different offensive, defensive and communicative activities, and is especially unusual because only a single type of light organ is involved. The multiplicity of functions suggest that the organ is like a flashlight, whose owner can exercise options in its use."

My first encounter with flashlight fish was incidental to an expedition mounted by the California Academy of Sciences in 1974 and 1975 to collect the rare coelacanth fish at Grande Comore Island in the Indian Ocean. Although my formal training at the Scripps Institution of Oceanography was in ichthyology, I had never seen a living or preserved anomalopid in the large fish collection at Scripps or in the enormous collections at Stanford University and at the California Academy of Sciences. Through the writings of the late E. Newton Harvey of Princeton University, an authority on bioluminescence, I was familiar with the strange and intensely bioluminescent fishes of the genera *Photoblepharon* and *Anomalops,* and I was therefore intrigued when Francis Debuissy, a French veterinarian and scuba diver stationed at Grande Comore, described a luminous fish called *le petit Peugeot* by the few French divers who had seen it. They were the intrepid men who ventured into the tropical Comoran waters at the time of the new moon. The fish were seen only on the darkest nights and were observed abundantly only at depths greater than 30 meters, where they swam above the reef with their

light organs exposed and were indeed reminiscent of a small, dark automobile on a country road at midnight.

Debuissy said rather wryly that he had only seen the fish, being unable to capture any of them because his battery light was too weak to immobilize them. (The flashlight fishes, like many other animals, tend to be transfixed by a bright light.) He felt sure that our newer and brighter diving lights would suffice, particularly when the moon was in its darkest phase. Subsequent nights proved him to be correct. I returned to San Francisco in March, 1975, with numerous living and preserved specimens of *Photoblepharon.*

In addition to the work that my colleagues and I have done, the number of investigations of flashlight fishes has increased significantly since the Arab-Israeli conflict of 1967. The connection between these seemingly unrelated events was explained to me by Morin, who in turn heard it from an Israeli ichthyologist with whom he had studied the population of flashlight fish in the Red Sea near the Heinz Steinitz Marine Biological Laboratory, which is on the Gulf of Eilat. During midnight patrols along the coastline of the Sinai Peninsula after the Six-Day War, Israeli soldiers had observed a faint green glowing mass beyond the coral reef. The soldiers, naturally assuming that they had encountered a team of enemy frogmen, responded by discharging explosives in the glowing shoals. To their surprise the result was a beach littered with the bodies of small, dark fish whose heads continued to blaze with a pair of green, glowing patches.

Since 1967 several groups of American and Israeli workers have returned to the Gulf of Eilat to study these little-known fish. They have discovered that during the daytime the flashlight fish live in the dark caves and recesses of the deep reef. On nights with little or no moonlight they venture out of the caves either alone or in small groups to forage along the bottom for small crustaceans and other plankton of the reef's edge. In the Red Sea they move into shallow water in compact aggregations, which give the appearance of a large green superorganism.

The first person to give a scientific name to the flashlight fishes was Boddaert. In 1781 he named a specimen from Indonesia *Sparus palpebratus,* meaning "the porgy with an eyelid." Boddaert proposed that the function of the large, unusual organs below the eyes was to shield the eyes from injury caused by the coral branches among which the fish lived. In 1803 the French naturalist Count Bernhard Lacépède suggested that the organ served to protect sensitive tissue against sunlight.

It was not until 1900 that the Dutch ichthyologist A. G. Vorderman recorded his observation of light being produced by the living fish. Taxonomists with little comprehension of this odd creature's characteristics subsequently classified it in various genera and families. Eventually the Dutch biologist Max Weber established the genus *Photoblepharon* (meaning eyelid light), in which one species of flashlight fish now resides. The recent discovery of a population of the species in the Red Sea

and my discovery of its presence in the Indian Ocean indicate that it is distributed from Indonesia to the Red Sea. It is likely that as night-diving biologists survey the intermediate areas more carefully the fish will be discovered to be widely distributed.

A second species from the Celebes Sea was named in 1856 by the Dutch ichthyologist Pieter Bleeker *Heterophthalmus* (meaning different eye) *katoptron* (meaning mirror, presumably referring to the luminous organ). The species was later placed in the new genus *Anomalops* because *Heterophthalmus* had already been adopted as the generic name

FLASHLIGHT FISH of the species *Photoblepharon palpebratus* is portrayed at about twice the actual size of the mature fish. The light organ under the eye is in the open position. The fish is not only small and dark but also reclusive, living in caves and recesses in deep water.

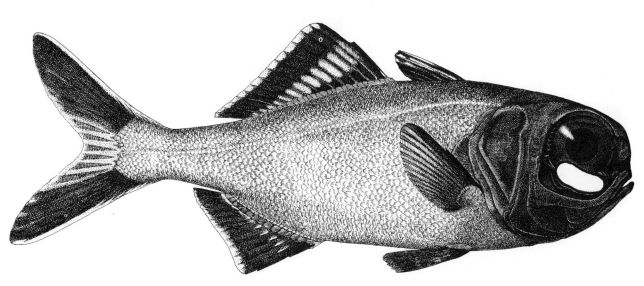

PORTRAIT OF ANOMALOPS KATOPTRON shows its anatomical differences from *Photoblepharon palpebratus.* The scale of the two drawings is the same. Specimens of *Anomalops* have been collected in Indonesia, in the Solomon Islands and in Japanese waters.

PHOTOBLEPHARON PALPEBRATUS were photographed at night at a depth of about 30 meters in the Indian Ocean by David C. Powell of the Steinhart Aquarium in San Francisco. The fish are active only on dark nights and in deep water, so that information about their habits has been difficult to obtain. Now the aquarium has several living specimens to observe.

ANOMALOPS KATOPTRON is the only other species of flashlight fish that biologists have observed alive. This species differs from *Photoblepharon* in the mechanism of obscuring its light: instead of raising a curtain, it rotates the light organ downward into a pocket. *Anomalops* also are more gregarious, forming schools of some 200 fish that feed near the surface at night.

PINECONE FISH (CLEIDOPUS GLORIAMARIS), also known as the port-and-starboard-light fish, is not a flashlight fish but has a somewhat similar light organ. The organ is fixed in position. Its orange surface acts as a filter to transform the bluish light produced by the organ to a blue-green. Studies at the Scripps Institution of Oceanography indicate that the light organs of each fish are colonized by a different clone of bacteria, as may be true of flashlight fishes.

for a beetle. This gregarious species behaves quite differently from *Photoblepharon* in that it forms schools of as many as 200 individuals feeding on plankton near the surface at night. *Anomalops* is like *Photoblepharon* in being reclusive during periods of daylight or bright moonlight, but its habitat during those times is not known. The species has been collected from several places in Indonesia and the Solomon Islands and is also known from five specimens taken in Japanese waters.

It is somewhat enigmatic that there appear to be two species of anomalopid fishes in the New World, each known from a single dead specimen. A Caribbean form, named *Kryptophanaron* (meaning hidden lantern) *alfredi* (in honor of Alfred Mitchell), was discovered floating on the surface off the coast of Jamaica in 1907 by Ulric Dahlgren of Princeton University. The American workers Charles F. Silvester and Henry W. Fowler described the specimen, which they subsequently lost, as a new genus and species. No other specimens of *K. alfredi* have been found.

In 1972 a small, dark fish with glowing patches under each eye was captured by a shrimp trawler in the relatively shallow water of the Gulf of California. The Mexican captain presented the specimen to W. Linn Montgomery, a graduate student in ichthyology at the University of California at Los Angeles, with the statement that he had not previously seen that species in 35 years of trawling for shrimp. Richard H. Rosenblatt of the Scripps Institution and Montgomery determined that the specimen represented a new species, which they named *Kryptophanaron harveyi* in honor of E. Newton Harvey.

The two forms from the New World are quite similar, suggesting that they have a common ancestry predating the formation of the Central American land bridge. That event, which is believed to have occurred between one and three million years ago, separated the aquatic populations of the Caribbean and the eastern Pacific and provided a basis for their subsequent differentiation into separate species.

When one considers the numerous collections of fishes that biologists and commercial fishermen have made in the Caribbean and the Gulf of California, it is puzzling that *Kryptophanaron alfredi* and *K. harveyi* are known only from single specimens. The rarity of these fishes must be attributable to the habitat they presumably prefer, namely reefs that are below the depths where most scuba divers go and rocky areas that are relatively inaccessible to collection with nets. I might add on the basis of personal experience that for sane biologists deep diving in tropical seas on dark nights with one's diving light turned off is rarely practiced and never enjoyed.

Anomalopids are distinctive in several ways. Their light organ produces what is perhaps the most intense light known to come from a multicellular luminescent organism. An anomalopid fish apparently employs the light not only for attracting prey, confusing predators and communicating with other members of the species but also as a flashlight to see what is in front of it.

The organ is cream-colored on its outer surface and black on its inner and upper surfaces. (If these surfaces were not black, the light would certainly blind the fish itself.) It is easy to experiment with the organ because it can be removed surgically without difficulty and will continue to glow for eight hours or more after removal.

The glow of the light organ in *Photoblepharon* and *Anomalops* comes from

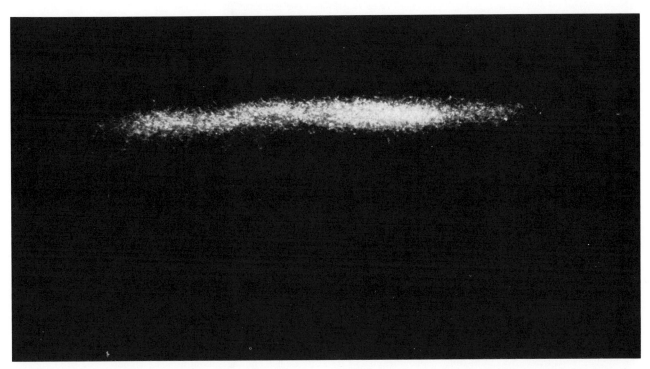

SCHOOLING FLASHLIGHT FISH were photographed in the Red Sea at night by the light of their own luminous organs. This school of *Photoblepharon palpebratus* contained about 30 fish. Since each individual turns the light off frequently in a blinking pattern by raising a curtain of skin over the organ containing the light-emitting bacteria, it is unlikely that all the lights were "on" in this group of fish.

DAYLIGHT VIEW of the place in the Red Sea where the school of flashlight fish was photographed shows that the far edge of the coral reef (about 12 meters from the camera) was the area where the fish congregated at night to feed on plankton in the shallow water above the reef. In other parts of the world *Photoblepharon* are seldom found abundantly except at depths of 30 meters (100 feet) or more.

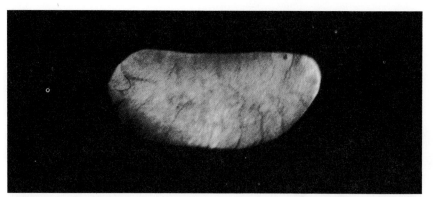

LIGHT ORGAN of a flashlight fish was removed surgically and photographed in its own light by means of a time exposure. The bacteria in the organ continue to emit their blue-green light for eight hours or more after the organ has been removed from the fish. This organ, which in the fish has a black inner surface, is from a living specimen of _Photoblepharon palpebratus._

bacteria that are symbiotic with these fishes. For the bacteria the light is a by-product of metabolism, as heat is a by-product of metabolism in warm-blooded animals. Enormous numbers of these bacteria (some 10 billion per milliliter of fluid in the organ) are packed into special nutritive compartments within the fishes' light organ.

In the Comores my colleague Michael D. Lagios and I had hoped to culture the bacteria in various nutritional mediums. Our thought was that it would be prudent to return to the Steinhart Aquarium with bacteria in culture as well as in the living fish; if the lights "went out," we could reinfect the organs and turn the lights back on. We were dismayed to find that we were unable to culture the bacteria. Later we learned that both Harvey and the Japanese workers Yata Haneda and F. I. Tsuji had failed in similar attempts. Haneda and Tsuji concluded that they were dealing not with normal bacteria but with prokaryotic (lacking a nucleus) cell-like organisms they termed "bacterioids."

A possible explanation for this perplexing failure of efforts to culture the bacteria has been advanced by Kenneth Nealson of the Scripps Institution. His hypothesis is that the light-generating microbe of _Photoblepharon_ is so specialized that it is an obligate symbiont: it cannot survive outside its host.

Nealson proposed that this bacterium, unlike more independent bacteria, can only partially metabolize the glucose of its host. Therefore in culture it probably generates toxic concentrations of pyruvic acid, a product of the incomplete oxidation of the sugar. The enzymes the host provides to break down the pyruvic acid are absent in the culture medium, and so the medium becomes too highly charged with pyruvic acid for the bacterium to survive. The phenomenon has been observed in similar symbionts (but not obligate ones). Fortunately the lights of the specimens at the Steinhart Aquarium have remained lit

since March, 1975. Indeed, at the time of writing the lights had become brighter rather than weaker as the fish grew accustomed to the conditions in the aquarium.

When I first dissected a specimen of _Photoblepharon_ to examine its sexual state and the contents of its stomach, I was surprised to discover a considerable amount of a fatty substance dispersed throughout the coelom: the space between the body wall and the digestive tract. This substance, which has the consistency of lard at room temperature, is probably an energy store. Presumably it provides energy for both the fish and the bacteria during the lunar periods when _Photoblepharon_ is less active than it is at other times and is fasting.

The low density of the fatty material would make the fish excessively buoyant if it were not for a compensating reduction in the volume of the gas bladder that is typically found in fishes as a means of controlling buoyancy. The gas bladder in related but nonluminous fishes is much larger. For biologists the reduced gas bladder of _Photoblepharon_ is a happy evolutionary circumstance, since it means that specimens can be brought up from deep water without the otherwise common problem of rupture of the bladder from expanding gas.

Flashlight fishes are also unusual among bioluminescent organisms in that their light is on more than it is off. To turn off its light _Photoblepharon_ simultaneously raises a black curtain over each light organ, completely blocking out the light. The rate of blinking varies with the water temperature and the conditions of the fish's environment. When live brine shrimp (a food for _Photoblepharon_) are added to an aquarium tank containing several of the fish, a rapid blinking sequence ensues, suggesting that the fish are somehow communicating the information to one another.

In using its light organ to deal with predators _Photoblepharon_ relies on a

sudden flash of the light that so startles the predator that _Photoblepharon_ gains time to escape. Moreover, our studies of the living fish in our aquarium have disclosed a most interesting adaptation that the species presumably employs to confuse predators. Each time a swimming _Photoblepharon_ changes direction, it turns off its light. Shortly afterward it turns the light on again, but from the viewpoint of a predator the position at which the light of the prey reappears is unpredictable.

This behavior is continuous. Presumably it makes _Photoblepharon_ difficult for a predator to track. Morin has observed the behavior in a natural setting, where he has recorded an average frequency of 75 blinks per minute with an average duration of 160 milliseconds for each blink. His term for the behavior is "the blink and run."

With photometric techniques Morin and his colleagues have identified certain basic patterns of the bioluminescent activity of _Photoblepharon_. The most obvious pattern is an infrequent blinking. (Most other bioluminescent organisms flash rather than blink.) This pattern is typical of undisturbed _Photoblepharon_ at night. They blink an average of 2.9 times per minute, and each blink has an average duration of 260 milliseconds.

A second pattern suggests that a more or less daily rhythm exists in the spontaneous blinking frequency of fish kept in continuous darkness. The blinking rate is much higher during the daytime hours than it is during the night, and the duration is somewhat longer. The average frequency is 37 blinks per minute and the average duration of the blinks is 800 milliseconds.

Several specimens of _Anomalops_ that arrived at the Steinhart Aquarium in a weakened condition, with their lights extinguished, provided an opportunity to test the hypothesis that flashlight fishes use their light for seeing. When live brine shrimp were put in the tank occupied by the specimens, the lightless fish largely failed to discover the shrimp. The shrimp were eaten immediately when a light approximately equal in intensity to the natural light of the fish was turned on in the room.

We tried without success to rekindle the natural light of these fish. Our first attempt was to reinfect the _Anomalops_ organ with a culture of bacteria from glowing specimens of _Photoblepharon_. It did no good to put light and dark fish in the same tank, even though tests of the water showed that many photobacteria were present. Attempts by Edward E. Miller of the Steinhart Aquarium staff to transfer bacteria from _Photoblepharon_ to _Anomalops_ with a hypodermic needle were also unsuccessful.

The failure of all these efforts suggests that the two species provide different environments for the bacteria. The answer

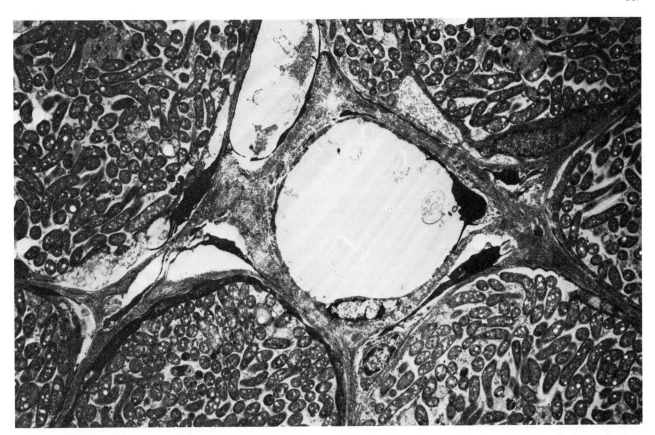

LIGHT-EMITTING BACTERIA in a specimen of *Photoblepharon palpebratus* appear in an electron micrograph made by Michael D. Lagios of Children's Hospital in San Francisco. The enlargement is about 5,700 diameters. The bacteria are the elliptical and round structures inside the compartments surrounding the central vessel, which provides nourishment to the bacteria from the blood of the fish.

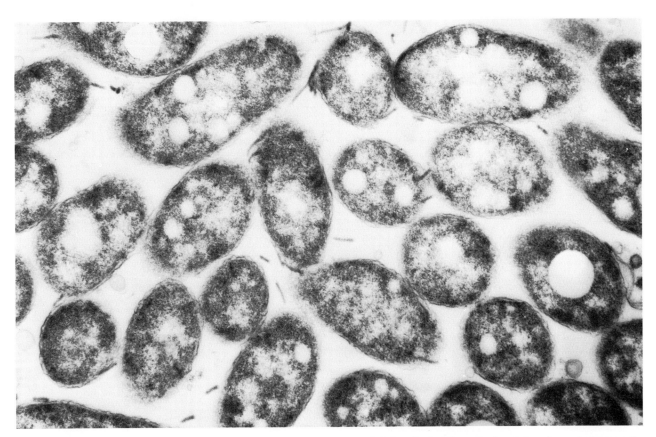

CLOSER VIEW of the light-emitting bacteria shows them at an enlargement of 28,500 diameters. The slender, dark projections are flagella; their function in this species is not clear. Inner circles are artifacts resulting from evaporation of granules by the electron beam.

may be more complicated, however, as Nealson and Edward Ruby discovered at the Scripps Institution in work with the bioluminescent pinecone fish. Their preliminary findings indicate that each pinecone fish is colonized by a different clone of bacteria and so is in effect a biological island.

Man has taken advantage of the fact that the flashlight fish lures prey with its light organ. Although the fish captures prey smaller than itself by attracting the prey with its lights, the fishermen of the Banda Islands in Indonesia learned that larger fishes are also attracted to the light. Harvey noted in 1922 that Banda fishermen removed the organ from the fish and attached it to their lines above the hook to act as a lure. The organ remained luminous for many hours as the symbiotic bacteria continued to glow. Other fishermen in Indonesia have learned to take advantage of these alluring properties without harming the flashlight fish. By suspending below

their canoe a perforated length of bamboo enclosing a dozen or more living *Photoblepharon* they can fish each night with reusable lures.

Although the light organ is essentially the same in all anomalopid fishes, the way in which it is operated differs considerably. *Anomalops* obscures its light by rotating the light organ downward into a darkened pocket. (The organ is hinged at the front.) *Photoblepharon* extinguishes its light by raising a black curtain over the organ. The curtain is much like an eyelid.

The reason for these differences may be indicated by the recently collected specimen of *Kryptophanaron*. Rosenblatt and Montgomery have noted that the light organ of the fish can be rotated but that a membrane is also associated with the organ. They propose that *Kryptophanaron* controls its light by both rotation and a shutter mechanism, with the rotational movement serving to in-

terrupt the light for fairly long periods and the shutter mechanism accomplishing rapid blinking. Rosenblatt and Montgomery believe both mechanisms may have been present in the ancestor of *Kryptophanaron* and the common ancestor of *Photoblepharon* and *Anomalops*.

If this supposition is correct, the membrane of *Kryptophanaron* or its ancestor provides the genesis for the shutter mechanism employed by *Photoblepharon*, a species that has retained the presumed ancestral habit of staying near the bottom. *Anomalops* has evolved in such a way that it leaves deeper water to feed on plankton near the shore at night. Behavioral accommodations accompanying this evolutionary change include the formation of large schools and a continuous rapid blinking. Both adaptations make it difficult for a predator to single out a lone prey.

Rosenblatt and Montgomery have speculated as follows about the change. "In clear water, near the surface, away

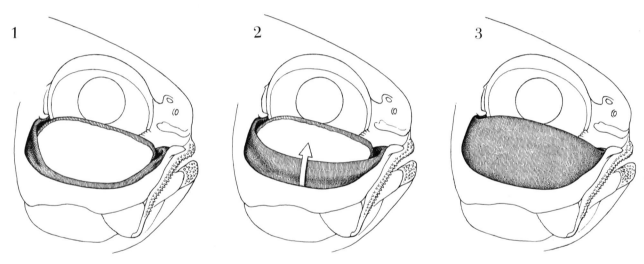

OBSCURATION OF LIGHT ORGAN by *Photoblepharon palpebratus* is depicted. The mechanism resembles an eyelid, except that in its normal position (*1*) it is folded below the light organ and that the fish raises it (*2 and 3*) in order to black out the light briefly.

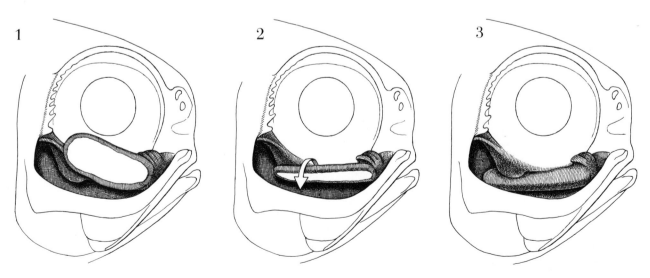

DIFFERENT MECHANISM serves to obscure the light of *Anomalops katoptron*. Its light organ is hinged at the front by a muscle (*1*). The fish employs the muscle (*2 and 3*) to rotate the organ downward into a pouch. Both species of fish normally blink several times per minute.

from obstructions of the reef, a predator would be able to see the luminous organ, and begin its rush, from a considerable distance. A rapid blink and short dash [as performed by *Photoblepharon*] would require only a small course correction on the part of the predator when the light came on. However, in *Anomalops* the light is occluded for a relatively long period, and all members of the school are blinking. This mechanism should be much more effective in lessening the opportunity of a predator to fixate visually on a single individual. Intermittent flashing in *Anomalops*, then, lessens exposure to predators caused by use of the light organ in feeding and maintaining the school."

Many interesting questions about the anomalopids remain to be explored. What is the behavior of *Kryptophanaron*, the fish that has yet to be observed alive? How do the bacteria in the light organs compare in the two species of *Kryptophanaron* and with *Photoblepharon* and *Anomalops?* Have the symbiotic bacteria evolved at a rate different from that of their hosts? How do larval flashlight fishes acquire their bacteria? Is the photobacterium carried within the fertilized egg, remaining dormant until the time when bioluminescence becomes important to the survival of the juvenile? The answers to these questions will require a combination of persistence and luck in observing the fishes in their natural environments and in experimenting with them in the aquarium.

Finally, it is appropriate to describe here a far-reaching experiment that has been suggested to me by Nealson and by J. Woodland Hastings of Harvard University for a beneficial application of our research on these rather esoteric fishes. Nealson and Hastings are impressed by the remarkable purity of the colonies of bacteria in the light organs of flashlight fishes, that is, by the absence of competition from other bacteria. They suggest that it testifies to the existence in the fish of an extremely effective immune system of the type that in other animals wards off foreign organisms invading the body.

Nealson and Hastings suggest that a genetic engineer might be able to replace certain unnecessary genetic information in the bacterium with information that controls operations useful to the human organism. An example would be an enzyme system that would trigger the production of insulin in a person who is diabetic. By linking such a system to a photobacterium the geneticist would have a visual measure of the purity of his culture. When the light is on, one can assume that the culture is free of contamination. I marvel at the suggestion. During my 50-meter dives in the Indian Ocean on moonless nights, I never suspected that the flashing lights I was following might lead to such an end.

13

Giant Clams

by C. M. Yonge
April 1975

The tridacnids, related to cockles, include the largest bivalve in evolutionary history. Their size is probably due to the fact that photosynthetic algae live in their tissues and nourish them

As the tide retreats from the upper surface of the Great Barrier Reef of Australia a remarkable vista materializes. Visible for miles among the coral heads are the occasional rounded tops of the giant clam *Tridacna gigas*. The presence of these huge bivalve mollusks, which can be as much as four feet long, two feet wide and two feet high, immediately raises two biological questions. One is: How did such a large bivalve come into existence? The other has to do with the fact that the waters of tropical reefs are notoriously poor in mineral nutrients and therefore in the plankton that support the animal life of the sea. How, when an adequate food supply must be of vital importance to a bivalve as large as *T. gigas,* can it inhabit such an impoverished environment?

The answers to these questions are related. Long before any tridacnids existed the first modern corals made their appearance, and massive coral reefs arose in the world's shallow tropical seas. The reef-building corals presumably lived, as they do today, in symbiosis with unicellular brown algae: the zooxanthellae, photosynthetic organisms that have now been identified as a species of dinoflagellate, *Gymnodinium microadriaticum,* in a resting stage. Among all the many thousands of molluscan species this kind of symbiosis is found in only seven instances: in the six species of tridacnid clams and in one fairly close relative. This fact suggests not only that the tridacnids evolved in the same kind of coral-reef environment they inhabit today but also that their survival in these nutrient-poor waters depends in no small part on the nourishment provided by the photosynthetic algae they harbor.

To trace the evolution of the tridacnids it will be useful to review the anatomy of bivalves in general. One can, for example, better appreciate the bizarre upside-down position of the tridacnid shell when it is compared with the normal orientation of the shell among other bivalves. Within the phylum Mollusca the bivalves form the class Bivalvia, so called because of their two-part shell. While a bivalve is still in its larval stage it becomes enclosed within a fleshy mantle that secretes the shell. This outer covering consists of the two calcified "valves" and a connecting elastic ligament, and it can be tightly closed by the contraction of the animal's adductor muscles. When the muscles relax, the shell gapes slightly, giving the animal access to the surrounding water.

The mouth of a bivalve, however, is never in direct contact with the external environment. Both the oxygen and the nutrients the animal requires come to it through paired, enormously enlarged gills. The distinctive molluscan gills were named ctenidia by early anatomists because of their comblike structure; *ktenos* is the Greek for "comb." The organ has a remarkable morphological potential, and in the bivalves it has been modified into a living sieve that endows the animals with the most efficient means of ciliary feeding known in the animal kingdom.

In its simplest form each of the paired gills of a bivalve consists of a main axis with lateral filaments on both sides that bear characteristic rows of cilia. Certain of the cilia create a powerful water current by their beating; others remove food particles and sediment from the passing water. The food particles, usually consisting of phytoplankton (plant plankton), are sorted, largely in terms of size, by ridged and ciliated palps before they enter the animal's mouth. At the same time other ciliated tracts accumulate the accompanying sediments for expulsion.

The bivalve's gills are enclosed in a respiratory cavity within the mantle. The water that passes through them enters the cavity through an inhalant opening and leaves through an exhalant one. Both openings are typically located at the posterior end of the animal. They are often modified into extensible and retractable siphons, and bivalves that are so equipped, such as the many kinds of burrowing clam, can penetrate deep into sand or mud. If the bivalve is a surface dweller such as a scallop or an oyster, its inhalant opening does not end in a narrow siphon but flares out. As the sole means of contact with the external environment the inhalant opening is fringed with receptive tentacles; in some bivalves, such as scallops and cockles, it is also surrounded by eyes.

The larvae of virtually all bivalves are free-swimming at first. Some adult bivalves remain mobile, others anchor themselves and others are anchored early in their life cycle and are mobile later. In all cases the same organ, the foot, plays a key role. An unattached bivalve such as a cockle uses its large, muscular foot to propel itself rapidly, sometimes with a leaping movement, to escape predators. In sedentary bivalves, such as mussels, special glands in the foot secrete an anchoring mass of fine threads known as the byssus, a term the Greeks applied to a closely woven cloth.

These are the generalizations concerning bivalves that are pertinent here. As a major class of mollusks the animals have exploited their potential with striking success. That is true not only in terms of the size of many bivalve populations but also in terms of the diversity of the bivalve form. Although the morphological theme is basically simple, the variations played on it have brought forth a large array of bivalve superfamilies, each exhibiting a particular struc-

SUNLIT GIANT CLAM is examined by a diver in the waters off the Eniwetok atoll in the Marshall Islands. This specimen is the species *Tridacna gigas*, largest of giant clams. Some specimens of *T. gigas* have exceeded four feet in length and 550 pounds in weight.

MIDDLE-SIZED GIANT, *Tridacna squamosa*, does not exceed 16 inches in length. Early in life it is attached to the reef by its byssal threads. As it grows larger its weight also helps to hold it in place. Its mantle tissue, which shows a spotted pattern in this specimen photographed at Eniwetok atoll, receives the sunlight that penetrates the water. In this tissue live most of the symbiotic algae.

MANTLE PATTERN of fine light stripes is displayed by another tridacnid, photographed in the waters of Malakal Harbor at Koror in the Palau Islands. This is an unanchored species, *Hippopus hippopus*; its mantle does not extend beyond the margins of its shell.

tural pattern. Moreover, within each pattern adaptation and specialization can run riot.

The Cardiacea, the superfamily that includes the tridacnids, provide an excellent example. In addition to the more specialized tridacnids this superfamily includes the cardids, or cockles, which number some 10 genera and a great many species. Beaches and shallow zones of sand and sandy mud the world over are the habitat of the cosmopolitan cockle, and the populations of some species reach astronomical numbers. More than one adaptive trend evident in the highly successful cockle family is suggestive of the pattern of tridacnid evolution. To understand the one family some knowledge of the other is needed.

One cockle trend is a tendency toward enlargement. The giant cockle of East African waters can be as large as a coconut; a New World cockle, *Cardium elatum,* which is common in the Gulf of California, is equally impressive. A second significant cockle trend is apparent in the typically globular shape of its conspicuously ribbed shell. The rotundity of the shell prevents the cockle from burrowing very far. At the same time the ribbing of the shell increases the stability of the animal within the surface layers of the sand it inhabits.

A third trend is apparent in the heart cockle, *Corculum cardissa,* of the tropical Pacific. The very thin shell of this cockle has been greatly compressed, not laterally but from end to end [*see illustration on next page*]. The animal works into the bottom with its anterior half undermost, so that its posterior half, with short siphons extended, remains exposed. The tropical sunlight that illuminates the shallows penetrates the translucent shell. Apart from the tridacnids the heart cockle is the only bivalve that harbors algae; these photosynthetic symbionts are found in the gills and other tissues of the animal that are exposed to the light.

The tridacnids also vary substantially in size and shape. Five of the six species in the family are in the genus *Tridacna;* the other is in the genus *Hippopus.* All are confined to the tropical waters of the Red Sea, the Indian Ocean and the western Pacific. *Tridacna crocea,* which can reach a length of six inches, is the smallest species and is also unique in habitat. Other tridacnids may grind shallow pockets into the limy rock where they take up residence, but *T. crocea,* by repeated contractions of its byssal muscles, bores so deep into the rock that the margins of its shell valves are all that remain visible [*see bottom illustration on page*

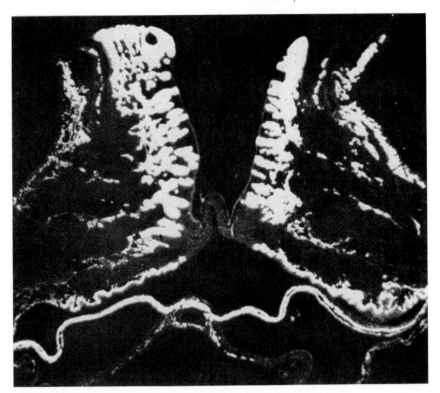

TRANSVERSE SECTION through the siphonal tissues of *Tridacna maxima* is seen in a radioautograph. The clam was exposed for an hour to carbon dioxide labeled with radioactive carbon atoms. Symbiotic algae, clustered in the siphonal tissues that are exposed to sunlight, incorporated the carbon dioxide in the course of photosynthesis. Radioactive carbon makes algae appear as bright areas. Radioautograph was made by Thomas F. Goreau.

127]. The species is found from the Nicobar Islands of the Indian Ocean eastward to Fiji in the Pacific.

Three tridacnid species are, relatively speaking, middle-sized. The length of *Tridacna maxima* does not exceed 14 inches; the upper limit for *Tridacna squamosa* and *Hippopus hippopus* is 16 inches. *T. maxima* is found from the Red Sea to Pitcairn Island; *T. squamosa,* from the Red Sea to Tonga. Both, like *T. crocea,* spend their life attached to the reef rock, often partly covered by growths of coral. *Hippopus,* however, remains attached only from the end of its larval stage until it has grown to about the size of a clenched fist. Its byssus threads then atrophy, and the currents that pile up sand in the lee of the reef roll the detached clam to the same shelter. The distribution of weight in *Hippopus* is such that the animal comes to rest in an upright position, and its siphonal tissues, extending up to the margins of the widely opened shell, catch the sun. *Hippopus* is found from the Nicobar Islands to Tonga.

Tridacna derasa is the next to largest in the family; its shell can be more than 20 inches long. Like *T. crocea* (and *T. gigas*) it is found between the Nicobars and Fiji; like *Hippopus* (and *T. gigas*) it

loses its byssal apparatus as its size increases and ends up as an unattached animal.

The term *Tridacna* is from the Greek *tridaknos,* "eaten at three bites"; this was a learned joke of the 18th-century naturalist Jean-Guillaume Bruguières, who named the genus. Although bivalves that range from six to 20 inches in shell length can scarcely be called small, the smaller five species are dwarfed by *T. gigas.* Even discounting the more fantastic stories about this species, the attested data are quite impressive enough. One specimen, collected in the Philippines early in this century and now on display at the American Museum of Natural History, weighs 579.5 pounds. The two empty valves of another *gigas,* four feet six inches long, weigh 507 pounds. In the 16th century the Republic of Venice sent a pair of *gigas* valves, three feet four inches long, to François I of France as a curiosity. They can still be seen in use as fonts for holy water at the church of St. Sulpice in Paris.

The habitat of the tridacnids, the upper surface or sandy lee of coral reefs, is covered by at most a few fathoms of exceptionally clear water. Thus the tissues that harbor the clams' symbiotic algae are exposed to intense tropical sunlight. Only when the animals are uncovered by the tides or are stimulated by the shadow of a predator do they withdraw their exposed tissue and close their shell valves. The sunlight is of course essential to the photosynthetic activity of the algae, but it is potentially lethal to the clam. The hazard of exposure to harmful wavelengths has been overcome by the evolution of a protective pigmentation. The pigments in the exposed tissue are contained in fixed cells known as iridophores. Mainly in the color range of blue to green or brown to yellow, they give rise to an almost infinite variety of patterns that are at their most vivid among the smaller species. The exposed tissues of *T. crocea* are invariably blue; those of *Hippopus* and *T. gigas* are usually olive green.

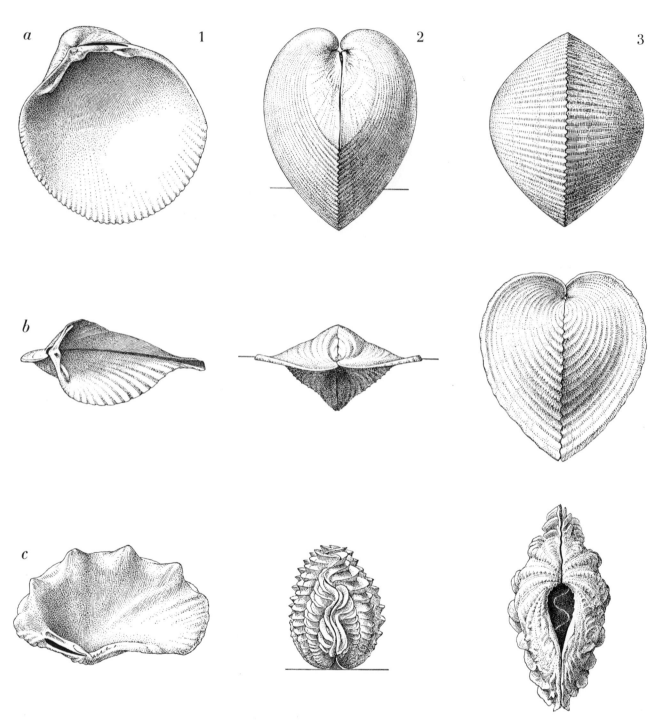

SHELLS OF THREE BIVALVES of the superfamily Cardiacea are shown in three views: (1) the inside of the right valve oriented as in life, (2) both valves together, rotated 90 degrees around the vertical axis from the first position, and (3) both turned 90 degrees around the horizontal axis from the second position, showing the shell from below as it lies in life. The lines suggest each animal's position in or on the sea floor. The first shell (*a*) is that of a giant American cockle, *Cardium elatum;* it is six inches long. The next (*b*), the heart cockle, *Corculum cardissa,* is 1.5 inches long. The last (*c*), *Tridacna maxima,* is five inches long; a mass of threads (*color*) emerging from its shell is the byssus that anchors the animal to the reef. The anchoring threads are secreted by glands in the foot of the clam. The muscular foot of the cockle also emerges on the underside, which is opposite hinge in this bivalve.

In a typical bivalve at rest the foot of the animal projects between the free margins of the shell valves; the margins are on the underside, the shell hinge being uppermost. If a tridacnid is to provide its algal partners with sunlight, however, it must rest in such a way that its exposed tissues face the sun. In all tridacnids the foot (together with the attaching byssus, when it is present) remains on the underside of the animal, but the margins of the shell valves, with their brilliantly colored tissues, are exposed at the top. The animals have achieved this paradoxical result in the course of their evolution by twisting themselves, so to speak, so that the hinge of the shell has moved from the top around to the underside, ending up next to the opening through which the byssal threads emerge [see illustration at right].

This process of rotation was the consequence of an extensive enlargement of the tissues that house the algal partners of the clam. It can most easily be visualized if one compares a cockle and a tridacnid, both animals being oriented in the basic foot-down attitude. The shell of the cockle is thus in the normal bivalve position, with its hinge uppermost. Imagine now that the cockle's shell is slowly rotated in relation to its enclosed body, while the body remains fixed by the foot. The rotation not only eventually brings the hinge around to the underside, next to the foot, but also stretches the soft tissue in the region between the inhalant lower siphon and the exhalant upper one [see illustration on next page].

The rotation, a process that probably required millions of years, may have been achieved by slow and gradual stages or by a few major mutations. The end result, however, was the same: an ancestral cockle had been transformed into a tridacnid. The siphonal tissues were now extended along the length of the upper surface, ready to cover the opening between the margins of the shell valves, and the shell hinge was now adjacent to the foot. The rotation had the further effect of squeezing out one of the ancestral cockle's two adductor muscles, so that, like scallops and oysters, although for entirely different reasons, tridacnids are single-muscle bivalves.

Rotation, in short, is the anatomical highlight of tridacnid evolution. To understand why the animal evolved at all, however, one must put the anatomical event in historical perspective. The cockles, the obvious ancestors of the tri-

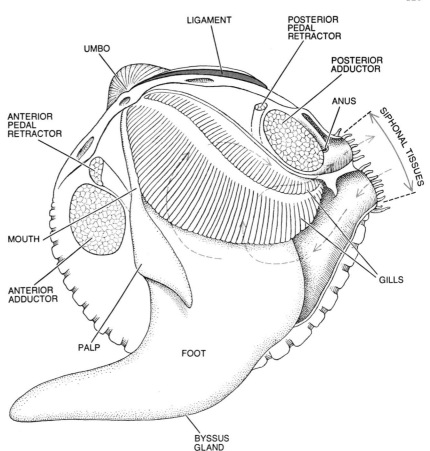

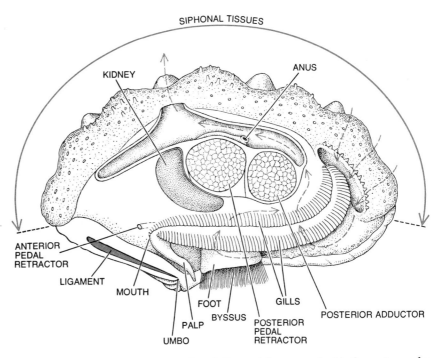

UPSIDE-DOWN ANATOMY of a tridacnid (bottom) is compared with the anatomy of a cockle (top); both bivalves are shown in a foot-down position. This orientation places the cockle's shell-hinge ligament (color) uppermost. The cockle's paired siphons are at right; broken arrows indicate inward and outward flow of water. The cockle's foot is extended beyond the gape of its shell. The byssal threads secreted by the foot of the tridacnid also extend beyond the shell, passing through the opening near the shell hinge. A great enlargement of the tridacnid's siphonal tissues (colored arrow) has been responsible for moving the shell hinge to the underside. A reconstruction of the process appears on the next page.

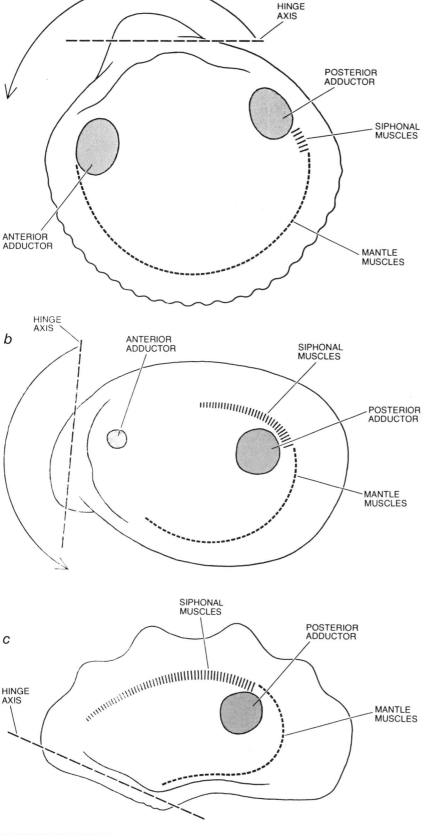

a

HINGE
AXIS

POSTERIOR
ADDUCTOR

SIPHONAL
MUSCLES

ANTERIOR
ADDUCTOR

MANTLE
MUSCLES

b

HINGE
AXIS

ANTERIOR
ADDUCTOR

SIPHONAL
MUSCLES

POSTERIOR
ADDUCTOR

MANTLE
MUSCLES

c

SIPHONAL
MUSCLES

POSTERIOR
ADDUCTOR

HINGE
AXIS

MANTLE
MUSCLES

POSSIBLE SEQUENCE in the evolution of the tridacnids is illustrated here in terms of a rotation of the shell. The process begins (*a*) with an ancestral cockle, shell hinge uppermost; rotation of the hinge axis is counterclockwise. With rotation half complete (*b*) on this hypothesis, stretching has extended the siphonal muscles of the tridacnid-to-be and may have reduced the anterior adductor muscle. Continuing rotation (*c*) carries the shell hinge to its ultimate position next to the foot. Enlargement of the siphonal muscles progresses, the anterior adductor is lost and what emerges is the basic tridacnid-family morphology.

dacnids, originated in the same period of the Mesozoic era that witnessed the rise of the modern corals; this period was the Triassic, which began some 225 million years ago. In the succeeding period, the Jurassic, the corals proliferated with remarkable success. It is believed it was at this time that modern corals first became separated into those that build reefs and those that do not. The corals that built reefs, living in symbiotic association with photosynthetic algae, were confined to the shallows of tropical seas; the corals that did not build reefs, living without symbionts, became widely distributed in all latitudes and down to great depths.

We may visualize the sandy shallows in the lee of these Mesozoic reefs as being the habitat of the cockles that were eventually to evolve into tridacnids. The heart cockle, the only other bivalve that is symbiotic with algae, evolved even more recently than the tridacnids, and so it can tell us nothing about the emergence of the tridacnids. Its existence nonetheless demonstrates that suitably illuminated surface-dwelling cockles can become the hosts of algae that are already prepared for life in the tissues of some host animal.

But how did the host-guest relation develop? The reefs that sheltered the evolving cockles, of course, harbored vast populations of algae living symbiotically in the corals. The relationship between the algae and the corals, however, can best be considered a kind of infection: an invasion by the plant cells that is certainly to their benefit. Corals are exclusively carnivorous animals; they will not accept plant food and so offer no threat to the algae. At the same time their metabolic processes provide the algae with elements such as phosphorus and nitrogen, which are essential for protein synthesis and yet are available only in trace amounts in the mineral-poor reef waters.

How the corals benefit from the association is less apparent. The algae multiply and eventually degenerate, but there is no evidence that the corals digest them. Some organic material does pass from the plant cells into the corals' tissues. The exact significance of the transfer remains uncertain, although it is clear that the algae assist in the process of coral calcification. In the final analysis, however, the algae are the organisms that gain the initial advantage from the association, and the "initiative" must therefore be attributed to them.

Was the first infection of the ancestral cockle by algae equally equivocal and

LARGEST GIANT, *Tridacna gigas,* is not anchored to the reef by byssal threads, but its weight keeps it in place. The clams are often partly concealed by growths of coral. This specimen was photographed at low tide on top of the Great Barrier Reef of Australia.

SMALLEST GIANT, *Tridacna crocea,* bores its way into the reef rock, grinding its shell inward by repeated contractions of the byssal muscles until only the edges of the shell are visible. This photograph made by author shows several *T. crocea* at low tide.

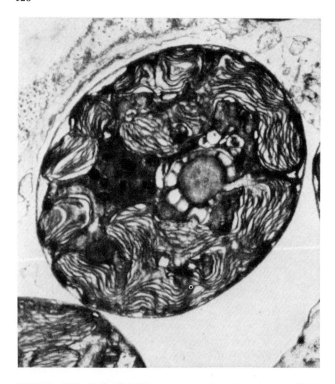

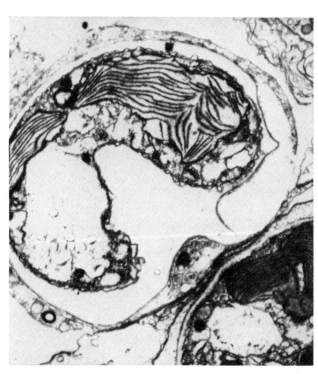

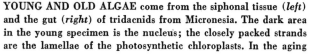

YOUNG AND OLD ALGAE come from the siphonal tissue (*left*) and the gut (*right*) of tridacnids from Micronesia. The dark area in the young specimen is the nucleus; the closely packed strands are the lamellae of the photosynthetic chloroplasts. In the aging alga the lamellae have spread apart and the nucleus is no longer visible; the symbiont is being digested by a scavenger blood cell. These electron micrographs were made by Peter V. Fankboner of Simon Fraser University; algae are magnified 10,000 diameters.

seemingly one-sided? We have no way of being certain, but to judge from modern tridacnids two conclusions seem tenable. First, the invading algae must already have been fully adapted for life within animal tissues. They could scarcely have been free-living forms, because the bivalves, supremely equipped for the collection and digestion of just such plants, would have promptly consumed them. Moreover, the algae do not invade all the tissues of a bivalve but are confined to those that are normally illuminated during the daylight hours.

These tissues are in the siphonal region, and as we have seen they have become greatly enlarged, in both length and width, in the course of tridacnid evolution. Thus they have offered the algae increasingly roomy accommodations. It is this fact of siphonal enlargement that leads to the second conclusion. Once algae became established in the sunlit part of the clam some advantage must have accrued to the host animal. Why else would the tissues have been stretched forward and extended laterally in a progressive series of changes occupying millions of years? The siphonal region has become so large that when one looks down through the clear reef water at an undisturbed tridacnid today, these strikingly pigmented tissues

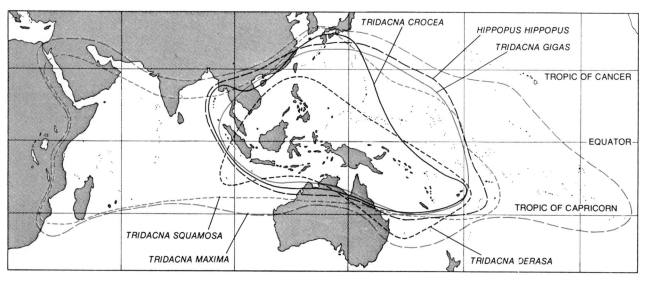

DISTRIBUTION OF TRIDACNIDS is shown on this map. Most widespread are *Tridacna maxima*, found from the Red Sea to Pitcairn Island, and *T. squamosa*. The most restricted are *T. gigas* and *T. crocea*. *Hippopus* and *T. derasa* are intermediate in distribution.

completely obscure the two great shells under them.

From study of the tissues the algae inhabit it is evident that the advantage accruing to the clam from the presence of its symbiotic partners is nutritional. The inhalant and exhalant siphons of the tridacnids are backward extensions of the mantle, the tissue that enfolds the body, adds new shell and thickens old shell as the animal grows. In the tridacnids, as in all other bivalves, the margin of the mantle has three parallel folds. The outermost fold is the one that secretes shell. The middle fold gives rise to sensory tentacles and in some instances eyes. The muscular inner fold controls the flow of water.

The inner fold is also the one that has become enormously extended. Here, apparently housed in the individual blood cells that occupy the blood spaces of the thickened tissue, the algal partners of the tridacnids congregate, multiplying most vigorously in the parts of the tissue that are best illuminated.

In other bivalves the blood cells often act as phagocytes, scavenging and digesting pathogens and other foreign matter. In tridacnids it has been observed that the tubules of the digestive gland are surrounded by blood spaces filled with cells that contain degenerating and apparently semidigested algae. These algae, however, never enter the gut. When this writer first studied these animals, he interpreted the condition of the degenerating algae as proof of a process whereby surplus symbionts were conveyed from the illuminated areas where they multiply into deeper tissues, there to be digested in the blood cells for the nutritional benefit of the clam.

(The residue after digestion would have been voided by way of the animal's enlarged kidneys.)

What stimulates the transport of the algae from sunlight to shadow was a mystery then and remains a mystery now. Recent studies with the electron microscope suggest, however, that what seemed to be semidigested algae may actually be senile algae. Perhaps that condition provides the stimulus for the transport. In any event tridacnids may not get as much nutriment in this way as was once considered probable.

What remains certain is that the photosynthetic algae release substantial quantities of organic matter into the clam's bloodstream. The principal product is a carbohydrate, glycerol. The release is readily demonstrated by exposing the clam to carbon dioxide labeled with radioactive carbon atoms. The algae in the mantle at once take up the carbon dioxide as a part of their photosynthetic activity. Almost immediately afterward, as radioautographs show, the labeled carbon is found in the glands of the foot where the byssus threads are secreted, in the pair of large mucus glands in the mantle cavity, in gill areas associated with food transport and, most interesting of all, in certain cells within the style sac. The style, a uniquely molluscan structure, plays an essential role in the digestive processes of these animals by liberating enzymes and mixing food in the stomach.

The evidence from radioautographs shows not only that certain carbohydrates produced by algal photosynthesis are available to the host animal but also that the carbohydrates are rapidly incorporated into the metabolic processes of the host. The tridacnids, however,

also need a supply of protein. This supply they obtain in part by ingesting all the plant plankton that is available. Their dependence on this normal source of bivalve nutrition is evident in their retention of the structures that all bivalves use for the purpose: feeding gills, selective palps and an elaborate digestive system.

As we have seen, however, the tropical seas in which the tridacnids live are notably poor in the mineral nutrients that support abundant planktonic life. As a result the proportion of the tridacnids' protein budget that comes from the uptake of plankton must necessarily be meager. Nonetheless, all six species in the family are large compared with the average bivalve, and *T. gigas* is not only the largest living bivalve but also the largest bivalve known to have evolved in the past 600 million years. (Some Mesozoic bivalves of the rudist group may have had larger shells, but the animal within was much smaller.) Moreover, there must be an upper limit to the size of any animal that feeds only by ciliary action, and that limit appears to have been far exceeded by *T. gigas*.

When the accumulated evidence is weighed, it is hard to escape the conclusion that the tridacnids get a significant part of their protein by digesting their surplus algal guests, whether or not the digested algae have become senile and degenerate along the way. Whatever uncertainty there is about the "how" of the process, it appears to be beyond question that the algae for whose environment and exposure to light the bodies of the tridacnids have become so strikingly adapted pay rent to their hosts by providing as nourishment both their photosynthetic products and themselves.

14 Corals and Coral Reefs

by Thomas F. Goreau, Nora I. Goreau
and Thomas J. Goreau
August 1979

*Tiny coral polyps, living in symbiosis with photosynthetic
algae, build huge limestone reefs that harbor more plant
and animal species than any other ecosystem on the earth*

Man's ability to alter the surface of the earth is rivaled among biological organisms only by colonies of tiny coral polyps, which over aeons of geologic time accrete massive reefs of limestone. True reef corals are limited in geographical distribution to the clear, warm, sunlit waters of the tropical oceans; they are found in the great reef tracts of the Indo-Pacific and the western Atlantic. Reefs are important land builders in tropical areas, forming entire chains of islands and altering the shoreline of continents.

There are three major types of coral reef. Fringing reefs grow in shallow water and border a coast closely or are separated from it by a narrow stretch of water. Barrier reefs also parallel a coast but are farther away from it, are larger and are continuous for greater distances; the best-known is the Great Barrier Reef off the northeastern coast of Australia, which forms an underwater rampart more than 2,000 kilometers long, as much as 145 kilometers wide and as much as 120 meters high. Atolls are rings of coral islands enclosing a central lagoon, and hundreds of them dot the South Pacific. Consisting of reefs several thousand meters across, many of them are formed on ancient volcanic cones that have subsided, with the rate of growth of the coral matching the rate of subsidence. This explanation of atolls was proposed by Charles Darwin during the voyage of the *Beagle* and was confirmed in the 1950's by Harry S. Ladd and Joshua I. Tracey of the U.S. Geological Survey when their extensive drilling programs on Pacific atolls hit volcanic rock hundreds of meters down.

Although tropical ocean waters are impoverished in nutrients, having low concentrations of dissolved nitrates, ammonia and phosphates, coral-reef environments have among the highest rates of photosynthetic carbon fixation, nitrogen fixation and limestone deposition of any ecosystem. The reef ecosystem also probably supports a larger number of animal and plant species than any other. The key to this prodigious productivity is the unique biology of corals, which plays a vital role in the structure, ecology and nutrient cycling of the reef community.

The Biology of Corals

Because corals are sessile they were for a long time thought to be plants. In Ovid's *Metamorphoses* he refers to the coral as an organism that is soft under water but hardens on contact with air. (What he was actually seeing was the death of the living tissue, which exposed the hard skeleton.) In 1723 the naturalist Jean André Peyssonel proposed to the French Academy of Sciences that corals are animals. His view was derided, and he subsequently abandoned scientific work. Since then, of course, he has been proved right. Corals belong to the large and varied phylum of coelenterates, which are simple multicellular animals. The phylum's name is from the Greek *koilos,* hollow, and *enteron,* gut, because the main body cavity of its members is the digestive cavity.

The closest relatives of the true corals are the sea anemones, which corals resemble in basic body structure and overall appearance. The soft coral polyp consists of three layers of cells and is basically a contractile sac crowned with a ring of six tentacles (or a multiple of six) surrounding a mouthlike opening. The tentacles have the specialized stinging cells called nematocysts, which discharge an arrowlike barb and a toxin that stuns animal prey such as microscopic crustaceans. From the mouth of the polyp the short muscular gullet descends into the stomach cavity and is connected to the body wall by six partitions (or a multiple of six), increasing the area of the digestive surface. The free edges of the partitions are extended into mesenterial filaments: convoluted tubes that can be extruded through the mouth or the body wall.

The size of the polyps is highly variable, from about one millimeter in diameter in some species to more than 20 centimeters in others. Each polyp can give rise to a large colony by asexual division, or budding. Corals also reproduce sexually, producing free-swimming larvae that settle and establish new colonies. The most striking feature of coral colonies is their ability to form a massive calcareous skeleton. Individual coral colonies weighing several hundred tons and large enough to fill a living room are common in many reefs. In most species the polyps are in individual skeletal cups, some extending their tentacles to feed by night and some partially withdrawing into the cups by day. In the contracted condition the polyps can resist drying or mechanical injury at low tide, when some of the colonies may be exposed. The skeletal cups consist of fan-shaped clusters of calcium carbonate crystals, which are arranged in patterns that are characteristic of each coral species.

A remarkable feature of all reef-building corals is their symbiosis with the unicellular algae known as zooxanthellae. The coral polyps contain large numbers of these algae within cells in the lining of their gut. The zooxanthellae are yellow-brown marine algae of the family *Dinophyceae,* to which many of the free-living dinoflagellate algae also belong. The algae live, conduct photosynthesis and divide within the cells of their coral host, and on this symbiosis rests the entire biological productivity of the coral-reef ecosystem.

Since the zooxanthellae of reef-building corals need light for photosynthesis, such corals grow only in ocean waters less than 100 meters deep. The corals also require warm waters (above 20 degrees Celsius) and do not tolerate low salinity or high turbidity. Where deeper colonies are shaded by a dense overgrowth of shallower ones, the deeper colonies maximize their light-gathering capacity by growing in ramifications like the branches of forest trees. In shallow water, where light is abundant but wave stress is high, the colonies deposit robust branching skeletons; in deeper water, where light is scarce, the colonies form horizontal platelike structures in which each polyp may harbor an increased number of zooxanthellae. Under highly adverse conditions such as prolonged darkness or freshwater flooding it is no longer advantageous for the coral polyps to maintain their zooxanthellae and they expel them from their tissues. Since the skeletal-growth rate of corals is dependent on their algal partners, true reef-building corals are almost never found outside the range of stable symbiosis.

Some coral species harbor no zooxanthellae; some of these species are found in crevices under the large structures erected by reef-building corals. Many of them are solitary cup corals such as *Astrangia,* which encrusts shells and rocks as far north as Cape Cod. Such corals can tolerate lower salinities, lower temperatures and greater depths: up to 6,000 meters in the deep sea. Even the deep, cold waters of the Norwegian fjords harbor great banks of *Lophohelia,* a colonial branching coral. Although these nonsymbiotic corals are distributed worldwide, their rate of growth is much lower than that of their symbiotic relatives, and they do not form massive reefs. In isolated instances colonies of these corals do contain symbiotic algae, but the algae do not appear to contribute significantly to the nutrition of their host.

The zooxanthellae are stored within individual membrane-bound cavities inside each of the cells in the stomach wall of the coral polyp. The feedback mechanism whereby the host regulates the number of its algal cells has not been determined, but there is little evidence that the corals "farm" and digest their

DIMES REEF off the Palau Islands in the western Pacific is shown in this aerial view. The shallow crest of the barrier reef is marked by the waves breaking over it. The Palau Islands, which are part of the Western Caroline group, are situated 1,060 miles southeast of Manila.

STAGHORN CORAL is a species commonly found in the shallow, sunlit waters of tropical coral reefs. Along the reef crest, where the corals must withstand the mechanical forces of waves, the staghorn colonies are robust and have short branches. In the sheltered waters behind the reef crest the colonies grow taller and have longer, slenderer branches, as is seen here. Corals always grow toward the light.

LIVING CORAL POLYPS were photographed at sunset, when they emerge to feed. By day they retract into their skeletal cups and so are able to withstand drying if the colony is exposed at low tide. The pol- yps, which reproduce sexually and by asexual division, coat the entire surface of the coral skeleton. They feed on small plankton animals, which they stun with stinging cells (nematocysts) on their tentacles.

algae. Instead the coral polyps seem to control the population of zooxanthellae by extruding the older and less metabolically active algae.

Robert K. Trench and his colleagues at the University of California at Santa Barbara have shown that specific strains of zooxanthellae are adapted to specific coral species. Some strains can live successfully in several different corals, and some corals are not discriminating about the lineage of their symbiotic algae. The fascinating problems presented by the symbiotic selectivity of corals are only beginning to be explored, and corals provide a valuable experimental system for the study of cellular interactions in general.

The Physiology of Coral Symbiosis

The modern study of the physiology of coral symbiosis began with a series of elegant experiments done by C. M. Yonge on the Great Barrier Reef Expedition of 1929. Yonge showed that symbiotic corals take up phosphates and ammonia from the surrounding seawater by day and release them at night. In order to study this phenomenon in greater detail two of us (Thomas F. Goreau and Nora I. Goreau) supplied carbon in the form of the radioactive isotope carbon 14 to reef corals. During the daylight hours the zooxanthellae assimilated the radioactive labeled carbon and photosynthetically fixed it into organic matter at a rate that was dependent on the intensity of the light. Some of this organic matter was then "leaked" from the algae to the coral host. Subsequent work by Trench and Leonard Muscatine of the University of California at Los Angeles and by David Smith of the University of Oxford showed that the leaked compounds include simple nutrients such as glycerol, glucose and amino acids. These compounds are utilized by the coral polyps in energy-yielding metabolic pathways or as building blocks in the manufacture of proteins, fats and carbohydrates.

It has long been known that the rates of metabolic reactions are strictly limited by the rates at which waste products are removed from the immediate environment. In higher animals the task is accomplished by specialized circulatory and excretory systems. These systems are absent in the anatomically simple coelenterates, which rely largely on the slow process of diffusion to remove soluble inorganic waste products such as carbon dioxide, phosphates, nitrates, sulfates and ammonia. The zooxanthellae, however, need for photosynthesis the very substances the coral polyp must get rid of, and they are believed to actively take them up from their host.

The photosynthetic demands of the zooxanthellae therefore result in the cycling of the coral's waste products into new organic matter. During the daylight

TWO SPECIES of the coral *Agaricia* growing side by side differ strikingly in shape and size. One type has whorled fronds, whereas the other has shinglelike plates. Such complex morphological differences are produced by subtle environmental gradients, such as the decline of ambient-light intensity with depth. The corals are at a depth of 43 meters off Jamaican coast.

hours the symbiotic algae produce more oxygen than the coral polyp can utilize for its respiration, and some of the carbon dioxide produced by the respiratory process is refixed by the algae into new organic matter. In order to estimate the efficiency of the internal carbon cycling in corals one of us (Thomas J. Goreau) determined the abundance in the coral tissue and skeleton of carbon 13, a rare but nonradioactive natural isotope, with respect to the abundance of the common natural isotope carbon 12.

For reasons it is not necessary to explain here, photosynthesis takes up carbon 12 slightly faster than it does carbon 13. Hence the organic matter synthesized by the zooxanthellae will have a relative preponderance of carbon 12, and a pool of carbon compounds enriched in carbon 13 will be left behind. It is from the compounds in this pool that the calcium carbonate coral skeleton is built. By determining the relative amounts of the two isotopes with a mass spectrometer it was estimated that about two-thirds of the carbon taken up in photosynthesis and calcification is recycled from the respiratory carbon dioxide of the coral polyp, with the rest being taken up from the seawater.

Organic matter leaked by zooxanthellae is only one of the three major sources of coral nutrition. Corals are efficient carnivores, immobilizing animal plankton with the stinging cells of their tentacles or trapping them on filaments of mucus that are then reingested. A polyp can detect a potential food item chemically, and it responds by extending its tentacles, by opening its mouth or

by extruding its mesenterial filaments. James Porter of the University of Georgia has analyzed the content of the coral stomach and found that the polyps feed mostly on tiny crustaceans and wormlike plankton that hide in the interstices of the reef by day and emerge at sunrise and sunset.

Studies with radioactively labeled compounds have also shown that corals are able to take up dissolved organic matter across their body wall. Since corals actively feed on plankton, take up nutrients from seawater and absorb chemicals released by their zooxanthellae, they fill several ecological roles simultaneously: primary producer, primary consumer, detritus feeder and carnivore. This complex food web reduces their dependence on any single food source, which might be subject to random variation as environmental conditions change.

Calcification in Corals

Growth in corals is achieved by an increase in the mass of the calcareous skeleton and the overlying living tissue. The skeleton of corals is composed entirely of aragonite, a fibrous crystalline form of calcium carbonate ($CaCO_3$); calcite, the commoner crystalline form of calcium carbonate, is absent. In the reef many algae also deposit aragonite or a more soluble form of calcite with a high magnesium content. Working in Bermuda, Heinz A. Lowenstam of the California Institute of Technology showed that some calcareous organisms tend to deposit the less soluble calcite in the cold seasons and the more soluble aragonite in the warm seasons, but the mechanisms by which organisms regulate the mineralogy of their skeleton are still unknown.

Coral polyps absorb calcium ions from seawater and transfer them by diffusion and by an active pumping mechanism to the site of calcification. Calcium ions are a major biochemical regulator of cell metabolism and must be kept at extremely low levels if the cells of a tissue are to function. Although coral tissues have a total calcium concentration similar to that of seawater, the concentration of free ions in them is much lower because most of the calcium is bound to membranes or to organic molecules. Lothar Böhm, working in our laboratory at the University of the West Indies in Jamaica, has shown that the calcium bound in these organic complexes turns over rapidly.

One of us (Nora I. Goreau), working in collaboration with Raymond Hayes of the Morehouse College School of Medicine in Atlanta, recently made detailed electron-microscope studies of coral polyps. In the course of these studies minute calcium carbonate crystals enclosed within membrane-bound vesicles were observed in the outer cell layer of the polyp. The crystals are extruded through the membrane to the coral skeleton, where they act as nuclei for continued crystal growth. This work may serve to clarify basic mechanisms of calcification in the cells of a variety of organisms, particularly because corals lack the hormonal controls over calcification that complicate these mechanisms in more advanced organisms.

The major obstacle to the study of the physiology of calcification in corals has been the difficulty of keeping corals alive and healthy in laboratory aquariums long enough to make accurate measurements of the calcium uptake. One of us (Thomas F. Goreau) circumvented the problem by measuring calcification in situ in the living coral reef. This was done by providing the coral with calcium in the form of the radioactive isotope calcium 45 and measuring the uptake of the radioactive calcium into the coral skeleton. The method is so sensitive that growth can be detected in corals that have been exposed to the radioactive calcium for only a few hours, which is what makes field studies practicable.

Such studies have shown that although reef-building corals grow under fairly uniform conditions of temperature, illumination and water circulation, there are very large differences in the growth rates of different species. The highest rates are invariably found in the branching corals, such as the West Indian elkhorn and staghorn corals. *Millepora* ("fire coral") is a close second, with the *Poritidae* ("finger corals") third. The massive corals grow more slowly. In the branching corals most of the growth takes place at the tips of the branches, and new branches develop almost anywhere on the older parts of the colony.

Factors Influencing Calcification

A crucial factor influencing the rate of calcification is the conversion of respiratory carbon dioxide (CO_2) into carbonic acid (H_2CO_3), which is in turn converted into bicarbonate (HCO_3^-) and carbonate (CO_3^{--}) ions. The enzyme responsible for the addition of water to carbon dioxide to form carbonic acid is carbonic anhydrase, which is present in high concentrations in corals. The subsequent formation of bicarbonate and carbonate ions is rapid and does not require catalysis by an enzyme. Drugs that inhibit carbonic anhydrase bring about a dramatic decline in the calcification rate.

The growth of the coral skeleton is on the average 14 times faster in sunlight than it is in darkness, and it can be decreased by drugs that block photosynthesis. Even daily variations in light intensity have a measurable effect on the

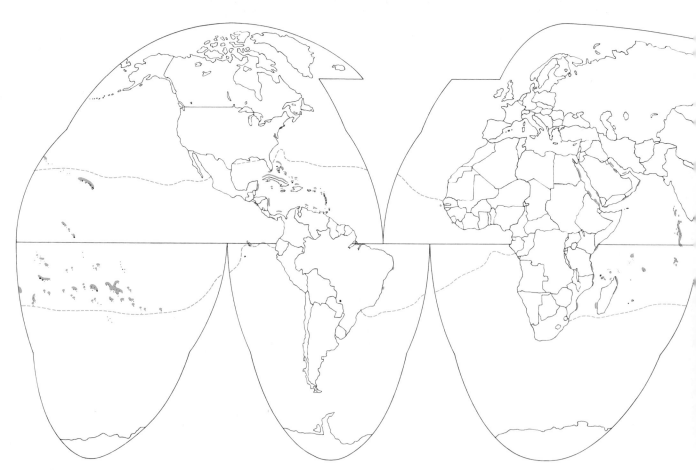

WORLD'S CORAL REEFS (*color*) **can be divided into three basic types: atolls, barrier reefs and fringing reefs. Reefs of the West Indies are primarily fringing ones. Reef-building corals are found only in sunlit tropical waters** (*broken lines*) **because their ability to rapidly**

calcification rate: the uptake of calcium is fastest at noon on a clear, sunny day, is reduced by 50 percent on a cloudy day and by nearly 90 percent in total darkness. The intensity of the ambient light also decreases with depth: the flux of light at a depth of 60 meters is only 4 percent the flux at the surface. As a result the rate at which calcium is deposited into the coral skeleton probably decreases rapidly with increasing depth.

The striking dependence of the growth of coral on the intensity of the ambient light is observed only when the zooxanthellae are present. If the symbiotic algae are removed (by keeping the coral colony in darkness for several months), the rate of calcification is low and is no longer affected by changes in light intensity, as is normally the case with nonsymbiotic corals. How do the zooxanthellae enhance the calcification rate? The answer seems to be that the fixation of carbon dioxide by the algae gives rise to an increase in the concentration of carbonate ions in the cells of the coral polyp through a series of linked chemical reactions, raising the pH of the fluid in the cells so that it is more alkaline. By precipitating its excess carbonate ions in the form of insoluble calcium carbonate the polyp is able to restore its pH to the normal level and at the same time build up its limestone

skeleton. The zooxanthellae may also stimulate calcification indirectly by increasing the amount of free energy available for the active transport of calcium ions to the site of calcification. The algae therefore work synergistically with carbonic anhydrase to enhance the formation of calcium carbonate. Calcification can proceed in the absence of algal photosynthesis but only at a greatly reduced rate.

The fact that calcification in corals is biologically controlled is further indicated by seasonal variations in the growth rate. These variations are reflected in measurements made by one of us (Thomas J. Goreau) of the concentration of the trace metal magnesium and of the heavy and light isotopes of carbon and oxygen in seasonal growth bands. Once the environmental and physiological influences that affect the growth of coral are better understood the variations in the composition of coral skeletons will provide a detailed chemical record of past environments much as tree-ring records do.

The synergistic effect of zooxanthellae on the calcification rate was clearly a decisive factor in the evolution of coral reefs. The development of enormous coral communities in the face of battering by heavy seas became possible only when the processes of calcium carbonate deposition became efficient enough for the rate of deposition to exceed the rate of loss through physical and biological attrition.

Reef Architecture

Coral polyps may not dominate the biomass (the total mass of living matter), the biological productivity or even the calcification in all parts of a coral reef. Nevertheless, the existence of many of the animal and plant communities of the reef is based on the ability of coral to build a massive wave-resistant structure. The dynamic interactions of the geological and biological processes that control the growth of coral reefs are well illustrated in the 150-mile fringing reef along the northern coast of Jamaica, which we have studied for the past 28 years.

The major structural feature of the living reef is a coral rampart that reaches almost to the surface of the water. It is made up of massive rounded coral heads and robust branching corals, which build a rigid, cavernous palisade of intergrown coral skeletons. Living on this framework are smaller and more fragile corals and large quantities of green and red calcareous algae. The biomass of these algae is small compared with that of corals, but their productivity and turnover are so high that the sand consisting of their skeletal remains makes up the bulk of the calcium carbonate deposited in the reef.

Hundreds of species of encrusting or-

ganisms live on top of the coral framework, binding the coral branches together with their thin growths. Innumerable fishes and invertebrates also hide in the nooks and crannies of the reef, some of them emerging only at night. In addition sessile organisms cover virtually all the available space on the underside of coral plates and on dead coral skeletons.

The crest of the reef runs parallel to the coast, in some places touching the shore and in others enclosing a sandy lagoon about five meters deep and up to a few hundred meters wide. This area is protected from the surf and is dotted with isolated coral heads. The lagoon is dominated by patches of calcareous algae and a community of bottom-living animals, notably sea urchins and sea cucumbers, which earn their keep by filtering organic matter out of the sediments or the overlying water. Many of these organisms graze on filamentous algae; if the grazers are removed from an area of the lagoon, a dense mat of algae forms after only a few days. The burrowing and churning activities of the grazers are important because they release nutrients created by the bacterial decomposition of organic matter buried in the sediments. Dense "lawns" of the sea grass *Thalassia* form special habitats harboring their own community of sea urchins, conchs and many other species.

Seaward of the reef crest is the fore reef, where corals blanket nearly the entire sea floor. The corals form massive buttresses separated by narrow sandy channels, down which passes a steady flow of fine sediment originating with the disintegration of dead corals, calcareous algae and other organisms. The channels resemble narrow winding canyons with vertical walls of solid coral growth. They may be as much as 10 meters deep, and some are completely roofed over with coral. This dramatic interdigitation of buttresses and channels dissipates wave energy and at the same time allows the free flow of sediments that would otherwise choke the growth of the coral.

Down from the buttress zone is a coral terrace, a slope of sand with isolated coral pinnacles, then another terrace and finally an almost vertical wall dropping into the darkness of the greater depths. The distribution of coral species and other animal communities in the reef is zoned by depth, a feature that enables paleontologists studying a section of an ancient reef now on dry land to accurately estimate the original depth of that section from the fossil animals associated with it. In water deeper than 100 meters few algae or symbiotic corals grow well because of the low light levels, and the fauna is dominated by animals that catch or filter the organic detritus sifting down from the reef above. The detritus feeders include the true sponges, the antipatharians ("precious corals") and the gorgonians (sea

accrete limestone skeletons depends on their symbiosis with algae known as zooxanthellae.

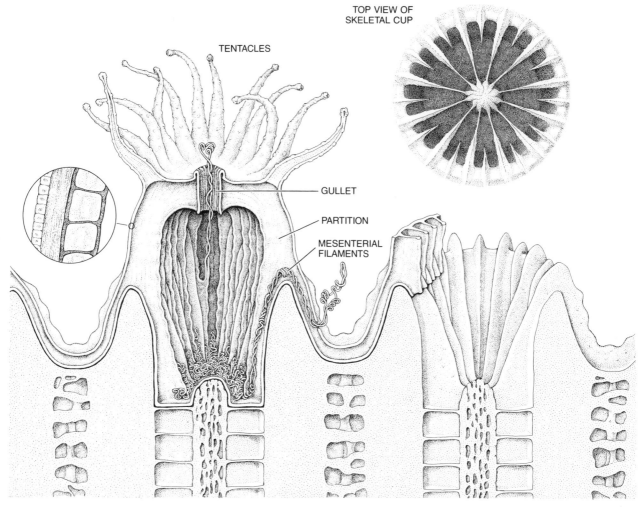

TENTACLES

TOP VIEW OF
SKELETAL CUP

GULLET

PARTITION

MESENTERIAL
FILAMENTS

ANATOMY OF THE CORAL POLYP is simple: the animal is basically a contractile sac made up of three tissue layers. The cylindrical body is topped by a central mouth surrounded by tentacles. From the mouth a muscular gullet descends into the central digestive cavity, which is connected to the body by a series of vertical partitions. The free edges of the partitions extend into mesenterial filaments. Cells in the lining of the digestive cavity harbor the symbiotic algae, which live, photosynthesize and divide within the host cells. The polyps sit in protective limestone cups consisting of a radial array of vertical plates, which interdigitate with the partitions of the polyp. Each polyp deposits new floors under itself as it grows upward. In the Tropics corals grow from one to 10 centimeters a year, depending on species.

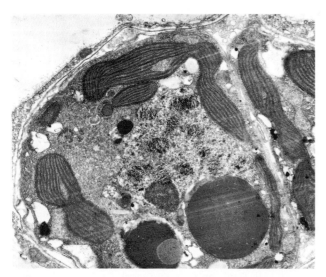

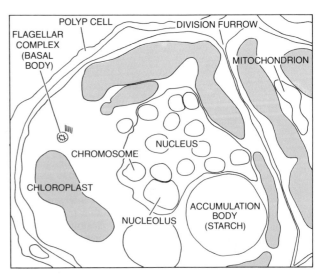

POLYP CELL · DIVISION FURROW

FLAGELLAR
COMPLEX
(BASAL
BODY)

MITOCHONDRION

NUCLEUS

CHROMOSOME

CHLOROPLAST

NUCLEOLUS

ACCUMULATION
BODY
(STARCH)

DIVIDING ALGAL CELL, or zooxanthella, is enlarged 13,250 diameters in the electron micrograph at the left. The striated sacs within the cells are sections through a single large chloroplast, where photosynthesis takes place; the other cell organelles are indicated on the map at the right. The zooxanthellae greatly increase the metabolic efficiency of the coral host by absorbing the waste products of coral respiration and recycling some of them into new organic matter. They also "leak" essential nutrients to the coral polyps and enhance the rate of calcification. The electron micrograph was provided by Robert K. Trench of the University of California at Santa Barbara.

fans). Also common here are the scle-rósponges, an ancient group that were major reef builders in the geological past but were long thought to have become extinct hundreds of millions of years ago. Our diving studies of the deep reefs of Jamaica showed them to be alive and well but displaced to deeper habitats by the faster-growing corals, which evolved later.

Reef Growth

The growth of the reef is the result of a dynamic relation between the upward extension of the coral framework and the flushing away of a much larger volume of fine-grained detritus. The export of sediment from the reef is largely accomplished by gravitational flow and creep, either into the lagoon or down the channels of the buttress zone into deep water. Unstable piles of coral may also grow until they topple under their own weight and slide away. When the lower Jamaican reef was explored in the research submarine *Nekton Gamma II* at depths of more than 200 meters, enormous piles of sediment and huge blocks of solid reef were observed at the base of the drop-off; they may have been dislodged by earthquakes. Such dislocations create fresh substrates for encrusting organisms and help to establish coral communities on the steep lower slopes, particularly the platelike whorled colonies of *Agaricia*.

Two other major processes influence the growth of the reef: biological erosion and submarine lithification. Many species of filamentous algae, fungi, sponges, sea worms, crustaceans and mollusks bore into coral skeletons, excavating holes by mechanical rasping or chemical dissolution. The commonest is the boring sponge *Cliona,* which saws out tiny chips of calcium carbonate; the chips are a major component of the fine sediments. *Cliona* can riddle a coral skeleton with holes without damaging the living coral polyps. In the deeper waters many corals grow in flat, thin sheets to maximize their light-gathering area and hence are quite susceptible to erosion by borers, which can cause the corals to break off and fall downslope. In some places, however, the coral is so overgrown with encrusting organisms that it remains in place even though it is no longer directly attached to the reef.

Counteracting the effects of biological erosion is submarine lithification: the deposition of a fine-grained carbonate cement in the pores and cavities of the coral skeleton. Sediments trapped in the reef framework are rapidly bound together by encrusting organisms and the calcareous cement. The origin of the cement is not yet clear; it may be an inorganic precipitate manufactured by bacteria that live in the crevices of the reef. Studies at the Discovery Bay Marine Laboratory in Jamaica done in conjunc-

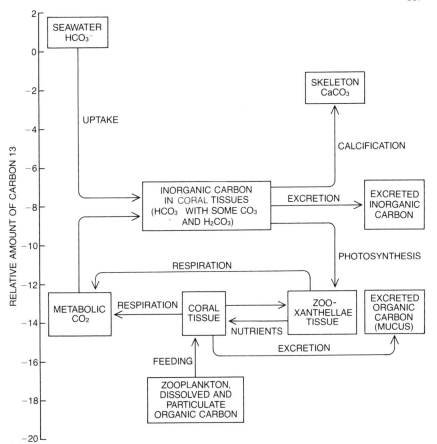

CYCLING OF CARBON among the zooxanthellae, the coral host and the environment is outlined. The various carbon pools are plotted on a vertical axis according to the ratio of the two stable isotopes of carbon: carbon 12 and carbon 13. The position of each pool is therefore an indication of the relative importance of the processes by which each pool gains and loses carbon and the extent to which these processes utilize one of the two isotopes preferentially. For example, because photosynthesis takes up carbon 12 faster than carbon 13 it leaves behind a pool enriched in carbon 13, from which the calcium carbonate of the coral skeleton is formed. About two-thirds of the carbon utilized in photosynthesis and calcification is recycled from respiratory carbon dioxide. Level of carbon 13 in coral tissue reflects the composition of its food sources.

tion with Lynton S. Land of the University of Texas at Austin showed that once the cement has hardened it is in turn bored and refilled; the filled holes are apparent when thin sections of the aggregate are examined under the microscope. Submarine lithification results in the outward accretion of the fore reef and stabilizes the steep profile of the drop-off wall. The growth of reefs is therefore the product of a dynamic balance among framework growth, sediment transport, bioerosion by borers, mechanical destruction and submarine lithification, with the relative importance of these factors varying from reef to reef.

The living reef is basically a veneer growing a few millimeters a year on top of a complex topography of superposed ancestral reefs. In Jamaica as much as nine meters of reef has built up since the present sea level stabilized some 5,000 years ago. The ancient reefs remain, providing a record of changes in sea level and of the uplift of land by the movements of tectonic plates.

The rise and fall of the sea level over

the past few million years has been caused by changes in the volume of water tied up in land glaciers and ice sheets during the Pleistocene ice ages. When ice sheets grew in the Northern Hemisphere, the sea level dropped and coral reefs were stranded above the waterline. Today fossil ridges and wave-cut notches mark the ancient sea level. A succession of stranded reefs are found in Jamaica, Barbados, New Guinea and on other coral coasts; these reefs were formed 80,000, 105,000, 125,000 and 200,000 years ago, when the climate was warmer and the sea level higher than it is today. Conversely, in Jamaica a series of drowned and overgrown ridges can be seen at 25, 40 and 60 meters below the present sea level. These drowned reefs were formed during periods of intensive glaciation 8,000, 11,000 and 14,000 years ago, when the sea level was considerably lower than it is today. The ancient reef is therefore a dimly visible palimpsest under the living reef, like a medieval manuscript that has been repeatedly erased and written over but shows faint traces of its history. Such features help in establishing the chronology of the Pleistocene ice ages and the volume of water added to the oceans by the melting of the ice.

Reef Ecology

The history of the modern Jamaican reef since the sea stabilized at its present level 5,000 years ago has not been long enough to establish a climax community: an ecosystem in equilibrium. This fact is evident from the almost haphazard development of reefs along any coral coast: some areas have well-developed reefs and others have only isolated patches of coral. Often there are no obvious environmental influences or catastrophic factors (such as earthquakes or tidal waves) that would explain such differences in development. It seems rather that chance variations in the settlement of free-swimming coral larvae and growth play a major role in determining the formation of reefs, and that there simply has not been enough time for corals to occupy all favorable habitats.

The role of chance in coral settlement is also reflected in the variability of the major species that fill the same structural roles in any reef. In some Jamaican reefs the dominant coral is the branching coral *Montastrea annularis,* but in similar habitats the same role is filled by the different species *Agaricia tenuifolia,* which forms colonies of identical shape, size and orientation. Hence in the creation of diversity in a coral reef historical variation is in many reefs just as significant as the approach to an ecological equilibrium where many specialized organisms coexist.

The many localized habitats and species in the reef give the reef community a wealth of interactions within and among species whose complexity can only be dimly grasped. An intuitive understanding of the major interactions can be gained only after years of field experience. Even then one may focus on so few components of the community that it is easy to miss the significant roles played by many obscure, unexamined or unknown organisms.

The intense competition for food and space in the reef habitat has given rise to a wide variety of survival strategies. For example, Jeremy Jackson and his students at Johns Hopkins University have shown that many encrusting organisms possess specific toxins for defensive or offensive purposes. Corals growing close together compete for space, and some species are able to extrude mesenterial filaments from the gut to kill the polyps of adjacent colonies. Judith Lang, working at Discovery Bay, has shown that among coral species there is a hierarchy of aggression such that slow-growing but aggressive corals can avoid being overgrown by faster-growing but less aggressive ones. This process may lead to an increased diversity of

species. In some instances, however, the result is precisely the opposite: James Porter has found that in the reefs on the Pacific coast of Panama the overwhelmingly dominant coral, *Pocillopora damicornis,* is both the fastest-growing and the most aggressive.

Grazing on algal and coral tissues by fishes, sea urchins and other animals has two important effects. Selective grazing may keep a few dominant species of algae from crowding out the more marginal species, so that a diversity of species are able to exist. Experiments in which grazers are excluded from an area of the reef usually result in choking densities of a few dominant algal species, which are rare under ordinary circumstances. Grazers that scrape tissues off hard substrates also create fresh surfaces where new algae can grow and the

larvae of sessile organisms can settle. Leslie S. Kaufman of Johns Hopkins has found that some fish species systematically kill patches of coral tissue so that "farms" of algae can grow on the bare coral skeleton. The fishes, which graze on the algae, chase any intruders on their territory, including much larger fishes and even human divers. How much damage to the reef is done by such biological space clearing compared with that done by slumping and storms is not known.

Much also remains to be learned about the nutrient and energy cycles of reefs. The richness of reef biological processes in the face of the poverty of dissolved nutrients in tropical surface waters is evidence that there is an efficient internal cycling of nutrients within the reef ecosystem, but the matter

ARCHITECTURE OF THE FRINGING REEF along the northern coast of Jamaica is depicted in these three-dimensional views. Several zones can be distinguished on the basis of reef structure, depth and the associated animal and plant communities. The reef crest extends to a depth of about 15 meters and comprises the shallow coral rampart and the surf zone (*a*). The fore reef extends from 15 meters to 30. This region is a medium-energy environment, with an ambient-light intensity about 25 percent that at the surface. The buttress zone (*b*), where coral

has yet to be investigated in enough detail. The major limiting nutrient in the oceans is generally thought to be nitrogen, and in coral reefs large amounts of the atmospheric nitrogen dissolved in the seawater are fixed in utilizable forms by filamentous blue-green algae. Another source of nitrates is the oxidation of ammonia by bacteria in the course of the decomposition of organic matter in the sediments of the reef lagoon. Recent work indicates that the oxidation of ammonia to nitrate is particularly intense in the fine-grained organic sediments trapped by the roots of sea-grass beds.

The coral reefs of the Atlantic, the Caribbean and the Indo-Pacific do not differ fundamentally in their structural forms, their habitats and the interactions of their species, even though the organisms occupying specific ecological roles vary greatly between oceans and even between individual reefs. Between the Pacific and the Caribbean, however, there is one significant difference: in the Pacific the active growth of coral goes down only to 60 meters, and in the Caribbean it goes down to 100 meters. The reduced range in depth of the Pacific corals may be due in part to periodic infestations by the crown-of-thorns starfish (*Acanthaster planci*), which feeds on coral by turning its stomach inside out, spreading it over the coral and digesting the coral tissues. Before the recent well-publicized outbreak of *Acanthaster* the organism was limited to deeper water and was rarely seen. Then an unexplained population explosion gave rise to a food shortage that forced the starfishes to move up to shallower water, where their destructive effects were readily apparent. The lower limit of reef growth in the Pacific may therefore be affected by periodic starfish grazing. Much remains to be done to prove the hypothesis, however, not least because many Pacific reefs also show signs of being more intensively eroded mechanically than Caribbean reefs.

These points illustrate some of the handicaps ecologists face in attempting to predict the stability of reef populations in response to environmental changes or the sensitivity of reef food networks to alterations in the abundance of particular species. Since coral reefs are localized centers of high biological productivity and their colorful fishes are a major source of food in tropical areas, many marine biologists view with alarm the spread of tourist resorts along coral coasts in many parts

a

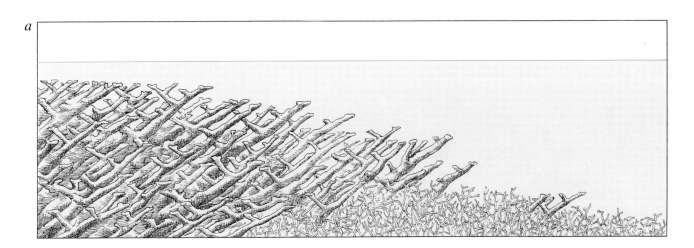

b

buttresses alternate with sandy canyons, serves to dissipate the mechanical energy of the waves and allows the flow down the reef of fine sediments, which would otherwise choke coral growth. The coral colonies are still varied but smaller in size, and much of the available space is occupied by sand-producing calcareous algae, sponges and large gorgonians (sea fans). The deep fore reef extends from 30 meters to 70. This zone has a steep topography and is poorly illuminated, with a light flux about 5 percent that at the surface. At depths below 30 meters coral growth becomes patchy, with a progressive reduction in number of species and size and density of colonies. There is also extensive transport of sediment from the shallow zones above. Beyond the deep fore reef the vertical wall drops off into darkness.

BUTTRESS CANYON between two walls of coral was photographed at a depth of about 12 meters off the northern coast of Jamaica. The wall at the right is covered by colonies of the coral *Monastrea annularia*. Shape of the colonies serves to maximize their light-gathering area.

DROP-OFF WALL of the Jamaican fringing reef is shown at a depth of about 40 meters. The steep fore-reef slope is covered by a dense growth of sponges, gorgonians and whip corals.

of the world. Such developments are almost always accompanied by increased dumping of sewage, by overfishing, by physical damage to the reef resulting from construction, dredging, dumping and landfills, and by destruction of the reef on a large scale to provide tourists with souvenirs and coffee-table curios. In many areas (such as Bermuda, the U.S. Virgin Islands and Hawaii) development and sewage outfalls have led to extensive eutrophication: the overgrowth and killing of the reef by thick mats of filamentous algae, which in turn support the growth of oxygen-consuming bacteria. The results, which are being intensively studied by Stephen V. Smith and his colleagues at the University of Hawaii, include an increased sensitivity of corals to bacterial diseases, the death of living coral and the resulting erosion of the reef, and the generation of foul-smelling hydrogen sulfide.

Breaching a Barrier

The proposal for digging a new canal at sea level across the Isthmus of Panama arouses further concerns about the viability of coral reefs and their intricately interwoven physical and biological resources. The large range of the tides on the Pacific side of the isthmus and the smaller range of the tides on the Caribbean side, together with the higher mean sea level on the Pacific side, would result in the effective movement through the canal of Pacific marine species into the Caribbean and the Atlantic. Since the reefs of the Caribbean and the Pacific have been evolutionarily isolated for millions of years, such a large-scale incursion of species into new habitats could enable certain species to multiply and spread unchecked, with ecological consequences similar to the explosive multiplication of the English rabbits introduced into Australia. For example, the crown-of-thorns starfish is common on the Pacific side of the isthmus but is not present on the Caribbean side, and its spread through the sea-level canal could decimate the corals of the Caribbean and the Atlantic. In addition poisonous sea snakes, unknown in the Atlantic, are common on the Pacific side of the isthmus. Peter Glynn and Ira Rubinoff of the Smithsonian Tropical Research Institute in Panama have warned that the sea-level canal could cause a greater perturbation in the natural environment than any previous engineering work.

The proposed sea-level canal illustrates not only the concerns of coral-reef biologists but also their ignorance, since it remains difficult to predict the deleterious effects of human activities in an environment as complex as the reef ecosystem. All the same it does not seem unduly alarmist to caution against taking the stability and productivity of the reef community for granted.

IV

MAKING SENSE
OF BEHAVIOR

IV MAKING SENSE OF BEHAVIOR

INTRODUCTION

Conrad Limbaugh's essay, "Cleaning Symbiosis," in the previous section, deals with a remarkable behavior that obviously has evolved independently among several marine groups (crustaceans, various kinds of fishes) and involves not only the cleaners' activities but also modifications of the behavior of the fish that are cleaned. The fish being cleaned refrain from chasing away the cleaners or eating them, and they actively adopt sites and postures that facilitate the cleaning interaction. I refer to this article here to juxtapose it to Evelyn Shaw's on fish schooling. Both phenomena challenge experimental analysis of what cues are at work, how behavior is established in each new generation, how complex group activities are coordinated and how their pace and range are set. Experimentation poses questions whose answers can be tested rigorously. But experimentation builds on rigorous prior observation. Limbaugh's article is about these observational foundations of experimentation; Shaw's report concentrates on experimental analysis itself. Thus, though their topics differ widely, these authors provide instructive essays about the methods of behavioral research as well as about the results of such research.

"The Schooling of Fishes," by Evelyn Shaw, moves quickly to an analytical dissection of this behavior. An ingenious device marks this investigation: Shaw has adapted the annular aquarium to her special use. She takes a developmental approach to behavior, seeking the circumstances that bear on the first appearance of schooling behavior in the life of the silversides. Her work suggests that schooling is controlled by a sequence of sensory and responsive interplays. Vision initiates schooling, but other perceptions (most likely involving the lateral line nerves) adjust inter-fish arrangements within the schools. This essay also underscores the contrast between the precise results of experimental ethology and the rougher texture, so to speak, of speculation about the adaptive value and evolutionary and phylogenetic history of a behavioral syndrome. "Adaptation" is, of course, inferred to be a central attribute of biological traits, yet it is a notoriously difficult one to pin down in procedures and data that lend themselves to those essential elements of rigorous scientific method—experimental test and confirmation or (especially) disproof. Phylogenetic statements are even less amenable to critical test and potential disproof, since lineages are unique historical events, raising the problems I mentioned in commenting on the reconstruction of events in "Plate Tectonics and the History of Life in the Oceans." This frustration has led some biologists virtually to dismiss whole domains of research as superficially descriptive. But the challenge persists. Our task is to develop investigative methods that can cope with the complex and apparently unmanageable issues in the *whole* biology of organisms, including the adaptive significance and historical background of their traits.

Like Evelyn Shaw's essay, William C. Leggett's report, "The Migrations of the Shad," combines an account of the life of an organism with one about the work of a biologist. The migratory routes and calendars of these fish were detected by the patient application of a simple technique, tagging fish and plotting the patterns of tag recovery. The resultant evidence first suggested and then substantiated a revised picture of the shad's seasonal movements, different from what earlier work had indicated. Novel and highly sophisticated equipment like ultrasonic transmitters now permit investigations beyond the reach of even the most ingenious earlier methods.

What emerges is a remarkable new look at the world of the shad—but still a look based mostly on correlations, not causal analyses, between the behavior of the fish and variations in conditions around them. Apparently, the fish stay in water of constant temperature, but they migrate to do so. Shad migrations reflect the seasonal sweep of these isotherms up and down the coastal waters. Selection of rivers for the spawning runs and the subsequent runs upstream involve a shift of sensitivities to break free from the offshore migratory pattern. Cues that are absent or inapplicable in the high seas (tidal shifts, changes of salinity, bottom configurations) dictate the local "decisions" that the spawning runs call for. It is rather like a driver leaving a freeway and negotiating local roads by the different kinds of signs that mark them. Tagging has revealed the extent and timing of the shad's migrations, and ultrasonic devices let us follow the fish in littoral and inland waters. But Leggett's caution is well taken, in light of the history of mistaken inferences about the movements of this fish. For these studies leave us with a certain nagging wonder whether we have hit upon the causative environmental factors generating and directing the shad's movements or merely upon correlative accompaniments to still other cues that the fish themselves directly respond to.

Field ethology has concentrated on vertebrates or on the complex behavior of some insects. We have studied the behavior of only a few marine invertebrates and of even fewer in their natural surroundings. Howard M. Feder's essay, "Escape Responses in Marine Invertebrates," reviews in particular the variety of reactions that predatory sea stars provoke. His own work and that of other West Coast zoologists have revealed behavioral patterns that clearly contribute to the welfare of the prey. But, as Feder notes, these behavioral patterns, relatively well known as they are at the descriptive level, still present many perplexing aspects. The selective factors that favor here one, there another, particular tactic of escape, have yet to be demonstrated. Nor has anyone exactly defined the provocative cue itself (is it really invariably saponins?) from all the cues that sea stars might present or their prey detect in the natural habitat's rough conditions. It is a most difficult challenge, to gain some sense of what another creature recognizes in its world and how. Feder's studies are a shrewd start on one aspect of this daunting endeavor in ethology, daunting because it seeks clues to the worlds that other animals "know."

Howard Feder discusses escape responses that animals use once they have been found by their predators. By contrast, Mary K. Wicksten, in her article, "Decorator Crabs," examines how these spider crabs often avoid such confrontations altogether. Inert surfaces in the sea are rapidly overgrown by fouling organisms. Such surfaces include not only rocks and hulls but also the remnant hard parts (shells, for example) of dead animals. Living animals usually have clean surfaces or a restricted and characteristic epibiota. Exceptions to this cleanliness call for explanation. In some cases (such as the tunicate *Microcosmos*) the surface fouling may reflect a passive arrangement or perhaps the release, as well, of attractants for the settling larvae of other creatures. Some sessile invertebrates that form tight intraspecific clumps, such as gooseneck barnacles or mussels or oysters, show larval settling responses to their own species' presence. But decorator crabs foster epibiotes in quite another way. They use their "nimble chelae" to prepare and apply materials to their bodies in particular ways and places. This active adornment may be supplemented by the sheer accumulation of debris, but even that is often controlled by the distribution on the crab of catch hooks that hold the material. Many decorator crabs are not only visually concealed but also camouflaged to touch and even to taste—an important consequence of their decoration, since many marine predators hunt by nonvisual senses. Behavior, then, is as crucial to effective concealment as it is to escape. Protective traits work only to the extent that the creatures put these traits to work. Again, adaptive traits are traits in action.

The articles on behavior that I have selected place behavior among other interactions of the organism with its habitat. In "Biological Clocks of the Tidal Zone," John D. Palmer provides a contrasting perspective, seeking behavioral cues within the animal's own genetic endowment. Many animals show rhythmic behavior that persists even in apparent isolation from the environmental changes the behavior seems to accommodate. Experimental isolation does diffuse the rhythm gradually, and drastic experimental manipulations can alter or obliterate it quickly. But then stimuli that are quite different from mere imitations of, say, solar or lunar tides can reinstate the rhythmic activity.

Rhythmic behavior may be driven by what Palmer calls an internal "horologe" or clock that is correlated with rhythmic environmental stimuli but functions independently of them. The clock ticks as a rapid-frequency (short-period) oscillator whose cycles accumulate to drive physiological or behavioral activities that are malleable by environmental factors. The oscillator is not integrally part of the activity that it paces; rather, it is "coupled" with that activity. Experimentally induced arhythmicity, by this model, is interpreted as an "uncoupling" of the activity from its clock, which continues to run "on time." Resumption of rhythmicity indicates a recoupling of the clock to its target activity. The oscillator-clock has not revealed its internal mechanics, despite enormous investigative efforts. But it remains the most prevalent model on which to construct hypotheses about the mechanics of rhythmic behavior. The clock is a perplexing biological entity, because it is unaffected by changes of temperature and other factors that alter the rate of chemically based biological reactions. Clearly, in rhythmic biological activity, we have "a capacity. . .[that is] the expression of a genetic potential," but it is one that so far has eluded attempts to get beyond an ingenious but troublesome model. Recently, J. L. Cloudsley-Thompson has reviewed the adaptive significance of physiological rhythmicities (J. L. Cloudsley-Thompson, *Biological Clocks* (London: Weidenfeld & Nicolson, 1980)). I recommend his essay as a timely supplement to Palmer's article.

The final article about behavioral biology deals with animals whose familiarity has been fairly forced upon us. In an introductory understatement in his article, "Dolphins," Bernd Würsig suggests, "the effort to position them firmly in the spectrum of animal intelligence is premature." Our attitudes toward cetaceans are extraordinarily complex. We imbue the great whales with a significance we reserve for the immense, majestic, awesome, and rare. Dolphins and porpoises are not huge, they are not so much majestic as astonishingly active and adept, and for the most part they are not very rare. If the great whales conjure images of Moby Dick, dolphins bring Flipper to mind; if whales are awesome, dolphins are merely charming. And yet, by their prominence in myths since ancient times, as a Christian emblem of love and dedication, and even as the title bestowed for almost five hundred years upon the eldest son of the King of France, dolphins have long figured in our own symbolic efforts to put our world in order. As Würsig cautions, a much more ordinary animal may be the actual creature behind this legendary façade. He reviews the behavioral traits by which dolphins have so impressed people. These traits are widespread among herd animals. Dolphins' antics could, in fact, be generated by highly developed imitative powers rather than by exceptional intelligence. Their spectacular leaps and other "play" could be fairly simple signaling behavior rather than an expression of exuberance. But should science disenchant us even about dolphins? What price explanation?

15

The Schooling of Fishes

by Evelyn Shaw
June 1962

What influences make a fish join others of the same species to form a school? The question is studied partly by observing the developing behavior of young schooling fishes in a special laboratory aquarium.

For sea gulls, fishermen and other predators the propensity of certain species of fish to assemble in large schools is a great convenience. A school of fish is something more, however, than a crowd of fish; it is a social organization to which the fish are bound by rigorously stereotyped behavior and even by anatomical specialization. Schooling fishes do not merely live in close proximity to their kind, as many other fishes do; they maintain, during most of their activities, a remarkably constant geometric orientation to their fellows, heading in the same direction, their bodies parallel and with virtually equal spacing from fish to fish. Swimming together, approaching, turning and fleeing together, all doing the same thing at the same time, they create the illusion of a huge single animal moving in a sinuous path through the water.

This peculiar social organization has no leaders. The fish traveling at the leading edge of a school frequently trade places with those behind. When the school turns abruptly to the right or left, the fish on that flank become the "leaders," and what was the leading edge becomes a flank. Except in the execution of such a turn and during feeding—when the school formation may break up completely—the fish swim parallel to one another. The distances between fish may vary as individuals swim along at different and changing speeds, particularly in a slower moving, loose school. When a school is startled, for example, by a predator or an observer, it closes ranks immediately and the fish-to-fish spacing becomes equal and fixed as the entire school takes flight.

Even in schools of as many as a million fish, all members are of a similar size. Speed increases with size and the fish of a species therefore tend to sort themselves out by size and by genera-

tion in the sea. Schools can take many shapes and usually have a third dimension, being a few fish or many fish deep. From above they may appear rectangular or elliptical or amorphous and changeable. Some species form schools of characteristic shape. The Atlantic menhaden, for example, can be easily recognized from the air because their schools move through the water like a giant amoeboid shadow, often changing course but never breaking apart.

The speed and synchronization of response, the parallel orientation and the constancy of spacing among members of a school inevitably suggest that their behavior is integrated by some central control system that makes each "think" of changing course at exactly the same moment. Of course, there is no such central control system. Nor is it possible to explain the simultaneity of the members' actions as response to external stimuli from the environment. From time to time the fish do respond, as other animals do, to such stimuli as food and change of light intensity. Environmental conditions, however, do not explain the high degree of synchronized parallel movement that the members of a school display moment after moment, day after day. In fact, the great stability of schools, persisting through the most varied environmental conditions, suggests that the school organization must be dominated by internal factors.

Schooling is easily enough explained as an instinct. The term implies a causal factor—saying, in effect, that fishes school because they have an instinct to school. This tautology does not explain much, even when it is amplified by the more sophisticated statement that the behavior is inborn, unlearned and characteristic of the species. Many animals exhibit clear-cut, species-specific pat-

terns of behavior, and it is useful to seek these out and compare them as they appear in related species. Such inquiry leaves equally interesting questions unanswered. In the present instance it does not explain what brings about the concerted action of the fish in a school. This requires, above all, study of the behavior as it unfolds in the developing organism. With growth and particularly with the maturation of the sensory system, the relation between the organism and its environment changes. The life history of the individual, however typical of its species, has a profound role in the molding of the behavior of the mature animal and holds the principal clues to the mechanism that governs its interaction with its social and physical environment. So far this approach to the schooling of fishes has only made the mystery more intriguing.

With progress on the question of how fishes school, one can also hope for some light on why fishes school. No other line of study has disclosed what function this highly organized social behavior serves in the perpetuation of the species that have adopted it.

In my own work at the Marine Biological Laboratory at Woods Hole, Mass., at the Woods Hole Oceanographic Institution, at the Bermuda Biological Station and at the Lerner Marine Laboratory on Bimini in the Bahamas, I have attempted to overcome the difficulty of study in the field by bringing fishes into the laboratory for observation and experiment. Life begins for most species of schooling fish in the plankton, where the eggs drift untended and abandoned by the school that laid and fertilized them in its passage. The eggs develop into embryos and the embryos into larvae, or "fry," which are capable of some feeble swimming movement. They grow, they ma-

SCHOOL OF HERRING was photographed by Ron Church near San Diego, Calif. The majority of herring caught in the Pacific Ocean are used to make fish oil and fish meal. This school, originally headed straight for the camera, has begun to turn to its right.

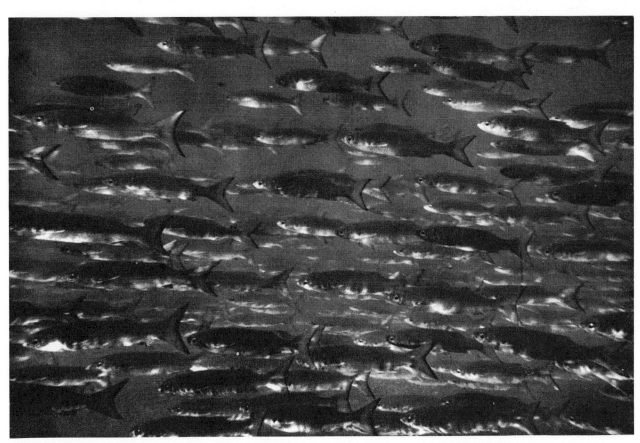

SCHOOL OF MULLET, which are common in the waters off Florida, was photographed there by Jerry Greenberg. A member of the order Mugiliformes, the mullet is an oceanic fish, and its distribution is primarily on both sides of the temperate South Atlantic.

ture and at some point during their early lives come together and form schools. One would like to be able to observe them during this epochal period. The only way to find the fry in the open oceans is to gather them in a plankton net, and the net necessarily disrupts the normal pattern of their behavior. My field studies have therefore been restricted to species that can be found as fry near the shore. But the fry are so tiny that crucial stages in the unfolding of their behavior in their natural habitat must go unseen.

In the waters around Cape Cod I have worked in particular with two species of *Menidia,* known commonly as whitebait, spearing or silversides. During late spring and early summer they spawn heavy eggs that adhere by sticky threads

to rocks and to the stems of marine grasses and algae. On hatching, when they are no more than five millimeters (about a quarter of an inch) in length, they become part of the plankton. In spite of patient search I have never observed fry this small in open waters. When they grow to seven millimeters or longer, they become easier to find in the plankton. I have seen fry seven to 10 millimeters long randomly aggregated in groups but not yet schooling or showing any sign of parallel orientation to one another. As the season progresses and as they grow from 11 to 12 millimeters in length, they can be observed forming schools for the first time, lining up in parallel, with 30 to 50 fry to the school. During the summer of 1960 my associates and I observed an estimated

10,000 of these tiny fishes in the plankton of the shallow waters near Woods Hole and collected many of them.

From these observations one could deduce that schooling begins when the fry reach a certain length. We could not tell, however, whether schooling develops gradually or happens suddenly. We accordingly proceeded to rear some 1,000 *Menidia* from the egg in the laboratory. For the study of these fry we set up a doughnut-shaped tank with a channel three inches wide, having observed that schools tend to break up when they approach the corners of a rectangular tank. We took care also to observe them in constant light and through a one-way mirror. We were reassured to find that under these condi-

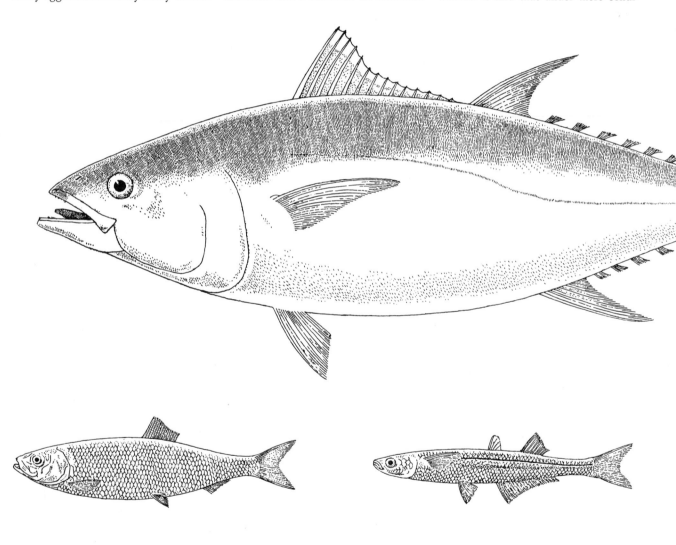

REPRESENTATIVE SCHOOLING FISH shown on these two pages are a tuna (*Thunnus thynnus*), at top left; a herring (*Clupea haren-* *gus*), at bottom left; a silverside (*Menidia menidia*), second from bottom left; a mackerel (*Scomber scombrus*), at top right; and a

tions schooling appeared in laboratory-reared fry when they grew to the same size as the smallest schooling fry observed in the sea.

The close-up and constant surveillance in the laboratory showed that schooling unfolds gradually in characteristic patterns of fish-to-fish approach and orientation. Newly hatched fry, five to seven millimeters in length, would approach the head, the tail or the side of other fry to within five millimeters and then dart away. At eight to nine millimeters in length, a fry would approach the tail of another fry and, when the two fry were one to three centimeters apart, they would swim on a parallel course for a second or two. If either fry approached the other head on at an angle, however, each would dart off rapidly in the opposite direction. At about nine millimeters in length the head-to-tail approach became predominant, and the fry would now swim on parallel courses for five or 10 seconds. When they reached a length of 10 to 10.5 millimeters, one fry would approach the tail of another and both fry would briefly vibrate their entire bodies. This curious behavior would terminate with the two fry swimming off in tandem, or in parallel, for 30 to 60 seconds, occasionally joined by three or four other fry in the formation of a recognizable little school. The number that would engage in this behavior increased to 10 or so when the fry reached a length of 11 to 12 millimeters. With the distances from fish to fish ranging from 10 to 35 millimeters, the school was a ragged one. By the time the fry had grown to 14 millimeters the fish-to-fish spacing became less variable, ranging from 10 to 15 millimeters, and there was less shifting about in the school.

Schooling behavior can therefore be described as developing initially from the interaction of two tiny fry. As they grow older and larger, the head-on approach gives way to the head-to-tail approach; the two fry tend to swim forward in parallel instead of fleeing from one another, and they are joined by increasing numbers of individuals in the formation of the incipient school.

At this point some speculation is in order, particularly if it suggests specific hypotheses for exploration by observation and experiment. During the head-on approach, one may suppose, each fry sees a changing visual pattern:

jack (*Caranx hippos*). The fish are not drawn to scale. Tuna have been known to reach a length of 14 feet. The mackerel averages 14 to 18 inches, and the jack about two feet. The silverside grows to six inches; the herring may reach a length of 12 inches.

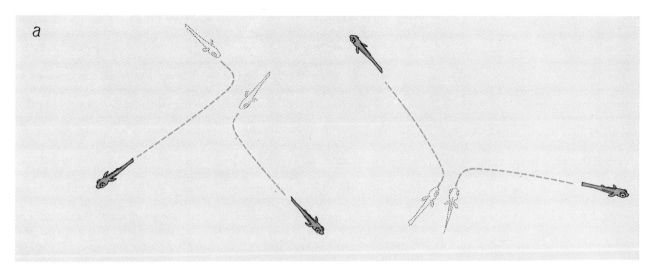

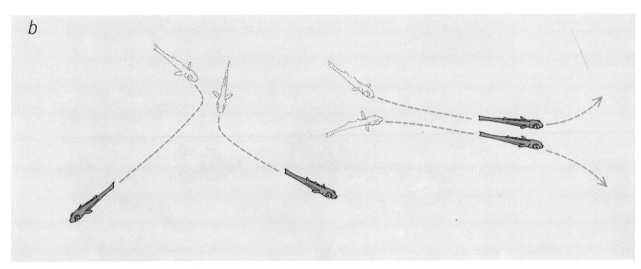

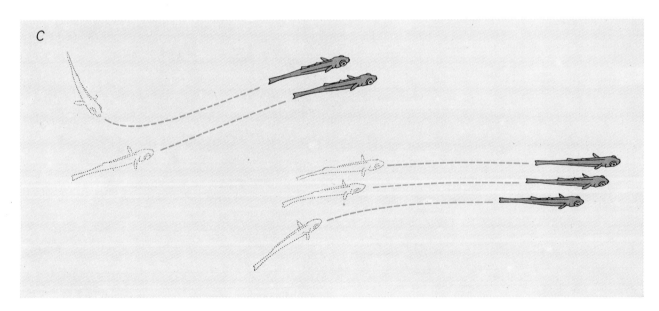

SCHOOLING ACTIVITY OF JUVENILE FISHES, or fry, develops as they grow. When newly hatched fry five to seven millimeters in length (*top*) approach the head, tail or side of other fry to within five millimeters, they dart away. At eight to nine millimeters (*middle*) two fry school momentarily if one has approached the tail of the other, but a side approach or one to the head still makes them dart away. As the fry grow from a length of about nine millimeters to 10.5 (*bottom*), the head-to-tail approach becomes predominant and two fry will school for five to 10 seconds; they later begin to school for short periods in threes and fours.

an oval mass (the head) and bright black spots (the eyes) coming closer and closer. The stimulus becomes too intense and each fry veers off in flight. The tail-on approach, in contrast, presents a quite different, although changing, pattern. This time it is a small silvery stripe and a transparent tail, swishing rhythmically and steadily moving away. The approaching fry follows. The leading fry may see, out of the rear edge of its eye, only a vague image of the follower. In each case the visual stimulus is moderate to weak in intensity, and the two fry swim forward together.

T. C. Schneirla of the American Museum of Natural History has postulated that, in general, mild stimuli attract and strong stimuli repel, and that most animals tend to approach the source of a mild stimulus and withdraw from the source of a strong one, even if they have had no prior experience with these conditions. Our fry had had considerable time to accumulate experiences of mutual encounter. We could not be certain, however, about the nature and impact of such experiences. A natural question therefore arose: Is such experience essential to the nature of schooling behavior? Or, to let the question suggest an experiment: Will fishes show schooling if they are taken away from their species-mates and raised in isolation? One must be cautious, however, in interpreting the results of such an experiment. On finding that a given behavioral trait appears in an animal that has been reared in isolation, some students of animal behavior are ready to conclude that the trait must be innate or instinctive and to close the book on further investigation at that point. Perhaps the pitfall lies in the word "isolation." No animal can grow up in a total vacuum of experience. In the case of the fry we proceeded to rear away from their species-mates, it was clear that each one had experience of itself (although we coated the bowls with paraffin so that the fry could not see their own reflection), of the water in its bowl, of the *Artemeia* shrimp on which it dined and of such stimuli as reached it from the world outside its bowl.

The mortality among the fry we isolated in this fashion proved to be extremely high. Only four out of 400 survived to schooling size in the first season and only nine out of 87 in the second. Apparently the fry need one another in the earliest larval stage, but we do not yet know why. The one noticeable difference between those reared in the community of their fellows and those reared alone seems to show up in the initiation of their feeding behavior. The fry in our laboratory communities began to feed two or three days after hatching, while they still carried a large yolky sac on their abdomen, whereas their siblings in isolation evidently starved to death. When we placed fry in isolation a week after hatching and after they had begun feeding, we secured a somewhat higher survival rate and, it turned out, a different and still enigmatic result when it came to observing the emergence of their schooling behavior.

As soon as the first four fry reared in isolation reached schooling size, we placed them in the company of schooling fish in community tanks. At first they showed disorientation; they bumped into their species-mates and occasionally swam away from the school. At the end of four hours, however, these fry could not be distinguished in behavior from the others. What this experiment showed is that fishes reared in isolation will soon join a school. It did not answer the question of whether or not schooling behavior would appear in fishes so reared.

With a more adequate supply of fry reared in isolation and in semi-isolation during the summer of 1960, we found that they would indeed form schools. The fry that had never had any contact with species-mates schooled within 10 minutes after being placed together in the test chamber. Those that had spent the first week after hatching in the company of species-mates also formed schools, but it took some of them at least 150 minutes to do so. What is more, we found that the shorter the time they had spent in isolation, the longer it took them to form a school. This suggests that their early experience with species-mates—at the period when the fry are still approaching one another at odd angles and darting away—may have set up some inhibitory process.

Although these experiments indicate that isolation in infancy does not keep these fishes from forming schools, the role of experience deserves further study. In this connection it should be added that schooling behavior was established

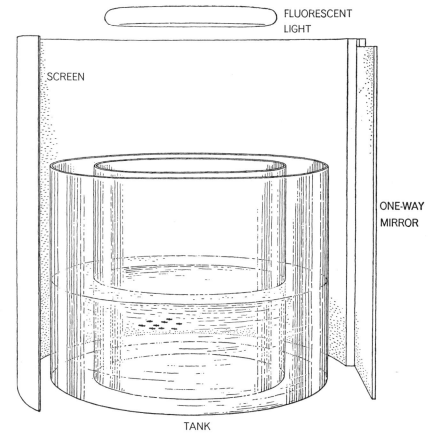

FLUORESCENT LIGHT

SCREEN

ONE-WAY MIRROR

TANK

FISH TANK used by the author to study the development of schooling behavior in *Menidia* fry is doughnut-shaped and has a three-inch-wide channel in which the fry can swim continuously without reversing direction. The tank is completely encircled by a screen (*here cut away*). The fry are observed either from above or through the one-way mirror.

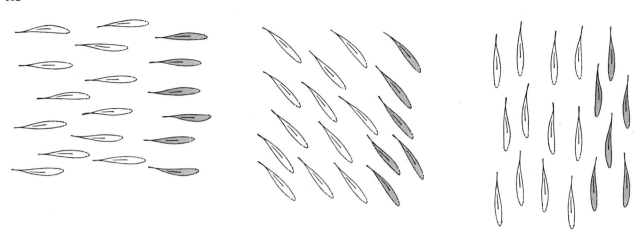

TURNING OF SCHOOL makes the relative position of the leading fish with respect to the school change. These fish (gray), which are originally at the leading edge (left), gradually shift around to the flank as the school turns (middle and right).

in our control communities when the fry were still a good deal smaller than the size at which we exposed our few precious isolates and semi-isolates to one another's company.

Another set of experiments with our laboratory fry produced evidence that the visual attraction of one fry for another develops in parallel with the emergence of schooling behavior. Very young fry showed no response at all to another fry swimming on the other side of a glass barrier. As the fry approached schooling age and size, however, they responded more and more actively to the visual image of the other fry. Finally they began to orient themselves in parallel to the fry on the other side of the barrier and were even observed to vibrate their bodies as they did so.

In a similar experiment with adult schooling fishes the visual attraction of one for another becomes readily apparent. Placed on each side of a glass partition, they swim toward each other immediately. In fact, fishes that cannot see cannot school. A fish blinded in one eye approaches and lines up with another fish on the side of the intact eye; a pair of fish blinded in different eyes swim at random when their sightless eyes are turned toward each other and school normally when they approach on the side with sight.

Just what visual cues are decisive in the mutual attraction of schooling fishes remains to be determined. Various experiments have shown that movement is important and that movement outweighs color and species, especially in attracting the initial approach. Albert E. Parr, now at the American Museum of Natural History, proposed in 1927 that fish-to-fish distances in schools might be explained by a balance of visual attraction and repulsion. According to Parr, the fish are repelled when they come too close together and attracted when they swim too far apart; the typical spacing in the school would thus represent the equilibrium of these two forces.

In a study of the schooling species around Cape Cod, Edward R. Baylor of the Woods Hole Oceanographic Institution and I found that many of these fishes are farsighted and that their retinas are therefore presented with a somewhat fuzzy image. The distribution of rod and cone cells in their retinas indicates, on the other hand, that their eyes may be well adapted for enhanced perception of contrast and so of motion against the hazy underwater background. This kind of vision would be highly adaptive in schooling behavior. Baylor and I also tried to modify the fish-to-fish schooling distance in pairs of fish by placing contact lenses over their eyes, but we observed no conclusive effects.

Although it appears that the visual apparatus is dominant in schooling behavior, there is also evidence that it does not serve as the exclusive channel of mutual attraction among fish. M. H. A. Keenleyside of the Fisheries Research Board of Canada has observed, for example, that Pristella, a species that sometimes schools, would respond to fish on the other side of a glass barrier by swimming back and forth along the barrier but would gradually lose interest, wandering away from the barrier more and more frequently and finally not returning at all. Sensory cues other than visual ones are most likely involved in establishing the parallel orientation and the fish-to-fish distances that give the school its ordered structure. It is difficult to determine which cues, because the experimenter cannot control for vision—a fish deprived of sight cannot make the initial approach so essential to the rest of the process.

Hearing, taste and smell have all been implicated, although inconclusively. James M. Moulton of Bowdoin College found that different schooling species produce different sounds, mostly of hydrodynamic origin, as the fish stream and veer through the water. Such sounds, in Moulton's opinion, may help to maintain the total school. There is no evidence, however, that sound helps to keep an individual fish oriented in position in the school. There is even less to be said for taste and smell, particularly in the case of oceanic fishes. Such odors as the fishes might produce would be diluted in the sea; although they might act on individuals at the trailing edge of a school, they could play little role in the behavior of those in the vanguard.

The one sensory system that would seem to be designed to play a role in the orientation of fish to fish is that associated with the lateral line—the nerve and its associated branches that are distributed over the head and run from head to tail along each side. It is thought that this organ is responsive to vibrations and water movements. Willem A. van Bergeijk and G. G. Harris at the Bell Telephone Laboratories have reported evidence indicating that the lateral line is sensitive particularly to "near field" motion of the water produced by propagated sound waves. Parallel orientation may well be facilitated by information about the movements of nearby fish picked up by the lateral line. The approach of one fish to another induced by visual attraction might also be checked by the increasing force of lateral-line perceptions of the movements of the same companion at closer range.

That schooling is a successful way of life can be judged from the fact that so many fishes have adopted it. Some 2,000 marine species school, and there is a major group—the Cypriniformes, consisting mainly of fresh-water fishes—that contains 2,000 more schooling species; among them are the common fresh-water minnows, or shiners, and the familiar characins of the tabletop aquarium. It is evident that these fishes must have converged on the schooling way of life by diverse evolutionary pathways. Of the marine fishes the best-known schooling orders are three that rank among the most numerous in the sea and constitute a vast portion of the world's fish supply. They are the Clupeiformes, the well-known herrings; the Mugiliformes, which include, in addition to the schooling mullets and our laboratory silversides, the solitary barracuda; the Perciformes, comprising the schooling jacks, pompanos, bluefishes, mackerels and tuna and the occasionally schooling snappers and grunts as well as numerous families of nonschoolers. Anatomically the Clupeiformes and the Mugiliformes are rather primitive fishes, whereas the Perciformes are advanced.

Although unrelated, these fishes do have significant features in common. Like many other schooling fishes, they are generally sleek and silvery. Significantly, they also have the same small and flattened pectoral fins actuated by musculature that does not permit much mobility. As C. M. Breder, Jr., of the American Museum of Natural History was the first to observe, these fishes cannot swim backward. When they make a pass at a bit of food and happen to miss it, they must come around on a wide turn for another attempt. This limitation on their maneuverability must nonetheless be an advantage in the maintenance of a school, because it tends to keep them all moving forward.

Since the schooling families include anatomically primitive as well as advanced forms, the evidence from living species does not show whether schooling is a primitive or an advanced adaptation. The fossil record is equally inconclusive on this score. Herrings are found in great number in Eocene deposits and one may reasonably speculate that they were schooling then. But fishes were evolving long before the Eocene, and it is impossible to determine one way or the other whether the fishes of those times schooled.

In spite of all the indications that schooling is an effective adaptation, no student of the subject has been able to show why it is so effective. Many advantages can be cited in favor of the behavior, but none seems critical to survival. It is said, for example, that the school creates for its predators, as it does for human observers, the illusion that it is a huge and formidable animal of some kind and so frightens off the predator. No real evidence supports this idea, and one can more plausibly see the school as providing easy prey. If the predator misses one fish, there is always another. In an experiment with goldfish, on the other hand, Carl Welty of Beloit College found that the fish consumed fewer *Daphnia* when they

SCHOOL OF ATLANTIC MENHADEN in Long Island Sound was photographed from the air by Jan Hahn of the Woods Hole Oceanographic Institution. The menhaden, which is a species of herring, forms schools containing as many as a million members.

were fed too many than they did if they were allowed a smaller number. Welty suggested that large numbers of prey might "confuse" the predator. This idea finds support in a mathematical analysis by Vernon E. Brock and Robert H. Riffenburgh of the University of Hawaii, which shows that a school cannot be decimated by attackers once it exceeds a certain number. But one must then ask: Why do some predators school?

Another rationalization for schooling holds that it facilitates the finding of food. When it comes to the search itself, however, only the fish on the school's periphery will be in a position to locate the food; the talents of those in the center of the school are wasted. Of course, once the food is sighted, all may partake. The young of many fishes travel in schools, and their social feeding seemingly promotes more rapid growth. As

our efforts to raise fry in isolation would indicate, the sight (or taste and smell) of other fish feeding induces fish to feed. Again, one must doubt that this advantage could account for the evolution of schooling behavior in so many different species.

Another advantage, often cited, has to do with the reproduction of the schooling species. When it is time to reproduce, there is no courtship behavior, no mate selection; as Parr observed some years ago, the males and females of schooling species are usually indistinguishable on casual inspection. The fishes simply shed their eggs and sperm in almost countless numbers into the plankton and leave the spawning site. This certainly enhances the probability of successful fertilization. In some of my collecting, however, I have found schools that were either all male or all female!

To the list of potential adaptive advantages I would like to add another one. Hydrodynamic considerations argue that schooling provides a more efficient way to move through the water. The exertion of each fish may be lessened because it can utilize the turbulence produced by the surrounding fish. Although the fish at the leading edge of the school may have to expend no less energy than solitary fish, the followers may receive enough assistance to help reduce their expenditure of energy. The attainment of maximum efficiency may dictate an optimum fish-to-fish distance in the school.

Study of the schooling of fishes has asked more questions than it has answered. But the questions have now begun to suggest fruitful programs of observation and experiment.

16

The Migrations
of the Shad

by William C. Leggett
March 1973

*The largest member of the herring family, much prized
as a food fish, moves between the sea and its river
spawning grounds with remarkable precision.
Its main guide appears to be temperature*

In many rivers, particularly those along the Atlantic Coast of North America, the annual "shad run" is a dramatic biological event. Within a space of a few weeks shad by the tens of thousands come in from the sea and move upriver to spawn. The timing of these migrations differs from river to river, being earliest in the south and moving progressively toward the north; in the St. Johns River of Florida, for example, the peak run is in January, whereas in the St. John River of Canada it is in June. What governs the timing of the movement of the shad into the rivers? The question is of both biological and practical interest: in North America the shad is an important seasonal food fish, and it is also highly prized by sport fishermen. The answer lies largely in the shad's preference for specific water temperatures in both the rivers and the sea. In these days of human intervention in such temperatures the question has added significance.

The American shad (*Alosa sapidissima*) is the largest member of the herring family, attaining an average length of from 40 to 50 centimeters (16 to 20 inches) and an average weight of from 1,400 to 2,200 grams (three to five pounds). Together with such fishes as the salmon, the sturgeon and the striped bass, the shad has the life cycle termed anadromous: the young, born in fresh water, remain there for a time (a matter of a few months for the shad) and then descend to the sea, where they remain until they attain sexual maturity and return to the home stream to spawn. By far the greatest part of a shad's life is spent at sea, since it takes from three to six years for the young fish to attain sexual maturity and since the adults that survive the annual spawning run return promptly to the sea.

In spite of the substantial commercial and recreational value of the shad, little was known of the fish's movements until recently. The state of knowledge at the turn of the century was indicated by C. H. Stevenson of the U.S. Fish Commission, who wrote: "It was formerly considered that the entire body of shad wintered in the south and started northward at the beginning of the year... sending a detachment up each successive stream, this division, by a singular method of selection, being the individuals that were bred in those respective streams.... But zoologists now recognize [that] the young shad hatched out in any particular river remain within a moderate distance off the mouth of that stream until the period occurs for their inland migration... entering the rivers as soon as the temperature of the water is suitable."

In the early part of the century a few

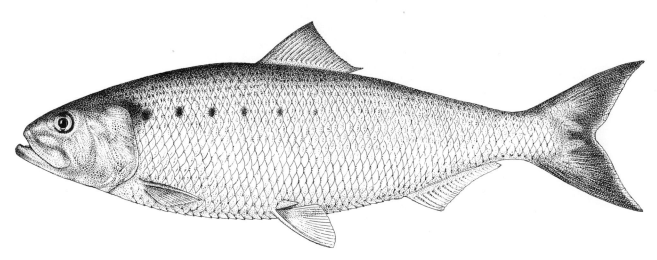

AMERICAN SHAD, which bears the formal name *Alosa sapidissima*, is the largest member of the herring family. An adult male is shown here. The fish attains an average length of from 40 to 50 centimeters, or 16 to 20 inches, and an average weight of from 1,400 to 2,200 grams, or three to five pounds. It was originally native to the Atlantic Coast of North America, but in 1871 the Sacramento River was stocked with shad from the East Coast, and the fish has since come to range widely along the Pacific Coast.

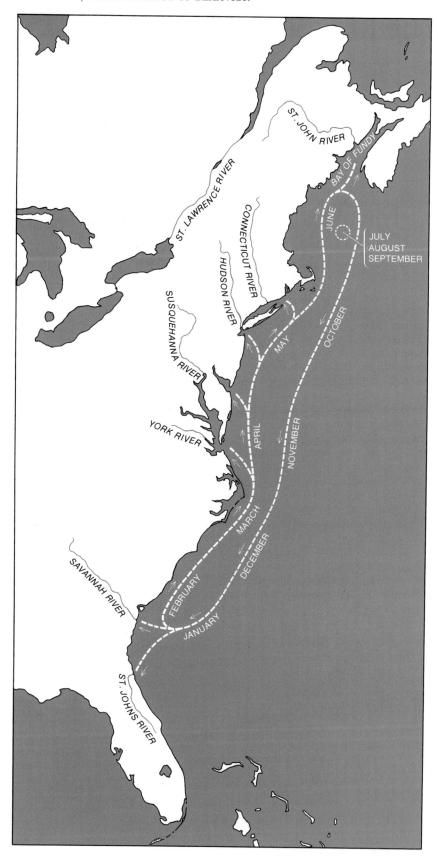

ATLANTIC MIGRATIONS of the shad are depicted according to the time of year when they are found or believed to be in each location. Also shown are several of the rivers that have shad runs on a large scale. The effect of the temperature dependence of the shad's movements at sea is to bring the fish near its home river in time for the spawning run.

shad were found (from tagging studies) to have moved over considerable distances at sea. Such movements were generally regarded as atypical. It was not until 1958 that Gerald B. Talbot and James E. Sykes of the U.S. Bureau of Commercial Fisheries showed long-range movement to be characteristic of the species.

Talbot and Sykes analyzed recoveries from more than 17,000 shad that had been tagged at several locations along the Atlantic Coast. They found that the distribution of recaptured fish exhibited a regular pattern in both time and space. The tag returns indicated that adult shad native to rivers from Chesapeake Bay to Connecticut migrated northward after spawning, congregating during July, August and September in the Gulf of Maine and the Bay of Fundy, together with shad from Canadian rivers.

In October and November the shad disappeared from the Gulf of Maine and the Bay of Fundy. Evidence of where they went was sparse. Tag returns indicated, however, that shad were off Massachusetts in October and November and between Long Island and North Carolina in February and March. On the basis of these returns Talbot and Sykes concluded that during the winter months shad move slowly southward to the Middle Atlantic region of the coast. In the spring they migrate in the direction of the river in which they were spawned, as indicated by the fact that tag recoveries were progressively closer to the home river as the season advanced.

The tag data taken as a whole showed that the shad migrated extensively, with annual movements exceeding 2,400 miles. The factors governing the movements, however, remained unknown until I began in 1965 a study of the population dynamics and migratory behavior of the shad native to the Connecticut River. The main purpose of the work was to assess the effect on the shad population of the discharge of heated water into the river from the nuclear generating station of the Connecticut Yankee Atomic Power Company at Haddam Neck, 15 miles above the mouth of the river. As so often happens, the work has produced a considerable body of new information quite unrelated to the problem at hand, including a clearer understanding of the migrations of shad.

As part of the study more than 32,000 adult shad have been tagged and released. More than 5,000 of them have been recovered in the river and close to 100 more were found at various places along the Atlantic Coast. As one might have predicted on the basis of the earlier evidence, the recoveries at sea ranged from North Carolina to New Brunswick. What was not anticipated was the remarkable precision in the timing of the movements as revealed by these recoveries.

In the Connecticut River from 1966 to 1972 the date when the first shad was captured varied by only five days, the earliest capture being on April 7 and the latest on April 12. By the last week of April large numbers of shad were entering the river to spawn, and this movement continued until the middle of June. Shad tagged and released in the river in April and May, while they were moving upriver to spawn, consistently reappeared in coastal waters off Rhode Island and Massachusetts in the last days of July and the first part of August. By late August they were in the Gulf of Maine and the Bay of Fundy, where they remained until late September or

early October. Annually there were no further recoveries until the second week of March, when tagged shad began appearing along the coast of North Carolina. By the first week of April they had moved north to the vicinity of Virginia and Chesapeake Bay, and by the second week of April they were off Delaware and New Jersey. Each year the timing of recoveries from specific locations was highly consistent, suggesting that some factor exerted a strong influence on the marine movements of the shad.

The first clue that this factor was water temperature came not from marine investigations but from observations of the movements of shad in fresh water. Although the idea that temperature was a factor affecting the timing of the spawning migrations was far from new, firm evidence in support of the idea had been lacking. I was able to compare records of the weekly catch of shad with

the mean weekly temperature of the Connecticut River over a period of 11 years. The comparison showed that few, if any, shad entered the river when the temperature of the water was below four degrees Celsius; above that temperature the number of shad increased steadily until the water was at about 13 degrees C. At higher temperatures the catch declined.

Further evidence was provided by records kept at a fish lift associated with a hydroelectric dam at Holyoke, Mass. The records showed how many shad were lifted over the dam each day and what the water temperature was. During 15 years the temperature at which the peak movement occurred was in a range of five degrees C., from 16.5 to 21.5 degrees. (The temperature is higher than it is during the peak migration at the mouth of the river because Holyoke is some 86 miles upriver and thus receives the main body of shad almost four weeks later.)

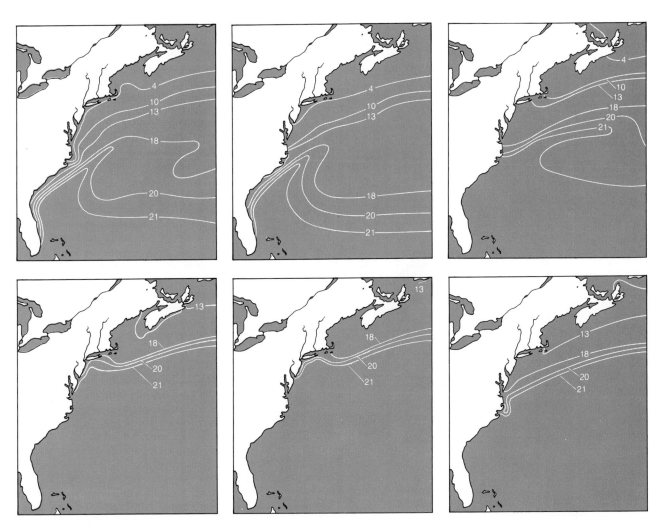

SEASONAL TEMPERATURES of the water along the Atlantic Coast of the U.S. are portrayed by means of isotherms, which show where a given temperature is found at a given time. The upper three maps are for January, March and May respectively and the lower ones are for July, August and October. Normally the seasonal distribution of the shad is closely associated with water temperature in the range from 13 to 18 degrees Celsius. The shad move northward in the spring and southward in the autumn within this range.

Although the data from the Connecticut River were strongly suggestive, I needed information from other rivers in order to make generalizations about the influence of temperature on the movements of shad. Accordingly I began in 1967 to assemble information from the York River of Virginia and the St. Johns River of Florida. Work had been done in the York River earlier by William H. Massmann and Anthony L. Pacheco, who at the time were associated with the Virginia Fisheries Laboratory. The data from the York River revealed a condition essentially identical with that in the Connecticut River.

In the St. Johns River the correlation between temperature and the timing of the migration was close but negative, that is, the shad moved in when the river cooled to their preferred range rather than when it warmed up. In the St. Johns River the water temperature seldom goes below 13 degrees C. even in the winter, and it is usually above 20 degrees C. from March through November. Few shad entered the river when the water temperature was above 20. As it fell below this temperature in November and December the number of shad increased. The peak movement occurred at the seasonal low of 13 to 16 degrees C., which was the same as the level that coincided with the peak movements in the rivers to the north.

These findings prompted my colleague Richard R. Whitney of the University of Washington to conduct a similar study on the Pacific Coast. He found that in 78 percent of the years from 1938 to 1969 the peak movement of shad at the Bonneville Dam on the Columbia River occurred in the temperature range from 16.5 to 19 degrees C. In 26 of the 32 years the temperature range within which 90 percent of the shad appeared at the Bonneville fish ladders varied by no more than four degrees. (Shad from Eastern rivers were stocked in the Sacramento River in 1871; by 1880 they ranged widely along the Pacific Coast.)

Having established conclusively that water temperature had a strong influence on the timing of the spawning runs of shad, Whitney and I turned to the question of whether or not temperature affected their movements at sea. We plotted the coastal recoveries from the Connecticut River taggings against the seasonal position of the portion of coastal waters where the temperature was between 13 and 18 degrees C. The correlation was again close. It is particularly noteworthy that during August and September the only area on the Atlantic Coast of North America having the preferred water temperature is the Gulf of Maine, which is precisely where the shad are at that time of the year. Moreover, the northern and southern limits of the range of the species in coastal water correspond closely to the range of acceptable temperature both in the rivers and in the ocean along the Atlantic Coast. It seems clear that the preference of the shad for a certain range of temperature leads to an orderly and precisely timed movement of the entire Atlantic Coast population southward in the fall and winter and northward in the spring and summer as the coastal water temperatures change with the seasons.

Examining the significance of this

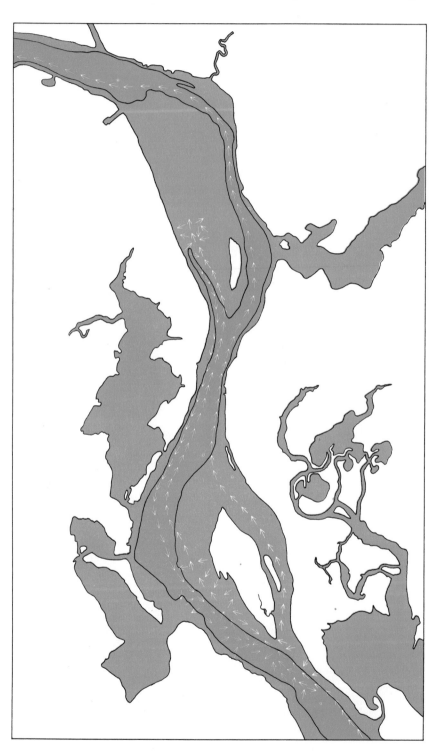

MEANDERING BEHAVIOR of shad at the interface of salt water and fresh water is typified by a single fish that was tracked in the Connecticut River by means of an ultrasonic transmitter that had been attached to the fish. Area is about six miles above river mouth. Shad on a run to spawn apparently meander while making physiological adjustments to fresh water.

temperature dependence, one is led to the view that it bears on the conditions needed to ensure maximum survival of eggs and larvae. In 1924 the Canadian biologist A. H. Leim published the results of studies showing that a minimum water temperature of 12 degrees C. was required for spawning but that optimum survival of shad eggs occurred at approximately 17 degrees. Subsequent experimental studies have shown that hatching and survival of shad are at a maximum when the water temperature is between 15.5 and 26.5 degrees C. The temperature-regulated oceanic migrations of shad bring individual populations to their home rivers at times when the river temperatures are approaching the level that is best for spawning.

W ater temperature appears also to affect the survival rate of adults. Studying returns from my tagging program in the Connecticut River, I found that shad entering the river to spawn early in the season, when temperatures were low, had a significantly higher rate of return to the river in subsequent years than shad tagged at higher temperatures. This finding seems to be related to the observation that in general the shad populations from rivers south of Chesapeake Bay lack repeat spawners and that north of the bay the proportion of repeat spawners increases with latitude.

To understand this phenomenon one must realize that shad do not eat while they are in fresh water. All the energy for the spawning run comes from stored reserves in the body. As a result a shad loses some 40 to 50 percent of its weight during its brief residence in fresh water. Since in shad as in most other cold-blooded animals the metabolic rate increases as the environmental temperature increases, shad migrating at 18 degrees C. would use more stored energy than shad migrating at eight degrees C. It seems likely that the additional energy required for migrating at higher temperatures is one of the reasons that shad spawning in southerly rivers do not survive to return another year. Brian Glebe of my laboratory at McGill University is investigating this aspect of the migrations of shad.

The absence of repeat spawning in the shad populations of Southern rivers would result in marked reductions in reproductive potential if reproductive adaptations were not developed to compensate for their loss. My studies of the reproductive ecology of the species indicate that Southern shad populations have adapted by earlier maturity and

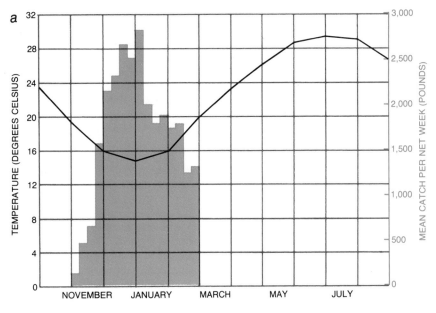

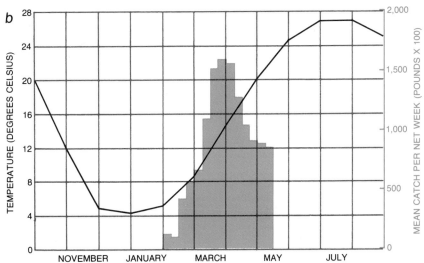

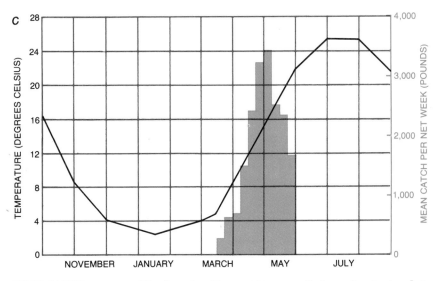

SHAD CATCH, represented by the colored bars, varies seasonally from river to river but in each case reaches a peak when the water temperature (*black curve*) is around 15 degrees C. In the St. Johns River of Florida (*a*) the peak catch is in January, when the river has cooled enough for the shad to enter for their spawning run. In the York River of Virginia (*b*) and the Connecticut River (*c*) the shad enter when the river has warmed up enough.

greater fecundity. In the St. Johns River the average age at maturity for female shad is 4.3 years and the average fecundity of females spawning for the first time is 412,000 eggs. In the Connecticut River the corresponding figures are 4.8 years and 263,000 eggs. Moreover, the variations in maturity and fecundity are gradual and in direct relation to corresponding gradual increases in the frequency of repeat spawning with latitude. The net effect of these changes in reproductive characteristics is to maintain a similar lifetime production of eggs in all the populations of the Atlantic Coast.

Our interest in the shad has not been limited to defining the path of their migrations and determining the environmental factors that govern the timing of their marine and freshwater movements. Since 1967 we have also been studying the migratory behavior of shad during the final stages of their approach to their home river from the sea and during their progress upriver. In these investigations we have employed ultrasonic tracking, which is a relatively new technique in fisheries research whereby the movements of individual fish can be followed from boats and shore stations by means of small ultrasonic transmitters attached to the fish.

Our work in Long Island Sound, which is under the direction of one of my graduate students, Julian J. Dodson, is aimed at determining how shad recognize that they are in the area of their home river and how they locate the river mouth. We have now tracked some 60 shad during their approach to the river from the sea; most of them were followed continuously for at least 24 hours. The observed movements of the fish were compared with data collected simultaneously on the temperature and depth of the water, the intensity of the light, wind and wave action, salinity, tidal currents and the location of the sun. So far only salinity, temperature and tidal currents appear to play a major part in directing the movements of shad in the final stages of the marine migration.

Shad enter Long Island Sound from the east, so that in order to reach the Connecticut River they must progress some distance westward. In doing so they exhibit a complex behavior. The open-water orientation of shad appears to be guided mainly by tidal currents: the fish tend to orient into the current. Since the current in Long Island Sound reverses direction approximately every

six hours because of the tide, the shad change their orientation by about 180 degrees every six hours too. We observed that they swim faster when oriented toward the west than when oriented to the east. Facing east against an incoming tide they swam at a speed about equaling the speed of the current and so made little net headway; facing west against an outgoing tide they exceeded the speed of the current and moved west.

It is noteworthy that the orientation was as precise at night as it was during the day, whereas the adjustment of speed was less precise at night. This observation suggests the existence of a compass sense to aid in orientation and the need for visual contact with the bottom or some other reference point in order to adjust swimming speed to the velocity of the current.

Reduced salinity at the surface and increased water temperature are among the more obvious indicators of the presence of water from the Connecticut River in Long Island Sound. Reduced salinity can occasionally be detected well east of the Connecticut River near the approaches to Long Island Sound from the Atlantic Ocean. Tracked shad were observed to respond to both indicators by swimming faster. It may be that the shad are actually responding to chemical substances characteristic of Connecticut River water. Indeed, this is the more likely hypothesis, inasmuch as changes in temperature and salinity alone would not be adequate clues for recognizing a specific river. Perhaps the detection of these identifying clues during the northward migration in the spring triggers the appropriate behavior pattern leading to the locating of the home river.

We have also carried out experiments in which we blinded some shad, blocked the olfactory capsule of others and performed both operations on still others. All three groups exhibited less ability in orientation than the unimpaired shad, the anosmic group being most seriously affected. Both the blind fish and the anosmic ones exhibited a reduced ability to find the river, and none of the shad that were both blind and anosmic homed on the river. The experiments indicate that the visual and olfactory systems figure importantly in the ability of the shad to orient and to home.

Robert A. Jones of the Connecticut Department of Environmental Protection and I have collaborated closely in the studies of the behavior of shad dur-

ing the river stage of the spawning migration. We have now followed the movements of some 230 individual shad in fresh water for periods of up to 100 hours. One phenomenon we observed was a distinct change in the behavior of the shad at the interface of salt water and fresh water. Because the Connecticut River has a narrow mouth and a large discharge, the salty part of the river is a well-defined wedge that extends only a few miles upstream. The upstream movements of shad in the saltwater section were quite direct. Near the upper edge of the saltwater intrusion, however, the movements of the fish were characterized by extensive meandering that continued for a day or two. During that time the fish meandered up and down the river as the wedge of salt water advanced and retreated with the tides, and their average swimming speed was about half what it was both before and after the period of meandering. At the end of this period they resumed their steady progress upriver, swimming now in water that was entirely fresh. We believe the meandering behavior is associated with physiological changes involved in the transition from salt water to fresh water. This hypothesis is supported by our finding that shad experience considerable osmotic stress when they are transferred quickly from salt water to fresh water.

The steady upriver migration of shad in fresh water reveals still another behavioral change. The Connecticut River is tidal in its lower 45 miles; as a result the direction of flow can reverse completely as much as 25 miles inland. Once the shad are in the river they abandon the strict countercurrent behavior characteristic of their marine movements and proceed upriver toward the spawning areas without regard to the direction of flow.

In fresh water the shad follow the natural and dredged channel of the river, making only occasional movements into shallow water. This orientation to channels results in remarkably consistent migration routes in any given area of the river. How the shad establish the location of the channel is unclear. The speed of the current is slightly higher in the channel than in the shallow areas, and it may be that the difference provides an orienting clue.

One of the major obstacles the shad confront as they move upriver is the series of nets (about 40 of them) put across the stream by commercial fishermen seeking shad. Our tracking studies showed that the shad have an impressive ability to detect and avoid the nets. Typically the shad would move close to a net before sensing its presence and then would swim parallel to it until they reached the end, where they would turn upstream again. This behavior was observed in both daylight and darkness, suggesting that senses other than vision may be involved.

So far we have not observed any effect of the "plume" of heated water from the atomic power plant on the shad we have tracked as they moved through that section of the river. Perhaps the reason is that at Haddam Neck the channel runs along the west bank of the river, whereas the thermal discharge is on the east bank, so that the shad, moving as usual in the channel, are about as far away from the plume as they could possibly be. Even if any are near the plume,

lack of knowledge about the depth at which shad swim leaves us uncertain about whether they swim under the plume or go through it without being affected. We intend to pursue the matter with equipment that shows how deep the fish is swimming and what the temperature of the water is at that point. In view of our finding that shad have a narrow range of preference for temperature, it is important to ascertain whether or not a permanent increase in the temperature of a river or a section of it resulting from the thermal plume would affect the shad run.

As is often the case in investigations of animal behavior, our findings have raised as many challenging questions as they have answered. Much remains to be learned about the path shad take during their southward migration in the fall and winter and of the conditions they encounter. We are also interested in knowing whether temperature is the only controlling factor in the timing of the ocean migrations or whether other factors such as food supplies that may be governed by seasonal cycles of temperature are also involved. Much remains to be learned about the sensory systems involved in orientation and homing and about the nature of the environmental clues that guide the ocean and river migrations. We must also gain a better understanding of the effects of man-made changes in the physical and chemical characteristics of rivers on the migratory and homing behavior of the shad. Fortunately recent advances in technology are providing new and powerful tools with which we can hope to attack these questions fruitfully.

Escape Responses in Marine Invertebrates

by Howard M. Feder
July 1972

*Limpets, snails, clams, scallops, sea urchins, and other
slow-moving sea creatures go into remarkable gyrations
when they are approached by a starfish. This lively behavior
enables them to deter predation*

One does not usually think of snails or other mollusks as being lively animals. Under certain conditions, however, some of these invertebrates exhibit a truly spectacular behavior. This was first noted many years ago by several marine biologists in Europe. Late in the 19th century Paul Schiemenz of Germany reported the curious response of a small sand-bottom snail (*Natica millipunctata*) to contact with a predaceous sea star, or starfish (*Marthasterias glacialis*). The snail would quickly extend a fold of its mantle tissue and slide it over its shell, thus dislodging any tube feet the starfish had planted on the shell. Other zoologists soon reported a variety of maneuvers used by mollusks to escape starfishes; some scallops were observed to begin violent swimming activities on being touched, and certain species of snails even executed a series of somersaults to shake off the predator.

These observations elicited little further research interest at the time. Then in the late 1940's Eugene C. Haderlie of the University of California at Berkeley reported on the "mild hysteria" and running behavior of the small mollusks known as limpets in response to contact with starfishes. This observation stimulated marine biologists to test a wide variety of mollusks for their reactions to

GREAT LEAP UPWARD is made by the cockle *Cardium echinatum* to escape from a preying sea star (*Asterias rubens*). The leap shown in the photograph on the opposite page is the last in a series of maneuvers executed by the cockle when it is touched by a sea star, beginning with extrusion of the tubular foot from the shell. The cockle continues jumping until it has managed to dislodge the starfish. The photograph was made by Holger Knudsen of the Marine Biological Laboratory at Helsingör in Denmark.

sea stars. In a summary of these studies Theodore H. Bullock of the University of California at Los Angeles concluded that the elaborate forms of escape behavior exhibited by marine invertebrates must represent a phenomenon of considerable ecological importance. He suggested that these escape activities were not simply generalized reactions to contact but were specialized and apparently effective responses each species had evolved to cope with its predators.

Currently there is a revival of interest in the investigation of this intriguing problem. The accumulation of reports from investigators around the world indicates that escape responses of prey species to predatory sea stars can probably be found wherever the investigator looks for them. How did the specific responses evolve? How effective are they in preserving the various species? What are the stimuli and mechanisms that produce the behavioral responses? These questions are being studied in a number of marine laboratories.

Sea stars inhabit most regions of the oceans and prey on many species of invertebrates. Most of the sessile victims (such as barnacles, mussels and other animals that are permanently attached to rocks or other substrates) manage to survive mainly by virtue of their high rate of reproduction. We are concerned in this article with the prey that have mobility: limpets, snails, clams, scallops, sea urchins, sea cucumbers, certain sea anemones and starfishes themselves (some species prey on other starfishes). All these animals have developed evasive maneuvers, some of them quite bizarre. Not all the responses depend on locomotion; in many instances the defense is a manipulative mechanism that enables the animal to remain in place while warding off the predator.

A survey of the escape behavior of such animals discloses several interesting points that invite detailed investigation. One is the variety of responses to the approach of the starfish. Another is the finding that related species show a great similarity in their escape behavior, even on opposite sides of the world. For instance, the running response that Haderlie observed in limpets in California is exhibited by limpets of the same genus in Europe as well as in North America, and the somersaulting movements characteristic of snails of the genus *Nassarius* are remarkably similar in related species in North America and in Europe. There are even more curious observations that need explanation. In some animals the supposed escape reaction initially includes a maneuver that would seem to make the animal more vulnerable to attack; for example, snails and limpets loosen their attachment to the underlying substrate immediately after contact. Furthermore, what are we to make of the finding that some of the animals that have highly efficient escape responses are apparently not attacked by starfishes at all?

Let us consider the varieties of escape behavior in some detail to see how effective they actually are for avoiding predation. Abe Margolin and I, working independently, closely examined the behavior of two different genera of limpets (*Diodora* and *Acmaea*) in the field and in the laboratory. (Margolin was at the Friday Harbor Laboratories of the University of Washington and I was then at the Hopkins Marine Station of Stanford University.) The two genera differ in their escape responses. When a limpet of the genus *Diodora* is touched by a species of starfish to which it is sensitive, the animal remains in place, raises its shell (in a movement that Bullock aptly called "mushrooming") and extends a

ESCAPE RESPONSE of a clam of the genus *Spisula* begins (*top*) when an arm of a sea star (*A. rubens*) touches the clam's siphon, which is protruding from the sand in this Knudsen photograph. The siphon is the small white object about a third of the distance in from the end of the right arm. On contact the clam stretches out its foot (*middle*) and thereby executes an escape jump (*bottom*).

fold of its mantle over the shell, thereby sweeping off the starfish's tube feet [*see illustrations on next page*]. The mushrooming reaction is so effective that *Diodora* is seldom attacked successfully by the two species of starfish whose intertidal distributions overlap the distributions of the limpets.

In contrast to *Diodora*, various species of *Acmaea* escape by running away, as Haderlie noted in his original observations of limpet behavior. Some species of *Acmaea* sense the approach of a predaceous starfish even before they are touched. After contact a limpet raises its shell in the mushrooming motion and rapidly moves away; its flight soon slows down, however, and the animal comes to a complete halt within a few minutes. In a laboratory aquarium, where room for flight is severely restricted, the limpets (and other organisms that depend on flight) are ultimately taken as food by predatory starfish in the same tank. In their coastal habitat, however, most species of *Acmaea* that can detect their predators are generally successful in evading capture. We have observed that the most important California intertidal starfish predator (*Pisaster ochraceus*) commonly feeds on only one species of *Acmaea*, known as the ribbed limpet; this limpet does not show an escape response. As an apparent consequence it thrives only in the high-tide region and its abundance declines sharply in the lower range where the starfish is active.

Of all the escape activities shown by animals attacked by starfishes perhaps the most spectacular are those exhibited by the snails of the genus *Nassarius*. These animals live on sediment bottoms and possess a strong foot that can flip the animal sideways or propel it into a somersault [*see illustration on page 167*]. Contact with a predatory starfish touches off a series of tumbling maneuvers by the snail, and the action is so violent that only a starfish considerably larger than the snail can remain attached to its prey. A large snail in New Zealand, *Struthiolaria papulosa*, displays a similarly striking performance. Robin Crump of the Orielton Field Centre in Wales, describing the behavior of *Struthiolaria*, reports that he has seen the animal perform as many as 50 consecutive somersaults in a four-minute encounter with a starfish. The European dog whelk (*Buccinum undatum*), a thick-shelled snail, also has a violent escape reaction. It rapidly rolls its shell counterclockwise and then with a twist of its foot throws the shell in the opposite direction, thereby effectively shaking off the starfish's tube feet.

The common black turban snail (*Tegula funebralis*) of intertidal waters in California has a varied repertory of responses after it has come in contact with a starfish. Contact causes it to flee, and if it is already in motion, it will speed up its travel from its normal two or three centimeters per minute to eight centimeters per minute. When an anterior surface of the snail is touched, it raises the front portion of its foot, makes a turn of about 90 degrees and moves off in the new direction. A touch on its side causes the animal to twist its shell away from the point of contact; when a posterior surface is touched, the snail tilts the shell over its head, twists about violently and moves off rapidly. On a steep slope the starfish contact will precipitate a tumble by the snail down the slope. In many areas black turban snails are apparently not fed on by sea stars in proportion to their abundance, and generally they are not attacked to any extent if sessile organisms such as barnacles or mussels are available.

The abalones (genus *Haliotis*) are another group of shellfish whose members exhibit very effective escape maneuvers. When touched by a starfish, an abalone raises its shell and whirls it violently from side to side; some species are capable of rapid locomotion, lifting the forward part of the foot and extending it far ahead for a leaping maneuver. The black abalones (*H. cracherodii*), dwelling in intertidal areas, seldom fall prey to starfishes. The predominantly subtidal red abalones (*H. rufescens*) are also usually successful in escaping the predators (except for the small young individuals, which can readily be taken by the large sunflower sea star *Pycnopodia helianthoides*).

Long before biologists began to notice the escape behavior of mollusks the Maori people of New Zealand were well acquainted with this behavior in abalones and put it to everyday use: they harvested abalones (a prized food) from inaccessible locations by deliberately touching them with a predatory sea star and then plucking the rapidly moving mollusks from the substrate. David Montgomery of California State Polytechnic College has discovered an instance in which an abalone, under attack by a starfish, liberates a substance that serves to alarm other members of its own species. When it is touched by a starfish, the animal exudes a cloudy fluid that causes other abalones in the vicinity to make escape movements typical of the species' reaction to contact with a sea-star predator. It was already known that some freshwater gastropods and tropical sea urchins discharge substances signaling fright or injury to members of their own species. The same phenomenon is well known in the world of the insects, some of which, notably certain ants, emit alarm pheromones [see "Pheromones," by Edward O. Wilson; SCIENTIFIC AMERICAN Offprint 157.

Even among the bivalve mollusks (clams, cockles, scallops), most of which would seem to have scant capability for rapid locomotion, one finds remarkable feats of agility. Typically a bivalve attacked by a sea star remains in place and slowly closes its valves (shells). Some bivalves show little sensitivity to the starfishes that prey on them; in fact, a few open their shells and continue to feed while an attacking starfish is manipulating them. Other bivalves, however, have evolved not only vigorous escape responses but also the ability to recognize their predators. Scallops can swim or jump about by rapidly opening and closing their shells; such movements, although uncontrolled as to direction, apparently are rather effective in protecting these mollusks from extensive predation. The cockles and a few species of clams use a different mechanism: when touched by a starfish, such a bivalve opens its shells wide, thrusts out its foot against the substrate on which it is resting and with a violent push of the foot makes a leap upward, often as far as 10 centimeters [*see illustrations on page 162 and on opposite page*].

The sea urchins, members of the phylum Echinodermata with a globular form studded with spines, possess two methods of protection against starfishes. They can run or use poisonous pinching structures, called pedicellariae, to grasp and remove the starfish's probing tube feet. In the latter case contact by the sea star generally results in a rapid retraction of the sea urchin's tube feet, a lowering of its spines and a gaping and erection of the pedicellariae. Significantly, a European species of urchin that customarily occupies exposed positions on rocks has sufficient speed to outdistance a pursuing starfish, whereas two slower European species of urchins typically inhabit protected niches among the rocks.

Another species, the purple urchin of North American coastal waters (*Strongylocentrotus purpuratus*), often aggregates in large intertidal groups, making up formidable beds of thousands of individuals. Although the urchins lie exposed to starfishes, the carpet of pinching pedicellariae they present undoubt-

EXTRUDED MANTLE is the response of the keyhole limpet *Diodora aspera* to contact with the sea star *Pisaster ochraceus*. Extrusion of the soft mantle can be seen most clearly in the limpet at right. The reaction, which is shown in more detail in the illustration at the bottom of this page, gives the limpet's shell a fleshy covering. The tube feet on the arm of the predatory sea star cannot grasp the covering firmly and may in fact be repelled by it.

edly deters attack [*see illustrations on page 169*]. Individuals of this urchin species also escape from predatory sea stars by means of a running response if the encounter between the two takes place on a uniform substrate. Recent work done on subtidal populations of the purple urchin off Point Loma, Calif., by Richard J. Rosenthal of the Scripps Institution of Oceanography and James R. Chess of the Tiburon Fisheries Laboratory of the National Marine Fisheries Service has shown that here the leather star (*Dermasterias imbricata*) is a major predator on urchins. This sea star has not been reported to prey on urchins

elsewhere along the northern Pacific coast. Rosenthal and Chess suggest that the unusually high level of predation observed off Point Loma represents a regional phenomenon. There seems to have been a population explosion among the sea urchins in an area practically devoid of the organisms typically fed on by the leather star elsewhere, and the sea star has apparently switched to an abundant but less "preferred" prey species.

Certain other echinoderms show escape reactions in response to a stimulus from predatory starfishes. A large sausage-shaped animal (*Parastichopus californicus*) known as a sea cucumber re-

sponds with awkward swimming motions produced by contractions of its longitudinal muscles. And among other invertebrates exhibiting sensitivity to the starfishes are some of the sea anemones, flower-like members of the phylum Coelenterata with a bush of tentacles topping a thick column. By bending the column anemones (*Stomphia* and *Actinostola*) make creeping or swimming movements to escape from a predator.

Several species of sun stars (*Solaster* and *Crossaster papposus*) feed intensively on stars of other genera and sometimes even attack smaller individuals of their own species; such small prey are readily captured and are ingested completely. All the prey species move away from the attacking sea stars after contact with them, and large stars may outdistance the pursuing predators. Karl P. Mauzey and his associates, who formerly worked at the University of Washington, noted that predatory sun stars can often counteract the rapid escape movements of their potential prey. The predator typically moves along alternately raising and lowering its leading arms. When the tube feet on one of the arms come in contact with another sea star, the arm drops onto the upper surface of the prey. The predator then raises the other arms facing the prey and "lurches" forward onto it. If the prey is the sun star *Solaster stimpsoni*, however, it may still avoid capture by curling its arms back over the upper surface of its body and using these arms to push upward and backward against the attacker. When a predatory sun star attacks a large sea star of the European genus *Asterias*, the captive often frees itself by throwing off its arms; the predator then feeds on these amputated members. Occasionally a sun star smaller than its prey has been seen clinging to the victim and feeding on its arm tips. In a laboratory tank the predator sometimes will cling tenaciously to a captive and feed in this way for several days.

So much for the observations of escape activities. Attempts to explain their development and their role in stabilizing animal communities in the sea begin with the examination of apparent anomalies. One of these has been the puzzling circumstance that many of the animals that show escape responses to starfishes in the laboratory are rarely, if ever, fed on by these predators in nature and in some cases apparently do not even inhabit the same waters.

Mauzey and his associates, whose investigations included free-diving observations of marine populations, suggest

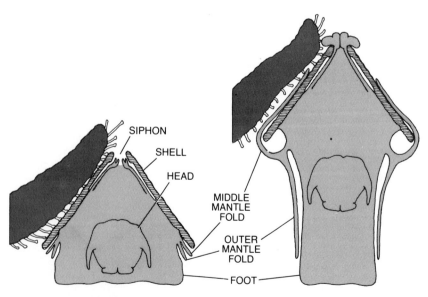

POSITIONS OF LIMPET are portrayed in the normal state (*left*) and when the animal is reacting to a sea star (*right*). The limpet is usually attached firmly to a rock by its foot. In the mantle response the animal stretches upward and then slides its mantle over its shell.

that the unmolested animals may actually owe their freedom from attack to the almost total effectiveness of escape reactions; such reactions may completely remove the potential prey from the predators' diet. Mauzey's group cites certain observations that tend to support this view.

One of the immune organisms is the large sea cucumber *Parastichopus californicus*. Because starfishes had never been seen to attack this animal in nature, observers supposed that the cucumber's reactions to starfishes in the laboratory did not actually represent escape responses related to this predator. Recently, however, intensive searching has turned up a few cases of successful predation on *Parastichopus* by the sunflower starfish, and it has also been found that all the other starfish species to which the sea cucumber responds with swimming movements feed on other species of sea cucumbers. Hence it appears likely that *Parastichopus'* activity was indeed developed as a defense against starfish predators and that the almost complete lack of predation on the large sea cucumber results from the effectiveness of the response. The other observations cited by Mauzey and his colleagues to support their hypothesis has to do with certain species of sea anemones. Under laboratory conditions these anemones react to contact with four species of starfish. There was reason to doubt that this reaction actually represented a predation-induced escape response, because in nature these anemones and starfishes had not been found to occur together in the same waters. In diving explorations Mauzey's group did, however, eventually find some instances of co-occurrence, and the starfishes there were feeding on the anemones. The four starfish species mentioned have also been observed to feed on other species of anemones and are almost the only sea stars with such a diet. It seems reasonable to conclude that the evasive behavior shown by the various anemone species is truly a response to the predatory sea stars.

What is the stimulus that initiates an escape reaction on the part of the prey of starfishes? It is not necessarily contact with the predator. Generally the most vigorous escape movements occur only after contact, but early in the investigation of the phenomenon Bullock and I discovered that many gastropods reacted to the approach of a starfish even before contact. This suggested that the animals responded to substances emanating from the predator. Investigation

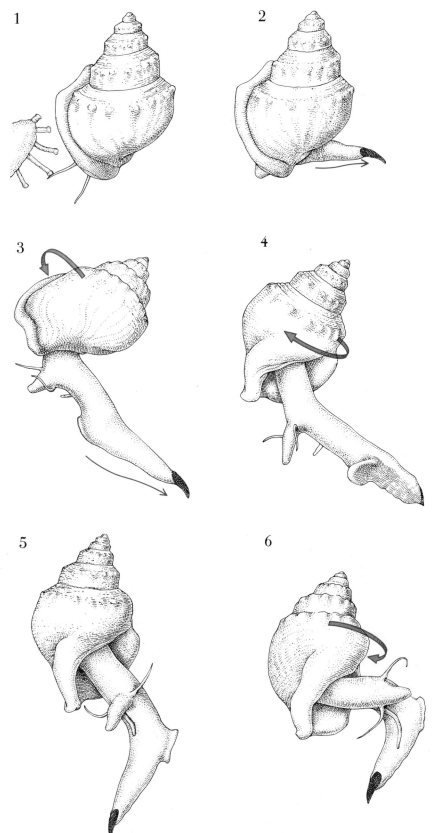

SEQUENCE OF MOVEMENTS in the rolling flight response of the New Zealand snail *Struthiolaria papulosa* following contact with a sea star (*Astrostole scabra*) is depicted. When a tube foot of the star touches a tentacle of the snail (1), the snail's foot extends to the side of the shell (2) and the shell is rolled by violent twisting of the body (3). Eventually the snail becomes inverted (4). Then the foot begins to curl under shell (5), soon attaining a position from which it can throw the shell in the opposite direction (6). Illustration is derived from a paper by Robin Crump of the Orielton Field Centre in Wales.

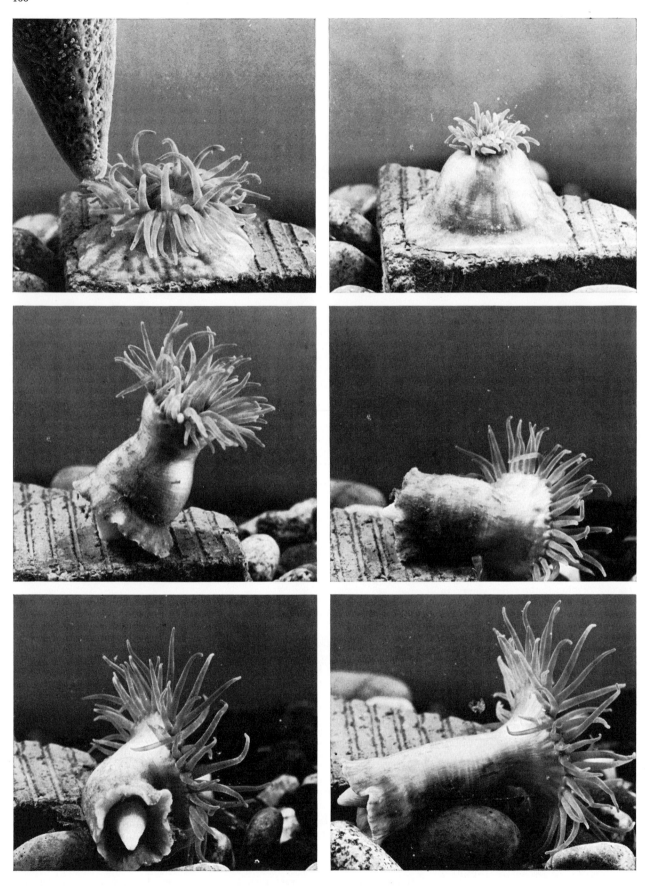

AVOIDANCE MOVEMENTS are made by the sea anemone *Stomphia coccinea* on contact with the sea star *Dermasterias imbricata*. At top left the star touches the anemone and the animal withdraws its tentacles (*top right*), meanwhile beginning to detach itself from the substrate. A few seconds later (*middle left*) the pedal disk has been detached from the substrate and a conelike structure is beginning to form at the bottom of the animal. Soon (*middle right*) the animal moves by flexing its body, and within about 10 seconds (*bottom left*) it has moved off the substrate. Finally (*bottom right*) the animal relaxes; it will resettle in about 15 minutes.

has since established that an active, irritating material is in fact continuously liberated by predatory starfishes. These substances are found in most tissues of the starfish; primarily, however, they are concentrated in the epithelial covering of the body, particularly in the tube feet.

Drippings exuded by starfishes and crude extracts from starfish tissues have been shown to be capable of stimulating escape responses in their prey. All the predatory starfishes of North America and Europe that have been examined possess active substances. The extracts are all relatively thermostable, weakly dialyzable, insoluble in fat solvents, attached to or readily absorbed by proteins and stable for many months when frozen or freeze-dried.

Sea-star drippings and crude extracts contain steroid saponins that are known to be toxic, and it has already been demonstrated in at least one case that saponin-like substances are responsible for a mollusk's avoidance reaction to a starfish; this was shown by Alexander M. Mackie, Reuben Lasker and Patrick T. Grant in the response of the dog whelk to the sea star *Marthasterias glacialis*. (Mackie, Lasker and Grant were working at the University of Aberdeen in the Fisheries Biochemical Research Unit of the British Natural Environment Research Council.) The similarity of the properties of the active materials obtained from the predatory starfishes so far examined suggests that perhaps all the predators contain saponin-like substances.

The indication that all predatory starfishes emit chemically similar irritating substances may explain the worldwide reactions to these animals by marine invertebrates. It could also account in part for the observation that often prey organisms respond with avoidance reactions even to species of starfish that they never normally encounter or that do not normally feed on them in nature. Mauzey and his colleagues point out that some of the anomalous responses might also be explained if one assumes that some of the chemical irritants originate in prey organisms and that the sea star which atypically induces avoidance responses has a diet similar to the diet of the predator.

In studies at laboratories in the U.S. and in Sweden, Montgomery, Jan Arvidsson and I have obtained evidence suggesting that most and possibly all echinoderms contain and exude substances that can induce avoidance reactions. Even nonpredatory echinoderms (sea urchins, sea cucumbers, and star-

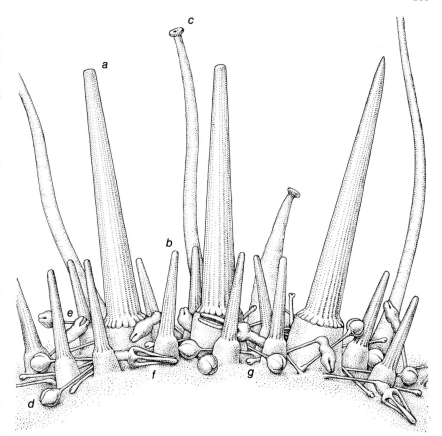

UNDISTURBED SURFACE of the sea urchin *Psammechinus miliaris* is depicted. The anatomical features include primary spines (*a*), secondary spines (*b*), tube feet (*c*) and several kinds of pedicellariae (*d–g*). The primary spines are about five millimeters long.

DEFENSIVE REACTION is manifested on the surface of *P. miliaris* when a tube foot of the sea star *Marthasterias glacialis* is held nearby with forceps. The spines bend away, and globiferous pedicellariae are raised and opened to close on the foot. Illustrations on this page are based on drawings by Kai Olsen of Marine Biological Laboratory at Helsingör.

fishes that are not predatory) contain irritating materials; however, they do not liberate any significant amount of it, either because their tissues contain very little of it or because it is held in a tightly bound form. In addition many nonpredatory echinoderms contain saponins. Therefore it is possible that steroid-like irritating substances are common metabolites in all echinoderms but that these substances are released in detectable amounts only by predatory sea stars.

How did it come about that predatory starfishes evolved the practice of exuding substances that alarm their prey? There is reason to believe that this behavior may have served originally as an effective prey-catching device. As we have noted, some species of limpets and abalones respond to a starfish's approach by raising their shells and loosening their attachment to the underlying substrate. Such a response, induced by an irritating toxic substance (which might also serve the sea stars as a repellent against predators), may initially have enabled the predator to carry off the prey more easily. The prey organisms in turn may have evolved devices for fleeing from or warding off the predator, presumably as responses to the chemical stimulus. Jefferson Gonor of Oregon State University has suggested that most of the escape maneuvers of marine snails, for example, are probably not new acquisitions but instead evolved from existing locomotory and righting movements that originally served functions other than defense.

The most novel and intriguing development, in Gonor's view, is the prey animals' acquisition of the ability to identify and recognize their specific predators. In most cases examined by our group and by other investigators a given starfish predator elicits strong responses only from the animals that serve as its prey. For example, D. Craig Edwards of the University of Massachusetts has shown that the olive snail responds strongly to a starfish species that preys on it but does not react significantly to another species of the same genus that it never encounters in nature. Similarly, the purple sea urchin shows almost no response to starfishes that rarely feed on it, and anemones do not swim away when they come in contact with starfishes that do not prey on coelenterates.

Further investigation of the chemical communications among all these animals should lead us to a clearer view of life on the sea bottom and at its teeming margins.

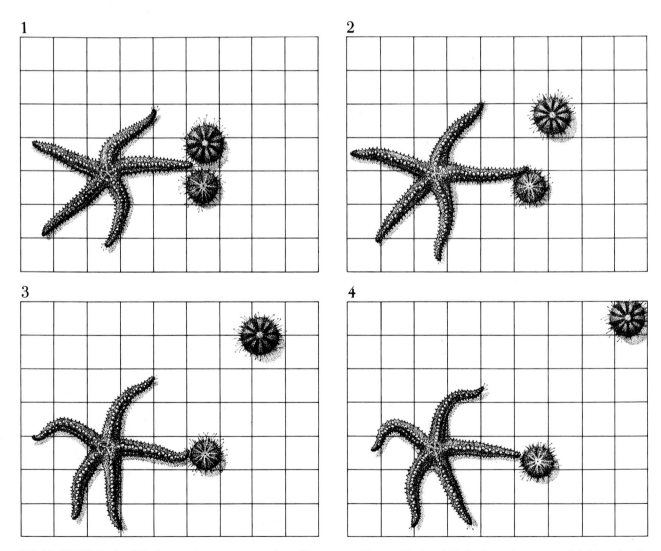

1

2

3

4

SEA URCHINS display differing reactions to a sea star depending on whether or not the urchin's natural habitat affords protection from sea stars. In a laboratory experiment the animals were placed on a grid with lines five centimeters apart. The urchin *Psam-* *mechinus miliaris*, which is from the protected habitat, barely moves during a period of 90 seconds, whereas the urchin *Strongylocentrotus droebachiensis*, which is from the unprotected habitat, moves away rapidly. The sea star in the experiment is *M. glacialis*.

Decorator Crabs

18

February 1980

*Many species of spider crabs can camouflage themselves
by patterns of behavior in which they select materials
from their environment and festoon them on their shell*

Animals that camouflage themselves are seen—or not seen!—throughout the natural world. Certain insects resemble a twig; certain lizards are able to change their coloring to match different backgrounds; certain mammals are spotted so that they blend into a pattern of light and shade. In the marine environment some of the best-hidden animals are crabs. Many of the thousands of species of crabs have lines, spots or patterns that serve to break up the outline of their carapace, or upper shell. A few species have a pair of modified back legs with which they pick up objects such as shells or sponges and hold them on or over their carapace. The crabs most adept at camouflage, however, are those that deliberately select bits of material from their environment and attach them to various parts of their shell. Called masking crabs or decorator crabs, these animals can become so heavily encrusted with strands of red or brown algae, pieces of fluffy branched bryozoans, segments of feathery brownish yellow hydroids and brightly colored sponges that an experienced underwater naturalist can sit on a decorator crab before realizing that it is there.

Crabs belong to the crustacean order Decapoda, and decorator crabs belong to the decapod family Majidae, commonly known as spider crabs. Spider crabs have a rounded or elongated body, four pairs of long, slender "walking" legs and a pair of front legs that are modified as chelae, or pincers. It is the chelae that gather fragments of decorating material and attach them to the minute hook-shaped setae, or bristles, found on various parts of the crab's shell.

Naturalists have known of spider crabs' habit of decorating their shell at least since the middle of the 19th century. In 1889 the Swedish naturalist Karl Aurivillius gave a detailed account of the decorating activities of many European species. Before the introduction of scuba diving equipment, however, all such studies were based on observations in aquariums of crabs that had been cap-

tured by dredging, a nonselective procedure that tended to injure the animals. In recent years it has been possible not only to stock aquariums with healthy crabs in a wide variety of species, sizes and stages of maturity but also to observe the animals in their natural habitat. With the application of modern techniques for underwater and slow-motion photography and scanning electron microscopy workers in this field have gained new perspectives on decorating behavior. I shall discuss here some of the most recent findings about the form, function and development of decorating in spider crabs.

Decorating is actually a complex chain of activities that begins with the acquisition of a piece of decorating material. For this task a spider crab relies on its nimble chelae, which are slender and forcepslike in females and immature males and heavier in mature males. The chelae can work separately to pick pieces of algae or other detritus directly off the bottom or together to twist off branches of bryozoans or hydroids and break them into pieces of suitable length. The fingers of the chelae can also snip off pieces of sponges or compound ascidians. (Orange or white compound ascidians, which are found on the shell of many decorator crabs, are fleshy organisms that grow in matted colonies on hard surfaces.)

Once a piece of decorating material of appropriate size has been obtained, one of the chelae conveys it to the oral field: the area near the front of the crab where the mouthparts are. The mouthparts manipulate the piece of material, rotating it repeatedly until its edges are rough. (A long, thin piece of material such as a strand of alga is roughened at one end only.)

To attach the roughened piece of material to its shell the crab takes the piece from its mouthparts with one of its chelae and rubs it against an area of the shell covered with hooked setae. Rows of these curved (and in some instances barbed) bristles are found on the crab's

rostrum (a projection in front of the eyes), on the back of its walking legs (and in some instances on the back of its chelae) and along the sides of its carapace, the distribution differing from species to species. The piece of material adheres to the shell because before the crab releases the piece from its mouthparts it turns the piece so that it is either parallel or perpendicular to the rows of hooked setae where it is to be attached. When the crab rubs the piece against the rows of setae with its chela, the piece is entangled or impaled.

A piece of decorating material that does not get quickly affixed is returned to the mouthparts for further manipulation. It has often been suggested that glands in the mouthparts of a spider crab manufacture a glue that this oral manipulation serves to apply. Those glands probably function in digestion rather than decoration, however; I have found that a spider crab can decorate itself normally after they are removed. On the other hand, if a crab's hooked setae are removed, the crab cannot decorate itself, although it will make repeated attempts to attach properly prepared material. (Other kinds of setae on the crab's shell function as tactile sensory structures and probably provide information about the position of the decorating materials.)

A few species of spider crabs are found on soft bottoms consisting of mud or sand, but most of them inhabit hard or rocky substrates anywhere from the high-tide mark on a coastline to the outer edge of the continental shelf. The crabs settle into such habitats after a larval stage in which they float free among other plankton at the surface. Little is known about the life cycle or molting patterns of spider crabs, but after settling on the bottom all the decorator crabs I have observed molted two or three times before they began to decorate themselves. It appears that many spider crabs continue to decorate themselves throughout their lives, although some, such as the male moss crab *Loxo-*

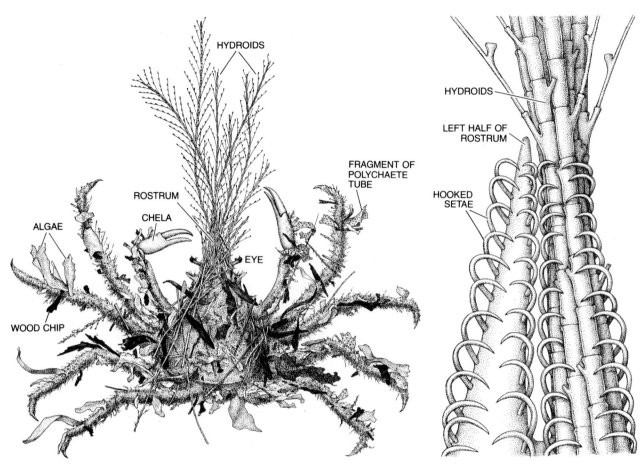

DECORATING MATERIALS on the carapace and legs of a crab of the species *Oregonia gracilis* in the drawing at the left include short strands of algae, segments of hydroids, wood chips and several sand-speckled fragments of tubes secreted by polychaete worms. The rostrum is decorated with long streamers of hydroids, and the entire surface is covered with a layer of filamentous algae. The chelae pick up or tear off a suitable fragment of material and carry it to the mouth-

parts, which manipulate the fragment until its edges are rough. Then one of the chelae rubs the fragment against an area on the crab's shell that is covered by rows of hooked setae. Depending on whether the fragment is held perpendicular to the rows of setae or parallel to them, it is either impaled or entangled. In enlarged view of crab's rostrum at right the decoration has been partly removed to show how in this species hooked setae are arranged in pairs of facing rows.

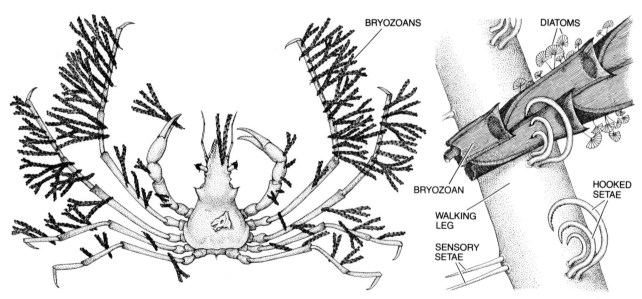

MORE SPARING DECORATION is found on the small California crab *Podochela hemphilli,* shown in the drawing at the left. In this species, which is found around docks and wharf pilings, much of the carapace and rostrum is usually bare, even though there are hooked setae on them. The walking legs, however, are often covered with small pieces of algae or, as is shown here, sections of branched bryozoans.

The enlarged view at the right shows how pieces of decorating material are impaled on the small clusters of barbed hooked setae that characterize the species. Also shown in this view are a number of diatoms, which often settle on pieces of branched bryozoans sticking out from the crab's walking legs, and straight setae, which function as tactile sensory organs, possibly for location of decorating materials.

rhynchus crispatus, stop decorating at about the time they reach maturity. (*L. crispatus,* one of the largest crabs found on the California coast, stops displaying decorating behavior when the width of its carapace reaches about four inches and the span of its legs about three feet. At this size it is presumably immune to attack by most predators.)

The materials spider crabs generally favor for decorating their shell are flexible, easily torn organisms such as sponges, algae, bryozoans and compound ascidians. The crabs make frequent use, however, of other types of decorating material, including pink or green coralline algae (algae stiffened by a coating of calcium carbonate), branches of brightly colored gorgonians (horny corals such as sea fans), tubes secreted by polychaete worms and even the leaves of land plants. The Caribbean spider crab *Stenocionops furcata* attaches the small striped sea anemone *Calliactis tricolor* to its carapace and legs. In aquariums spider crabs stripped of their decoration have been known to decorate themselves with torn sea pansies, strips of paper, chips of cement from the aquarium wall and even bits of hamburger.

Just as the materials of decoration can differ from species to species so can the decorating style. Compare two crabs that decorate intensively. A juvenile crab of the species *Loxorhynchus crispatus* covers its entire carapace with a thick layer of algae, bryozoans and other organisms; *Oregonia gracilis,* a crab found along the northwestern U.S. coast, covers its carapace and walking legs but also anchors at its rostrum long streamers of algae, bryozoans and hydroids. Among the crabs that decorate themselves more sparingly is the small crab *Podochela hemphilli,* which is often found around California docks and wharf pilings. This crab generally leaves much of its carapace and rostrum uncovered but attaches small bits of algae or bryozoans so that they stick out from its walking legs. The European spider crab *Inachus scorpio* is reported to be even more selective, decorating only its first pair of walking legs.

Some of the spider crabs that do little active decoration get covered passively, through the natural accumulation of detritus in their hooked setae or through the attachment of encrusting organisms to their shell. For example, although the sharp-nosed crab *Scyra acutifrons* of the Pacific coast of the U.S. often decorates its own rostrum, much of its carapace is covered by sponges or ascidians that settle there on their own. And as the tiny crab *Pitho picteti* moves across a bottom of sand or rocky rubble in the Gulf of California its hooked setae generally collect a covering of sand and debris. Moreover, small sea anemones and bits of sponges and compound ascidians, however they come to decorate a crab's shell, may remain alive after they have become attached and grow out from the original area of attachment. Many spider crabs of the genus *Pelia,* including the California crab *Pelia tumida,* tend to decorate themselves with small bits of sponges that regenerate and spread out to cover the entire carapace.

Although decorator crab species have been identified in all living subfamilies of the Majidae, many spider crabs do not decorate at all. For example, as I have mentioned, *Loxorhynchus crispatus* does not decorate after it reaches a certain size. Other large spider crabs (species with a leg span of two and a half feet or more) that do not decorate themselves as adults include the California sheep crab *Loxorhynchus grandis,* the Caribbean giant crab *Mithrax spinosissimus,* the Pan-American crab *Maiopsis panamensis* and the Japanese giant crab *Macrocheira kaempferi.* (With a leg span of up to 14 feet, *M. kaempferi* is the world's largest crab.)

Size is not the only factor that may keep a spider crab from decorating itself. For example, species with a knobby shell that already matches the substrate they inhabit generally do not decorate themselves. Hence the lumpy spider crabs of the tropical genus *Mithrax,* which live on a hard, rocky bottom, tend to remain totally free of decoration except for a few encrusting organisms that settle on their own. Spider crabs that live among the stipes, or stemlike parts, of giant algae tend not to decorate themselves, possibly because as they climb through the algae any decorating material would get knocked off. Thus if the brown kelp crab *Pugettia producta,* which is generally found among the giant brown kelps on the California coast, decorates itself at all, it attaches only bits of kelp to its rostrum. The purple kelp crab *Taliepus nuttalli* does not display any form of decorating behavior. Finally, spider crabs that inhabit soft substrates, such as the tanner crabs of the genus *Chionoecetes,* rarely decorate themselves, probably because of the absence of decorating materials on bottoms of sand or mud.

It is important to note that although a motionless, well-decorated crab may be difficult to see in its natural habitat, in some instances the match may be far from exact. I have seen a spider crab encrusted with white ascidians sitting in a field of dark red algae and one decorated with orange ascidians, yellow hydroids, red algae and blue sponges resting on a bed of bright pink coralline algae. *Loxorhynchus crispatus* can usually be detected by looking for patches of dead, bleached bryozoans. Moreover, if in aquariums the Caribbean decorator crab *Microphrys bicornutus* is placed against a background whose color contrasts with the crab's covering, the crab makes no effort at adjustment. In a similar situation a small crab of the species *Loxorhynchus crispatus* will intersperse material from the new environment among the bits of decoration already accumulated.

These examples probably reflect the fact that crabs living on reefs characterized by a diverse biota would, as they move from place to place, have great difficulty trying to match each different background. More generally, it appears that decorating serves not to make a spider crab invisible but rather to make it look less like a crab. To accomplish this end the crab must expend metabolic energy that could be devoted to other important activities such as feeding or mating. What does the animal get in return, that is, what is the function of decorating behavior?

To begin with, in a number of species the camouflage provided by decorating activities appears to facilitate the capture of prey. For example, in aquariums fishes and other swimming prey have been observed coming close to motionless crabs of the well-decorated European species *Hyas araneus* (a lyre crab) and *Macropodia rostrata* and being captured by them. And in tanks in my laboratory at the University of Southern California another lyre crab, *Hyas lyratus,* has proved to be a voracious predator on smaller crabs, which its heavy decoration often enables it to approach unnoticed. Most spider crabs, however, feed on carrion, on eyeless organisms such as echinoderms (sand dollars, sea urchins, starfishes and so on), on slow-moving organisms such as mollusks and polychaete worms, and on sessile plants and animals such as algae or ascidians. There is obviously no need of camouflage in capturing these organisms, and for the spider crabs that feed on such prey the main function of decoration is probably to provide protection against predators.

The animals that prey on decorator crabs include the European lobster *Homarus gammarus,* the sea otter *Enhydra lutris* and a wide variety of fishes, octopuses, starfishes, spiny lobsters, other crabs and even sufficiently adventurous human beings. During the daylight hours, however, most decorator crabs are effectively camouflaged by their shell decoration and their lack of movement. The crabs spend the day lying pressed against the bottom with their chelae folded under them. The only movement is that of the antennae, which extend out from the crabs' ventral surface, and they are generally hidden from view by overhanging decoration. Hence in aquariums well-decorated specimens of *Loxorhynchus crispatus, Pelia tumida* and *Pugettia richi* (a kelp crab found on

HEAVILY DECORATED CRAB *Oregonia gracilis,* **covered by bits of whitish yellow sponges and fluffy branched bryozoans, is barely visible at the center of this photograph. Like all decorating species this crab belongs to the family Majidae. familiarly known as the spider crabs. Spider crabs have a pair of chelae (pincerlike front legs) that in decorating species serve to gather material and attach it to hooked setae: minute curved bristles found on the sides of the carapace (the upper shell), on the rostrum (a sometimes forked projection in front of the eyes), on the back of the long, slender walking legs and sometimes on the back of chelae themselves. (The distribution of the hooked setae differs from species to species.) During the day spider crabs lie on bottom with chelae folded under them, as is the case here.**

the Pacific coast of the U.S.) are rarely eaten by predators. Similarly, although sea otters off Pacific Grove, Calif., frequently make a meal of the lightly decorated kelp crab *Pugettia producta,* they have rarely been known to feed on crabs of the species *L. crispatus,* which are also found in abundance there.

A decorator crab's camouflage is not its only form of protection; spider crabs display several types of behavior other than decoration that help them to avoid predators. For example, although many studies of spider crab behavior have emphasized the slow, deliberate movements that characterize the behavior of these animals during the day, at night they can be seen walking openly across sandy bottoms, climbing kelp or foraging on reefs. A spider crab that is threatened by a predator will run, display its chelae, pinch the predator or autotomize, or allow to break off, a limb in the predator's grip. A crab that is fleeing a predator will often drop off an underwater ledge with its legs spread and land on the bottom below. (The European crab *Macropodia rostrata* is the only spider crab known to swim.)

Furthermore, the materials a spider crab attaches to its shell do not protect the crab merely by hiding it. The sponges and the compound ascidians found on the shells of many decorator crabs may be noxious or even toxic to the crab's predators. Hydroids and sea anemones often possess nematocysts, specialized stinging cells that can make a predator drop a captured crab. In aquariums when three fishes that feed on crabs (the cabezon, the lavender sculpin and the California scorpion fish) captured small crabs of the species *Loxorhynchus crispatus* and *Pelia tumida,* they promptly spat them out.

Moreover, to a prowling octopus a furry, well-decorated crab might not feel like a crab at all. Two small specimens of *Loxorhynchus crispatus* in an aquarium with 15 octopuses survived unharmed while crabs of several other species in the aquarium were captured and eaten. One of the two decorator crabs even molted and redecorated itself with bits of the octopuses' crab prey. (Crabs are most vulnerable to attack when they molt or when their shell is damaged. When *L. crispatus, L. grandis* and the New Zealand crab *Notomithrax*

ursus molt, however, they immediately remove bits of decorating material from the abandoned shell and attach them to the new one. Molting proceeds at night or in the early morning, when it is difficult to see even an undecorated crab. Within a day the crab has a hard and completely decorated new shell.)

The complicated chain of activities that make up decorating behavior in spider crabs is probably the result of a long evolutionary process. Unfortunately there are not many fossil crabs available to aid in the reconstruction of the process, but consideration of the different behavior patterns displayed by modern species of decorator crabs suggests a possible evolutionary sequence. To begin with, it is interesting to note that the first maneuvers of decorating—picking up materials and conveying them to the mouthparts—are similar to the activities of feeding. It seems likely that the evolution of decorating behavior began with early crabs that picked morsels of food out of detritus. Some may have resembled the modern spider crab *Pyromaia tuberculata,* which picks edible particles out of detritus on the

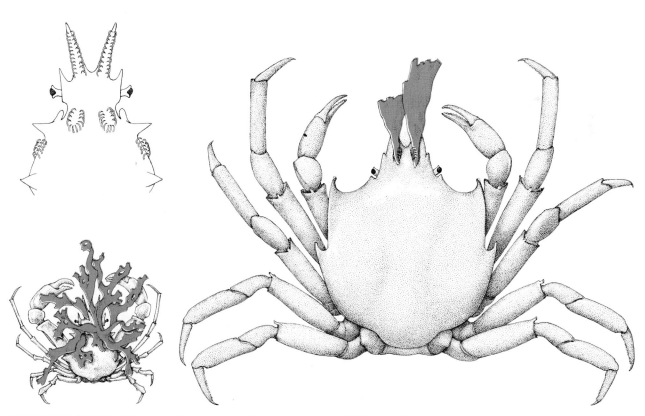

SOME KELP CRABS of the genus *Pugettia* seem to have lost decorating behavior. Small species such as *Pugettia dalli* (*left*), which live among low-growing algae or on the holdfasts (rootlike parts) of larger kelps, decorate the sides of the carapace and the rostrum, generally with pieces of kelp (*color*). The distribution of hooked setae on the shell of such kelp crabs is shown in the enlarged view at the upper left. The larger kelp crab *Pugettia producta* (*right*), which lives among the stipes (stemlike parts) of giant kelps, decorates only the rostrum, lacking the hooked setae along the edges of the carapace and behind the eyes that characterize the smaller species. *P. producta* is probably advanced species that lost most of decorating habit as it moved from smaller to larger kelps. Crabs shown here are drawn to same scale.

176

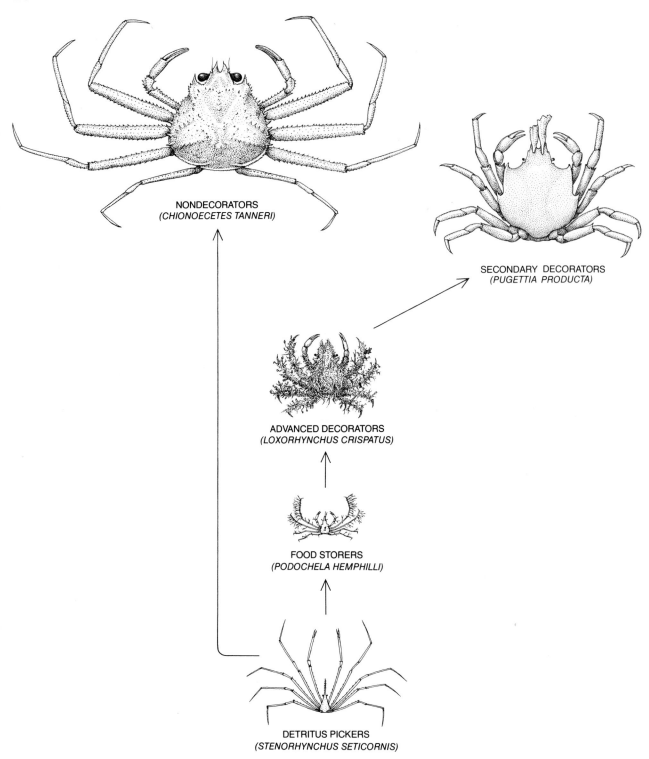

NONDECORATORS
(CHIONOECETES TANNERI)

SECONDARY DECORATORS
(PUGETTIA PRODUCTA)

ADVANCED DECORATORS
(LOXORHYNCHUS CRISPATUS)

FOOD STORERS
(PODOCHELA HEMPHILLI)

DETRITUS PICKERS
(STENORHYNCHUS SETICORNIS)

EVOLUTION OF DECORATING BEHAVIOR in spider crabs is suggested by behavior patterns in living species. Because the first activities in decorating (finding a suitable piece of material and conveying it to the mouthparts) are identical with activities in feeding, decorating probably began with early spider crabs that picked edible particles out of detritus. Some of these crabs may have resembled the living species *Pyromaia tuberculata* **(not shown), which feeds on debris it finds on the bottom. Others may have resembled the living arrow crab** *Stenorhynchus seticornis,* **which picks food particles out of debris that collects on its hooked setae. Eventually some of these early spider crabs may have taken to storing food on their hooked setae in the manner of living decorator crabs such as the species** *Podochela hemphilli.* **The habit of attaching food to the shell may have given a selective advantage to the early crabs by hiding them from predators, and some of them may have made the transition to attaching nonedi-** ble materials for camouflage alone. The most heavily decorated living crabs, such as the moss crab *Loxorhynchus crispatus,* **seldom eat anything they have attached to their shell and often decorate themselves with materials that few marine animals eat, including noxious sponges and stinging hydroids. The specimen of** *L. crispatus* **shown here is a juvenile; adult males of the species, one of the largest found on the California coast, stop decorating when they reach maturity. The same is true of both the males and the females of several other large species of decorator crabs. Among the other living species of spider crabs that decorate sparingly or not at all some, such as the kelp crab** *Pugettia producta,* **may have lost much of the decorating habit in response to environmental pressures (*see illustration on page 6*); others, such as the tanner crab** *Chionoecetes tanneri* **(whose soft-bottom habitat lacks decorating materials), may never have acquired the decorating habit. All spider crabs shown here are drawn to same scale.**

bottom. Others may have resembled the arrow crabs of the genus *Stenorhynchus* and the small California spider crab *Erileptus spinosus*, which perch on gorgonians or other elevated places and by extending their legs into a current gather detritus from which they later pick edible particles.

A later stage in the evolutionary process may have been the storage of uneaten food on the setae. The spider crab *Podochela hemphilli* will eat bits of algae it has removed from its shell. And in aquariums in my laboratory the tropical decorator crab *Camposcia retusa* attached to the hooked setae of its first walking legs bits of chopped fish, which it later ate. The attachment of uneaten food to the setae of primitive crabs may have given them a selective advantage by hiding them from predators.

Species such as *Loxorhynchus crispatus* and *Pelia tumida* generally do not eat anything they have attached to their shell. Indeed, these species often decorate themselves with noxious sponges, ascidians, sea anemones and other things few marine animals eat. This suggests that at some point the early food storers must have switched from attaching to themselves edible materials to attaching inedible ones providing only camouflage.

The decorating patterns of the kelp crabs of the genus *Pugettia* suggest a further evolutionary sequence away from decoration. The small species *Pugettia richi*, *P. hubbsi*, *P. dalli* and *P. gracilis*, which live among low-growing algae or on the holdfasts (the rootlike parts) of larger kelps, decorate their rostrum and the sides of their carapace. The larger crab *P. producta*, which lives among the stipes of giant kelps, decorates only its rostrum. *P. producta*'s lack of lateral hooked setae, the square shape of its carapace and its large size all suggest it is an advanced species that may have lost the decorating habit as it moved from small algae to large kelps. Similarly *P. venetiae*, which inhabits sandy bottoms, does not decorate itself at all. It may have lost the behavior pattern as it moved to a habitat where decorating materials were scarce.

In studying the relations among different species of birds and mammals behavioral comparisons have frequently proved to be of great value. It is possible that the application of similar methods to the study of decorating behavior would also be revealing. A deeper understanding of the evolutionary history of the spider crabs could be gained by comparing in different species the sculpturing and distribution of the hooked setae, the patterns of attachment of decorating materials, the types of material chosen, the degree of decoration achieved and the variations in decorating behavior according to size, sex or age. At present information about the natural history of most species of spider crabs is scant. In order to make meaningful, quantitative evaluations of their decorating behavior much close observation remains to be done.

Biological Clocks
of the Tidal Zone

19

by John D. Palmer
February 1975

*Endogenous clocks set to the rhythm of the solar day
are known throughout the biological world. Many organisms
that live along the shore also have a clock set
to the rhythm of the lunar day*

In the sands between the tide marks on the north shore of Cape Cod lives a microscopic golden brown alga, the diatom *Hantzschia virgata*. The protoplasm of this single-celled plant is encased in an elongated glassy cell wall perforated in places by pores and slits. Through some of the end pores is exuded a mucuslike substance that serves to slowly jet-propel the diatom through its subterranean habitat. During each daytime low tide the tiny motile organism glides up through the interstices between the grains of sand to the surface. There it remains throughout the ebb tide, its photosynthetic machinery bathed in sunlight. In midsummer the diatoms are so abundant that in spite of their microscopic size they form a prominent golden brown carpet over the beach. Moments before they are inundated by the returning tide they move down into the comparative safety of the sand.

A fascinating aspect of this vertical-migration behavior becomes more apparent when sand bearing the diatoms is transferred from the north shore of Cape Cod to the Marine Biological Laboratory at Woods Hole on the south shore. The samples are placed in an incubator where the temperature is held constant and the light is left on continuously. In this new environment, which lacks days, nights and tidal changes, the diatoms continue their periodic excursions up to the surface of the sand in virtual synchrony with the diatoms 27 miles away. Their movements in the laboratory are sufficiently punctual so that when we plan a collecting trip to Cape Cod Bay, we sometimes observe the diatoms in the incubator instead of consulting tide tables. Since the rhythm of the diatoms persists in the absence of the environmental periodicities that would be ex-

pected to govern such behavior, it seems that within the plants there is a biological clock that directs the temporal aspects of their lives.

This account is not just another amusing anecdote about a rare occurrence in nature. Clock-controlled rhythms are displayed by most inhabitants of the tidal zone. The rhythms are characterized by the repetition of some behavioral or physiological event, such as a flurry of activity, synchronized with a particular phase of the tide. Since there are two tides each lunar day (a lunar day is 24.8 hours in length, the interval between successive moonrises), the rhythms are called bimodal lunar-day rhythms, in contrast to the unimodal solar-day rhythms of organisms geared to the 24-hour solar day. The biological clocks related to both the lunar-day and the solar-day rhythms are apparently important as an aid to survival in that they give advance warning of the regular changes in certain periodic aspects of the environment, such as nightfall or the return of the flood tide. Under unchanging conditions in the laboratory the clocks continue to function, and thus biological rhythms persist for a considerable length of time.

The fiddler crab, a common denizen of mud flats and sand flats on North American coastlines, emerges from its burrow at low tide. It scurries sideways around the flat eating detritus. The males feign battles with one another and try to entice females into their bachelor burrow with awkward beckoning movements of their enormous fiddle claw. With each flood tide all the crabs retreat back into their burrow, where they sit out the deluge.

In the laboratory quantifying the locomotor behavior of fiddler crabs is quite

simple. Single crabs are placed in plastic boxes (the kind in which fishing lures are bought, and therefore a common commodity in Woods Hole). The boxes are balanced on a knife-edge fulcrum, and as the incarcerated crab moves between ends of this improvised actograph the box teeters, closing a microswitch that causes a deflection of a pen on a chart recorder. The actograph is placed in the unchanging environment of an incubator, and the crab is allowed to perform spontaneously for days. In these monotonous surroundings the crab's clock continues to operate and dictates almost, but not entirely, the same ambulatory pattern found in nature. The difference is slight but significant: in the laboratory the period of the bimodal lunar-day rhythm is slightly longer or slightly shorter than the period displayed in nature. This change in periodicity when an organism is placed in constant conditions is a property of almost all clock-controlled biological rhythms. Since tidal rhythms follow the lunar day, they are called circalunadian (about a lunar day).

The rhythms in some species of fiddler crab will persist for as long as five weeks in the laboratory, but more often they are damped out rather quickly. The crabs must occasionally be exposed to periodic immersion in seawater if their tidal rhythm is to be maintained. Even in nature whenever small populations of fiddler crabs become established along the margins of pools not subject to tides, they lose their tidal rhythm and display only a solar-day rhythm. When the crabs are returned to a tidal flat, they quickly reestablish a lunar-day rhythm that will then persist for some time even after the animals are removed from the tidal location.

Living side by side on the same flats

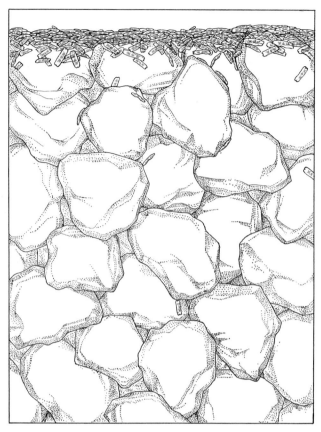

MIGRATORY BEHAVIOR of *Hantzschia virgata* is depicted in vertical section. Diatoms of this species normally reside about a millimeter below the surface of the sand (*left*). Each diatom has two X-shaped chloroplasts, which are the site of photosynthesis.

During daytime low tides the organisms are propelled upward to the surface by mucus that is forced through pores at the end of their elongated glassy cell wall (*right*). The diatoms remain in the sunlight until moments before sand is inundated by returning tide.

with the fiddler crab are the green crab and the penultimate-hour crab. Both of these crustaceans display tidal rhythms, but they differ from the fiddler crab in that their activity is synchronized with the times of high water. The rhythms of the two species will persist in constant laboratory conditions for about a week before being damped out. In the case of the green crab it has been found that

animals that have lost their rhythm in the laboratory need not be subjected to the tides to reestablish the rhythm. Instead it can be reinstated by cooling the crabs to a temperature of four degrees Celsius (39 degrees Fahrenheit) for six hours.

This technique has also been used to demonstrate that tidal rhythms are not learned or otherwise impressed on crabs

by the tides themselves. Barbara Williams, working with Ernst Naylor at University College of Swansea, had the perseverance and the rare skill necessary to raise green crabs in the laboratory from eggs through several larval stages to adults. During the entire maturation process the crabs were exposed only to the alternating day-night changes in the laboratory. When the crabs were large

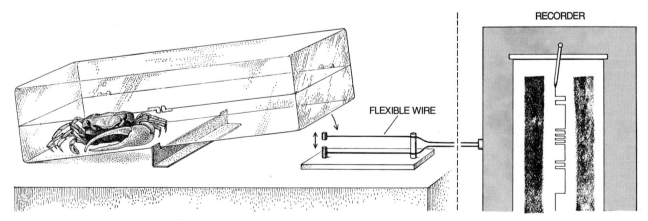

DAILY ACTIVITY of a fiddler crab is recorded by an actograph that consists of a plastic box balanced on a knife-edge fulcrum. When the crab moves to the near end of the box, the box tilts and

closes a switch that causes an excursion of a pen on the recorder. The number of daily back-and-forth movements of the crab is determined by counting the number of excursions made by the pen.

enough for their activity to be studied, it was found that their locomotor activity was limited to the daylight hours. Williams then gave the crabs one 15-hour cold treatment and recorded their subsequent locomotor behavior. A distinct tidal component appeared in their activity. Since a single 15-hour cold spell could not have provided the crabs with any information about the 12.4-hour cycle of tides, it is reasonable to conclude that the clock that measures the tidal frequency is innate, and that it merely needs to be activated by some environmental stimulus for its first expression.

Under natural conditions both the penultimate-hour crab and the green crab display in their locomotor activity a clear-cut solar-day rhythm as well as a lunar-day one. In the penultimate-hour crab the solar rhythm appears as a broad peak of activity spanning the hours of darkness. In the green crab the solar rhythm is represented not as an individual peak but as a decrease in the amount of activity at the crest of the daytime tide. The combination of solar-day and lunar-day rhythms is rather common in intertidal organisms, and it raises the question of whether such organisms have a solar-day clock for one rhythmic component and a separate lunar-day clock for the other, or whether a single horologe drives both rhythms. A single-clock mechanism might be regarded as analogous to the kind of wristwatch worn by surf fishermen, in which a single movement is transmitted to present on the dial both the time of day and the time of the tide.

Processes other than locomotor activity are also controlled by the crab's biological clock. Color-change rhythms have been investigated in the fiddler crab, the green crab and the penultimate-hour crab by Frank A. Brown, Jr., Marguerite Webb and Milton Fingerman at the Marine Biological Laboratory and by B. L. Powell of Trinity College in Dublin. Within the hypodermis of these crabs are star-shaped chromatophores that contain granules of dark pigment. When the pigment granules are tightly aggregated in the center of these cells, the coloration of the crabs is light. When the granules are evenly dispersed throughout the extensions of the cells, the coloration is dark. All three species of crab blanch during the night and darken during the daylight hours, even when they are placed in constant conditions in a laboratory. The color-change pattern of the fiddler crab has

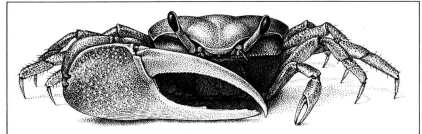

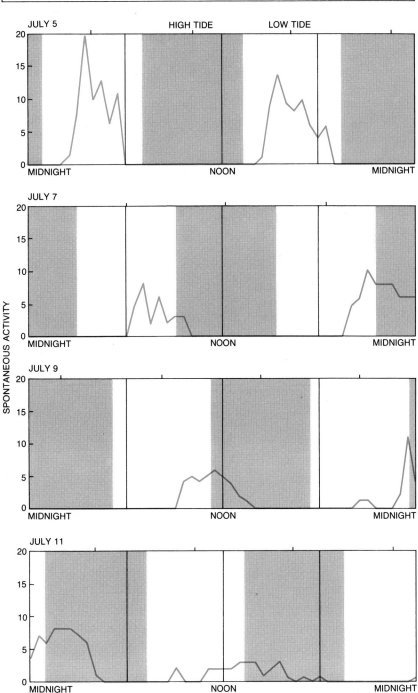

FIDDLER CRAB (*top*) is an inhabitant of tidal mud flats and sand flats. At high tide it remains quiescent in its burrow; at low tide it emerges to look for food. When a fiddler crab is taken from its natural habitat and put in an incubator where the light and temperature are constant, its periods of peak activity initially correspond to the times of low tide at its home location (*top curve*). The period of the crab's rhythm then begins to lengthen, and by the end of a week it is no longer synchronous with the times of the tide (*bottom curve*).

also been found to have a tidal component that gives rise to additional darkening at times of low tide.

The eyes of crabs are mounted on movable stalks. The stalks also house a neuroendocrine unit called the X-organ sinus-gland complex, which secretes a hormone that causes the pigments to disperse within the chromatophores. Powell found that the removal of the eye stalks from the green crab (and thus the X-organ sinus-gland complex as well) destroyed the color-change rhythm of the crab. Furthermore, Powell showed

that the rhythm can be restored in a stalkless crab by implanting in it the stalk glands from another crab. These findings strongly suggest that in the green crab the eye stalks are the site of the clock that controls the color-change rhythm. In the penultimate-hour crab and the fiddler crab, however, removal of the eye stalks only reduces the amplitude of the color-change rhythms.

The neuroendocrine system of the eye stalks also exerts some control over the locomotor-activity rhythm of the green crab. When the eye stalks of green crabs

were removed in experiments conducted by Naylor and Williams, all locomotor activity ceased, and it returned only gradually over the next six days. Since in green crabs the tidal rhythm normally vanishes after about a week in constant laboratory conditions, it was not surprising to find a lack of rhythm in the stalkless crabs when they resumed their activity. Attempts to reinstate the rhythm, however, by immersing the crabs in cold water were unsuccessful. On the other hand, crabs from which only the retinas were removed, not the stalks, returned to

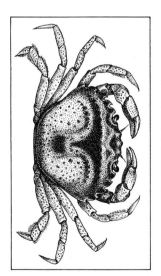
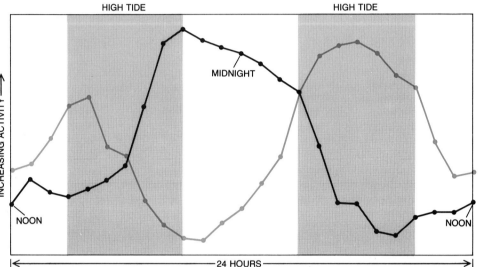

PENULTIMATE-HOUR CRAB lives in close proximity to the fiddler crab on tidal flats. (The name of the crab is derived from the fact that the activity of a newly caught animal peaks one hour before midnight.) The black curve shows the mean solar-day activ-

ity of 30 penultimate-hour crabs over a period of a month, which is expressed as a broad peak of activity during the hours of darkness. The penultimate-hour crab also has a mean lunar-day activity rhythm (colored curve) that corresponds to the times of high tide.

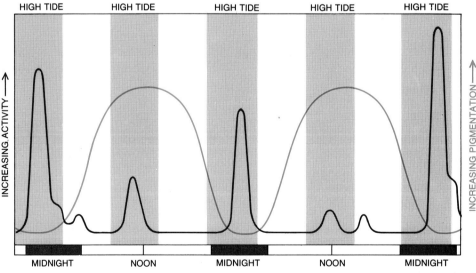

GREEN CRAB, which also lives on tidal flats, has a basic pattern of locomotor activity that corresponds to the times of high tide. Its activity greatly decreases when the high tide comes during the hours of daylight. This pattern of activity continues when the crab is placed under constant conditions in the laboratory (black curve).

In addition the green crab displays a solar-day rhythm in its body color, which blanches at night and darkens during the daylight hours. The color-change rhythm (colored curve), which persists when the green crab is kept under constant conditions, is thought to be controlled by a neuroendocrine system in crab's eye stalks.

their normal rhythm when they were immersed in cold water.

In further experiments Naylor and Williams found that subjecting the entire crab to a cold dip was not needed. If an arrhythmic crab is placed in water with a temperature of 15 degrees C. (59 degrees F.) but tethered so that its eye stalks protrude above the water, the tidal rhythm of the crab can be reinstated simply by dripping iced seawater onto its eye stalks for a brief period. Finally, Naylor and Williams made extracts from the eye stalks of green crabs in the quiescent phase of their locomotor-activity rhythm and injected the extracts into active stalkless crabs. The injection caused a significant reduction in the level of activity of the stalkless crabs, showing that there is an inhibitor substance that is periodically liberated from the stalk glands.

I have conducted similar experiments with the penultimate-hour crab. When the eye stalks were removed from these crabs, all locomotor activity stopped. I made eye-stalk extracts from rhythmic crabs during either the active phase or the quiescent phase of their locomotor activity and injected the extracts in various concentrations into crabs that had become arrhythmic because of long-term storage in constant conditions. No consistent alterations in the activity levels of the recipients were observed.

Since the neuroendocrine glands in the eye stalks of the penultimate-hour crab did not appear to be involved in the control of the crab's locomotor-activity rhythm, I carried out an experiment to determine if there was a chemical messenger coming from somewhere else in the crab's body. I joined two crabs, one strongly rhythmic and the other arrhythmic, by cutting small openings in their dorsal exoskeleton and cementing the openings together with sealing wax. In crabs most of the blood is not confined to vessels but flows freely through the spaces between organs; therefore when two crabs are joined, the blood of one mixes freely with that of the other.

I also capitalized on an anatomical peculiarity of crabs, the process called autotomy. When a crab is attacked, the attacker usually grabs one of the animal's 10 legs. The crab's defense is to cast off the leg and dash away before the predator can grab another. The leg separates from the body at a predetermined breaking point. Excessive loss of blood from the open stump is prevented by a self-sealing mechanism, and the sacrificed leg is regenerated during successive molts.

Taking advantage of this self-amputa-

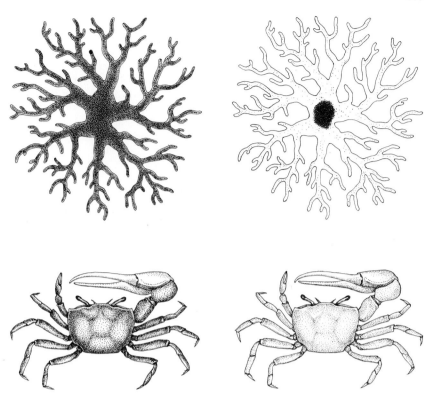

CHANGE IN COLOR of the fiddler crab is the result of aggregation or dispersal of pigment granules in cells in the crab's hypodermis that are called chromatophores (*shown greatly enlarged at top*). In some chromatophores there are only dark pigment granules; in others there are white or orange granules. The fiddler crab blanches at night and darkens in daylight. Tidal component in the rhythm produces additional darkening at times of low tide.

tion mechanism, just before I joined two crabs I made the rhythmic crab cast off all its legs. The joined crabs therefore consisted of an ambulatory arrhythmic crab on the bottom and a rhythmic amputee upside down on the top [*see illustration on next page*]. Any rhythmic locomotor activity recorded thereafter would have been the activity of the legged member, signifying that some substance in the blood of the legless crab had induced the rhythm. In 47 fusion-pair experiments not one rhythm was found. On the other hand, a rhythm was always displayed when two rhythmic crabs were joined in control experiments, indicating that the fusion procedure was not responsible for the lack of expressed rhythmicity.

It is clear, then, that whereas the endocrine system is involved in rhythms of color change and locomotion in the green crab, it is not necessarily involved in such rhythms in other crabs. Nor is it mandatory that an endocrine or a neural mechanism form the basis of any physiological rhythm, since a single-cell level of organization such as that found in the diatom *Hantzschia* is sufficient for the expression of all the known properties of clock-controlled rhythms.

The capacity for rhythmicity is not learned or impressed on organisms by the environment; it is the expression of a genetic potential. Heredity also determines whether the crab will be active at high tide or at low tide. This is not to say, however, that the environment does not play a significant role in the overt manifestation of a rhythm. It is the schedule of the tides on a particular stretch of coastline that determines the hour-to-hour settings of the rhythm. (The relation between a biological clock and the environment is similar to that between a pendulum clock and its owner. The rate at which the pendulum clock runs is determined by the escapement mechanism and the pendulum, but the owner can set the time to any hour by moving the hands on the face of the clock.) Thus a green crab will soon be active at high tide and a fiddler crab at low tide even when they are transported to an unfamiliar beach on a different ocean.

In the sand high on the beaches of southern California lives the sand hopper *Excirolana*. At the peak of each high tide, when the waters flood the habitat of this tiny isopod, it emerges from the sand to swim and feed in the breaking waves. Two or three hours later, when the tide turns, it burrows into the sand

and awaits the return of the next flood tide. James T. Enright of the Scripps Institution of Oceanography found that when he kept sand hoppers in a jar of seawater in constant conditions, they swam actively during the times corresponding to peak tide and remained in repose at the bottom of the jar at other times.

In southern California the tidal pattern changes greatly with the phases of the moon. Over a single month the tides change from one crest per lunar day to two per lunar day. Furthermore, during the transitions from one tide per day to two tides and back to one tide the height of consecutive tidal peaks also changes. L. A. Klapow, who was then working at the University of California at San Diego, showed that the pattern of the tides at the time sand hoppers were collected was reflected in the form of the activity rhythm the animals displayed in the laboratory [*see illustration on opposite page*]. In separate experiments Klapow and Enright also demonstrated that it is the pounding waves and the swirling waters that determine the pattern of the sand hopper's rhythm.

It therefore seems that inhabitants of beaches exposed to the open sea have their activity patterns shaped by the action of the surf. Intertidal organisms that live in protected bays are not normally exposed to a pounding surf and so we must look elsewhere for the elements that help to set their rhythms. The possibilities are numerous, including periodic inundation and periodic changes in temperature, hydrostatic pressure, the

chemical composition of the water or the availability of oxygen. Of all these possibilities only two have been shown to play an important role. As a clear-cut example I shall cite another study of the green crab by the prolific team of Naylor and Williams.

One of their most surprising findings was that the principal feature of the tide, the periodic inundation of the shoreline, was not itself an important agent in synchronizing the locomotor-activity rhythm to the tides. This fact was demonstrated by bringing crabs into the laboratory and subjecting them for five days to 6.2 hours of immersion in seawater followed by 6.2 hours of exposure to air. The immersion in seawater was timed to correspond to low tide at the crabs' home beach, in effect reversing the animals' tidal schedule. The temperature of both the water and the air was held constant at 19 degrees C. After this treatment the crabs were placed in actographs, and their locomotor-activity patterns were measured for the next three days at the same constant temperature. The treatment did not rephase the crabs' rhythm. The procedure was repeated, but this time the air temperature was maintained at a level 11 degrees higher than the water temperature. Five days of this treatment did rephase the crabs' rhythm, and the change persisted in constant conditions.

The final version of the same experiment omitted the seawater-immersion portions of the cycle. Crabs were exposed to air at 13 degrees C. for 6.2 hours and then to air at 24 degrees for

the same length of time. Complete and persistent synchronization resulted. Recently I have educed the same behavioral changes in fiddler crabs and penultimate-hour crabs. It is therefore the drop in temperature brought by the flood tides that plays an important role in setting the phase of the crabs' rhythm.

Hydrostatic pressure is the other environmental force that is known to synchronize organismic rhythms to local tides. In one experiment arrhythmic crabs were exposed for five days to a cycle of high pressure (1.6 atmospheres) for 6.2 hours followed by 6.2 hours at normal sea-level pressure. The crabs responded with an increase in activity during the high-pressure periods, and this periodicity persisted when the crabs were kept in constant conditions.

So far we have no firsthand knowledge of how the living horologe actually works. In the search for the elusive timing mechanism, however, several of its properties have been elucidated.

When rhythms in biological processes such as oxidative metabolism, photosynthesis and the like were first discovered, and it was found that these rhythms would persist without external stimuli, the controlling clock was thought to be simply some oscillatory step in the chain of chemical reactions underlying the process. As a result early attempts at locating the clock consisted in dissecting the chain of relevant reactions in the hope that the oscillatory segment could be identified. The rhythmic component was not found, and subsequent experiments showed that it probably does not exist. In fact, the clock is now known to be quite distinct from the process it makes rhythmic.

One of the many observations leading to this conclusion was conducted with the green crab. When the body temperature of the crab was lowered to 10 degrees C., all locomotor activity stopped for the duration of the chilling. When the body temperature was allowed to return to a normal level, activity resumed, and the locomotor rhythm was in exact phase with that of control crabs that had not been chilled. Clearly the crab's clock had continued to run accurately even when no rhythm was being expressed. This finding shows that the clock and the processes it causes to be rhythmic are separate and must be joined to each other in such a way that they can be uncoupled from each other and recoupled.

The disengagement of the coupling between the clock and the driven process may also be responsible for the even-

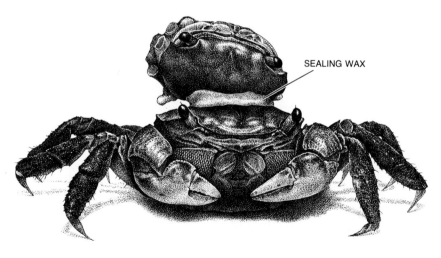

PENULTIMATE-HOUR CRABS ARE JOINED parabiotically so that their blood can continuously mix. Small openings are cut in the dorsal exoskeleton of each crab, and the openings are cemented with sealing wax. The top crab has cast off all its legs through autotomy, the process by which crabs shed a leg when it is seized by a predator. Before the union the locomotor activity of the top crab was synchronous with the tides. The activity of the bottom crab was arrhythmic. In experiments with fused crabs no rhythmic locomotor behavior was found, indicating blood does not contain a chemical messenger that induces such behavior.

SEALING WAX

tual loss of overt rhythmicity in animals that are maintained in constant conditions in the laboratory. Speaking somewhat teleologically, when intertidal animals are taken away from their tidal environment, there is no longer any pressure for them to maintain a tidal rhythm. Their life processes are emancipated from the clock and become arrhythmic.

The validity of the notion of a coupling between a clock and vital processes is enhanced by the fact that rhythms once lost by crabs, either in the laboratory or in nontidal natural habitats, can be reinstated by a single short-duration stimulus such as being chilled. Since the treatment provides no information about tidal intervals, the simplest interpretation is that the stimulus recouples the clock, which had continued to run, to the processes governing locomotor activity, causing such activity to become rhythmic again.

As we have seen, the vertical migration rhythm of the diatom *Hantzschia* demonstrates that a biological clock needs only the level of organization characteristic of a single cell to express itself. Two other unicellular organisms provide even better examples. The marine dinoflagellate *Gonyaulax* is known to simultaneously display different rhythms in four processes: photosynthesis, luminescence (it glows at night), irritability and cell division. Five different rhythms have been detected in the single-celled green alga *Acetabularia,* and all the rhythms persist even when the nucleus of the cell has been removed by microsurgery. There is evidence that in multicellular plants and animals the clock is also to be found in single cells. When organisms are subdivided and the parts are kept alive in tissue culture, the cells continue their original rhythm. Indeed, the plausible place to look for the living clock is within the single cell, where one would expect to find it in the form of some physiochemical entity. In spite of intensive investigation, however, neither the clock nor any of its components have been located. The search has nonetheless revealed two unusual aspects of the horologe: the rate at which it runs is almost completely insensitive to temperature, and the rate also is not affected by a wide variety of potentially disruptive chemical agents.

In general increasing the temperature increases the rate at which chemical reactions proceed. One would expect that the living horologe, with its chemical clockwork, would be accelerated in a similar way. To test this assumption we subjected groups of crabs to increasingly higher constant temperatures in the lab-

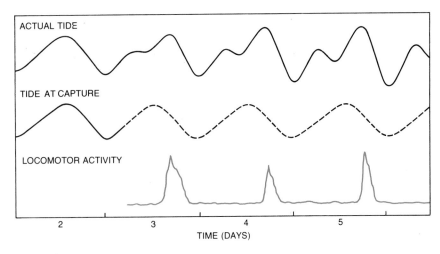
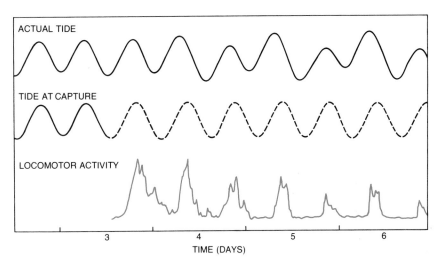
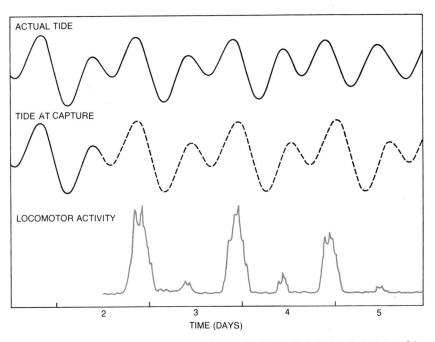

PATTERN OF ACTIVITY OF THE SAND HOPPER (also called the beach flea) found in the beach sands of southern California is adapted to the peculiarities of the tides of the region. The tides alternate every month between one peak per lunar day and two peaks per lunar day. When sand hoppers are kept in constant laboratory conditions, their pattern of activity (*colored curves*) tends to mimic the form of last tidal pattern to which they were exposed (*broken curves*) rather than the form of the actual tidal pattern (*black curves*).

oratory and observed their rhythms over periods of several days. Since the period of an expressed biological rhythm is believed to closely mimic the driving frequency of the clock, a temperature-induced change in the rhythm is assumed to indicate a change in the frequency of the clock. The usual result obtained in such experiments is that there is no change in period at all. If any change is recorded, it is only a fraction of what one would expect from a chemical system.

Attempts to disrupt the rhythms of organisms with chemical substances such as inhibitors of protein synthesis, stimulants, metabolic inhibitors and narcotizing agents have proved to be almost equally futile. Out of hundreds of substances tested only four—deuterium oxide, ethyl alcohol, valinomycin and lithium ions—have been found to alter the period of a rhythm. In view of the variety and number of substances screened, it appears that biological clocks, unlike most other pacemaker systems in organisms, are virtually immune to chemical manipulation.

From a pragmatic point of view these insensitivities might have been predicted. Certainly one of the most important

attributes of any clock—living or man-made—is accuracy, and a clock whose rate of running is altered by changes in the temperature or chemistry of its environment would not meet this requirement. In fact, if the clock responded to every change in ambient temperature, it would not be a clock at all but rather a thermometer that signaled ambient temperatures by the rate at which it ran.

The accuracy of biological clocks is even more amazing when one takes into account the fact that precision must be maintained during cell division, when presumably not only the cell but also the clock is replicated. The ease with which this replication is accomplished has been demonstrated in a study of the single-celled protozoan *Paramecium* by Audrey Barnett of the University of Maryland. In the strain of paramecium she worked with the sex of each animal changes from one mating type to another and back again each day. Barnett placed in constant darkness eight paramecia whose clocks had been set to regular day-night cycles. The cells in the population divided 2.2 times per day, and at the end of six days they had given rise to slightly more than 121,000 cells. On the seventh day the sex-reversal behavior of the en-

tire population was examined and was found to be rhythmic. The phase of the rhythm was close to that of control cells that had remained under the regular day-night cycle. Since only the original eight cells had been subjected to a day-night cycle, it appears that each of the original cell clocks had been replicated time and time again with very little loss in accuracy. An alternative interpretation is that each cell contains many clocks, some of which are replicating themselves while others are still coupled to cell processes, causing them to be rhythmic.

We still know very little about the mechanism of living horologes in the tidal zone, and the properties of clocks that have been elucidated in some ways compound the problem. The continued search for such mechanisms is nonetheless a worthwhile endeavor because clock-controlled rhythms are found not only in intertidal organisms but also throughout the kingdom of life. One may hope that the continued effort will eventually lead to discoveries that will enable us to perceive the fundamental principle that underlies the operation of all biological clocks.

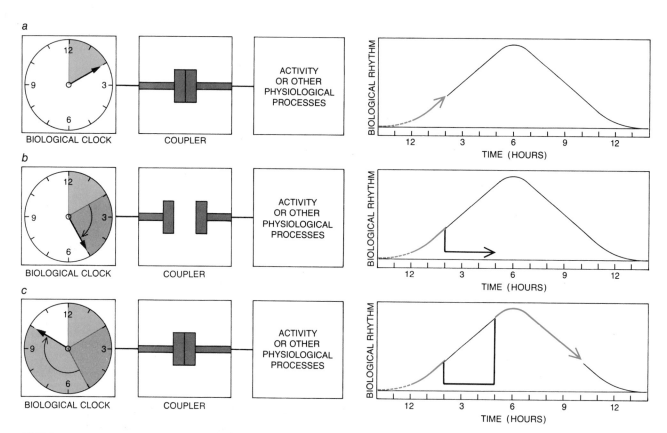

COUPLER that has not yet been identified is believed to join biological clocks and biological processes. When the coupler is engaged (*a*), the biological rhythm is expressed. The disengagement of the coupler (*b*) may be responsible for the loss of rhythmic behavior of organisms that are kept under constant laboratory conditions. The clock continues to function, however, and when recoupling occurs (*c*), the rhythm takes up not at the point where it left off but at point that corresponds to "time" determined by the clock.

Dolphins

by Bernd Würsig
March 1979

*These descendants of land mammals that took to sea
have a large brain, learn quickly and exhibit a rich
vocal repertory. Yet lack of evidence leaves open the
question of how intelligent they are.*

Because dolphins have big brains, are quick to learn tricks devised by human trainers and exhibit a rich repertory of vocal signals, they are widely reputed to have a level of intelligence unmatched by any other animal and perhaps even the equal of man's. On the basis of observations of dolphins in several of their natural habitats I believe the effort to position them firmly in the spectrum of animal intelligence is premature. In the present state of marine technology it is simply impossible for human observers to spend more than brief and isolated moments with an animal that lives in the ocean and moves rapidly over great distances. When a great deal more is known about the behavior of dolphins, the question of their intelligence will answer itself. At present the most one can say is that dolphins are gregarious herding animals, comparable in their individual and social behavior to more easily observed herding and flocking mammals on land.

Dolphins evolved at least 50 million years ago from land mammals that may have resembled the even-toed ungulates of today such as cattle, pigs and buffaloes. After taking to the sea dolphins became progressively better adapted to life in the water: their ancestral fur was replaced by a thick coat of blubbery fat, they became sleek and streamlined, they lost all but internal remnants of their hind limbs and grew a powerful tail, their forelimbs were modified into steering paddles and apparently as a further aid in steering and stabilization many species evolved a dorsal fin.

More than 30 species of dolphins can be identified. They belong to the suborder Odontoceti, or toothed whales, of the order Cetacea. (The quite similar porpoise, which is often confused with the dolphin, is also a member of the whale family; it is distinguished from the dolphin mainly by having a less beaklike snout and also by its laterally compressed, spadelike teeth. It should be noted, however, that many American students of marine mammals refer to all small odontocete cetaceans as porpoises, regardless of their physical charac-

teristics.) A few dolphin species live in fresh water, but most species have an ocean habitat. The freshwater species travel in small groups or are nearly solitary, whereas the ocean species (such as the Pacific spotted dolphin) may congregate in aggregations of several thousand. Such numbers are reminiscent of buffalo herds in North America and grazing animals on the Serengeti plains of Africa; one wonders about behavioral and ecological similarities between the dolphins and their distant terrestrial relatives.

In general highly social mammals have complex social signals and a rich behavioral repertoire. Hence they can interact with other members of their group in sophisticated ways. Examples include signals of aggression that are useful in establishing and maintaining dominance hierarchies, signals for courtship, warning sounds or movements at the approach of a potential attacker and many other signals that contribute to the functioning of the group and its individual members.

The effort to accumulate data on the behavior and social systems of dolphins is made difficult by the fact that most of their communication goes on below the waves. It is extremely difficult to approach a group of dolphins in a boat and to stay with them long enough to begin to understand their social system. All one sees is a group of dorsal fins as the animals surface to breathe, and then they are lost to view as they move on underwater. Even in the rare circumstances when the water is calm and clear and the dolphins can be seen for more than a few minutes, the proximity of a boat may disturb them, so that it becomes difficult to separate what is natural in their behavior from what is unnatural, being merely a reaction to the boat.

As a result of these problems and others most of the early observations involved dolphins in captivity. Although captivity must be an awkward situation for animals that are accustomed to a life with few physical boundaries, the captive dolphins nonetheless yielded useful

data simply because observers could watch them for extended periods of time. The best-known work of this kind was done by Margaret C. Tavolga of Marineland of Florida. She observed a group of 12 bottlenose dolphins (*Tursiops truncatus*) in a large tank for a total period of about five years.

Tavolga found that the group had a definite dominance hierarchy. The one adult male, which was the largest animal in the group, was more aggressive and less fearful than any of the females, subadult males or young dolphins. In general the larger animals were dominant over the smaller ones.

Similar data were obtained by Gregory Bateson at the Oceanic Institute of Hawaii. He found a dominance hierarchy in a group consisting of two spotted dolphins (*Stenella attenuata*) and five spinner dolphins (*S. longirostris*). The largest male threatened other dolphins (by lunging at them or showing his teeth) but was never threatened himself. The second-ranking dolphin, also a male, threatened the animals below him, and so on down the line. Bateson's findings showed, as the findings of other workers have shown, that the hierarchy is not as strict as it is among some other mammals. For example, the lowest-ranking male was still able to mate with a female without being challenged by the largest male.

If the dominance hierarchy is not necessarily related to the access of the males to the females (as it is, for example, in the harem of the elephant seal), one wonders what its function is in wild dolphin populations. Kenneth S. Norris and Thomas P. Dohl of the University of California at Santa Cruz have speculated that the function may be to organize the members of the group to deal with a variety of situations. For example, threats and chases by the larger dolphins could cause the smaller females and young animals to be herded into the center of the group, where they would be better protected from such potential predators as sharks and killer whales. Norris believes he has seen such struc-

THREE STAGES OF A LEAP are demonstrated (unintentionally) in the Sea of Cortés (the Gulf of California) off the coast of Mexico by three dolphins of the species *Delphinus delphis*. The repertory of leaps, spins and somersaults executed by dolphins is richly varied.

HERD OF DOLPHINS of the species *Delphinus delphis* was photographed leaping in the Atlantic. Dolphins are social mammals that sometimes congregate in quite large herds, as they are doing here, but more often they are found in subgroups of perhaps 20. Dolphins are also air-breathing animals, so that their leaps serve in part to enable them to breathe. The animals also leap during play and hunting.

turing in spinner dolphins and spotted dolphins.

For such a system to evolve it is helpful and perhaps necessary that the animals possessing it be genetically related. W. D. Hamilton of the Imperial College of Science and Technology and other biologists have argued that closely related animals should tend to protect one another more than they protect distant relatives, since close relatives share more genes. If a mother saves her offspring by herding it inside the group while putting herself in at least some danger, her apparent altruism will be adaptive for the group because a significant proportion of her genes will be preserved.

If dolphins herd group members in this way, it is likely that at least some of the animals in a group are related. Dolphins have also been seen to help an ailing member of the group reach the surface to breathe and to protect a group member from predators or other dangers. These patterns of behavior have often been cited as evidence of human-like altruism and of great intelligence. It appears more likely that they represent an outgrowth of an evolved tendency to help related individuals.

Unfortunately the degrees of consanguinity in a wild dolphin population are not known. Only recently has any idea at all of social structure among dolphins been gained, and the knowledge is complicated by the existence of many different species and many more separate populations within species. From terrestrial mammals it has been learned that the social system represents in part an adaptation to the population's habitat. For example, wolves that live mainly on deer tend to travel in small packs, whereas wolves that hunt moose are found in larger and more highly organized societies. This difference is apparently related to the need for a coordinated effort by several animals to successfully bring down the larger prey.

A few excellent studies that make comparisons possible among dolphin species and populations have been done recently. Norris and Dohl studied Hawaiian spinner dolphins from sea cliffs and from underwater. They found that these highly social mammals traveled in schools averaging 25 members. The structure of the school varied during the day in a predictable manner. In the morning the dolphins moved slowly and in tight groups, with individuals almost touching. They appeared to be resting. Later they became increasingly active, swimming faster, with individuals leaping clear of the water in the spins, somersaults and other displays for which dolphins in oceanariums are famous.

At such times the schools became more spread out, with animals often as much as 20 meters apart. Moreover, groups tended to join, so that 50 or more members might constitute the expanded school, with all the animals moving in the same direction. As night approached, the school moved several kilometers away from the shore, entering deeper water and beginning deep dives in order to feed on fish several hundred meters below the surface. The array consisted of many widely spaced groups within an area with a diameter of several kilometers. Because Norris and Dohl were able to recognize some individuals, they found that a given small group of individuals tended to stay together but often shifted as a unit from one school to another.

Norris and Dohl suggested that dolphins may form close groups while they are resting so that they can employ the combined sensory abilities of all the individuals in the school to scan the environment and to detect potential danger. It is well known that dolphins can scan the water by echo location over much greater distances than would be possible by eyesight. (In echo location a dolphin projects high-frequency sounds in short pulses, much as bats do. The sounds bounce off objects, and the echoes give back information on the distance, size, shape and even texture of the object.) Norris and Dohl's hypothesis about combined sensory abilities, particularly during rest, relies in part on the ability of dolphins to get information from echo location. Presumably each dolphin

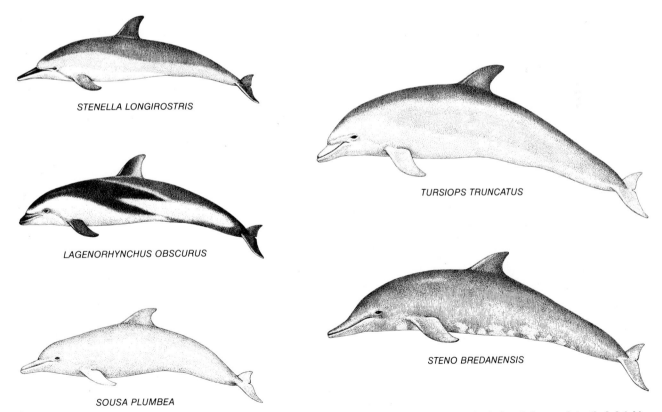

STENELLA LONGIROSTRIS

LAGENORHYNCHUS OBSCURUS

SOUSA PLUMBEA

TURSIOPS TRUNCATUS

STENO BREDANENSIS

FIVE DOLPHIN SPECIES are portrayed to indicate their differences in shape and marking. They are the bottlenose dolphin (*Tursiops truncatus*), the spinner dolphin (*Stenella longirostris*), the South Atlantic dusky dolphin (*Lagenorhynchus obscurus*), the Indo-Pacific humpback dolphin (*Sousa plumbea*) and the rough-toothed dolphin (*Steno bredanensis*). All of them are drawn to the same scale. About 25 other species of dolphins are known. Porpoises, which are quite similar, have a blunter snout and are ordinarily shorter and fatter.

190

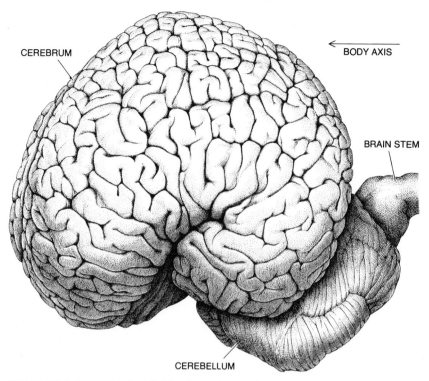

CEREBRUM

BODY AXIS

BRAIN STEM

CEREBELLUM

BRAIN OF DOLPHIN is depicted for the species *Tursiops truncatus,* **the bottlenose dolphin. A dolphin brain weighs about 3.5 pounds (1.59 kilograms); a typical human brain weighs about three pounds (1.36 kilograms) but is larger than the dolphin's in proportion to body weight.**

in a closely organized school can hear the echo-location sounds made by other members of the group. Therefore even though any given individual might not make many sounds, much information about the environment would be rapidly and efficiently disseminated to all. It is also probable that a resting group swims

close to shore in order to be in shallow water that is not frequented by large deepwater sharks.

During periods of alertness the spacing of the spinner dolphins increases and the animals do a great deal of leaping. This activity may in part be play, as many people have suggested, but it may

also represent a form of communication. A leap is usually followed by a loud slap or splash as the dolphin enters the water. Such sounds travel fairly long distances underwater and may signify the presence of the leaper to others. Indeed, groups of some dolphin species sometimes converge on an active, leaping school from a distance of several kilometers. Nighttime feeding in deep water is also attended by much leaping and loud splashing. At this time the members of a group are quite widely separated, and the assumption once again is that leaping may serve to communicate location and possibly information such as the number of other dolphins nearby and what they are doing.

Graham Saayman and C. K. Tayler of the Port Elizabeth Museum in South Africa studied Indian Ocean bottlenose dolphins (*Tursiops aduncus*), and my wife Melany Würsig and I studied South Atlantic dusky dolphins (*Lagenorhynchus obscurus*). Both species have habitats similar to the habitat of the Hawaiian spinner dolphin. The habitats are coastal-pelagic, meaning that all three populations can often be seen and studied from the shore but that they also move far from the shore, usually to feed. All three populations exhibited similar patterns of behavior and movement. It therefore seems reasonable to say that the habitat of these marine mammals is largely responsible for their way of life.

This assertion can be examined by studying dolphins in a different environment. Three such environments can be found: deep-ocean, coastal and freshwater. The most thoroughly examined populations have been coastal ones. Susan H. Shane of Texas A & M University observed Atlantic bottlenose dolphins off Texas; A. Blair Irvine, Randall S. Wells and Michael Scott of the University of Florida observed them off Florida; Melany Würsig and I observed them off Argentina, and Saayman and Tayler observed Indo-Pacific humpback dolphins (of the genus *Sousa*) off South Africa. Again major similarities were evident. They can be summarized by a description of our study of bottlenose dolphins off Argentina.

The bottlenose dolphin is a coastal species in many parts of its worldwide range. Hence it can be observed readily from the shore. For a period of 21 months my wife and I observed a school of bottlenose dolphins that passed close to the shore (always within a kilometer) in water less than 40 meters deep. We studied them by observations from coastal cliffs and from a small rubber boat, by underwater recordings of their sounds, by photographing their dorsal fins in order to recognize individuals and by tracking their movements with a surveyor's transit on the shore. (With a transit one can determine precisely the

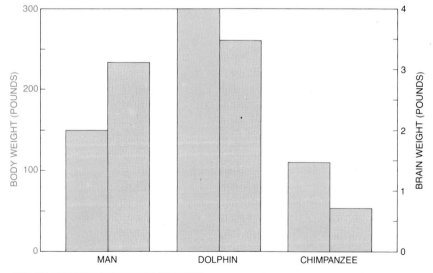

BRAIN WEIGHT AND BODY WEIGHT are compared for man, the dolphin and the chimpanzee. Man's brain weighs about 2 percent of his body weight, the chimpanzee's less than 1 percent and the dolphin's slightly more than 1 percent. Although rough ideas of comparable intelligence can often be obtained by comparing the ratio of brain weight to body weight, other factors such as the length of the body, the manner in which limbs are used and the complexity of the brain must also be considered. It is not possible to say on the basis of a simple comparison of brain weight to body weight that one kind of organism is more intelligent than another.

location, movement and speed of wild animals. The method was developed by Roger Payne of the New York Zoological Society and has proved to be useful in studies of the coastal movements of marine mammals.)

The school ranged in size from eight to 22 dolphins; the average was 15. Some individuals stayed together consistently; others were at times absent. Thus the population was made up of subunits of shifting membership.

Eventually our sightings were all of animals we had identified previously. We concluded that we knew all or nearly all 60 or so dolphins that at one time or another were part of the group. Although all the animals in the larger group interacted to some degree, we never saw the entire group together at one time. It is therefore likely that other subgroups, also composed of individuals known to us, traveled near other shores. Apparently they cover a large range; we once sighted a subgroup of known animals more than 300 kilometers south of our base. It is probable that the size of a subgroup represents an optimum balance among the number of animals needed for defense against predators, the number needed for efficient feeding and the number best suited for social interaction, reproduction and the survival of young.

In the fall, winter and spring the bottlenose dolphins seemed to feed on schools of anchovies at about midday in water from 15 to 35 meters deep. At such times they advanced as a spread-out school, each animal being separated from the next one by as much as 25 meters. After several minutes in this formation they began to dive and mill around in one area. From comparison with the behavior of dusky dolphins, which in this region feed almost exclusively on anchovies, we deduced that the bottlenose dolphins were herding schooling anchovies to the ocean surface and feeding on them there. The spread formation before feeding probably serves the purpose of acoustically scanning as large an area as possible in the search for anchovies.

This cooperation among members of a school is quite different from what the dolphins do in the summer. Anchovies are not present in the waters off Argentina during the summer, and so the dolphins feed mainly on large solitary fishes living among rocks near the shore. At such times the dolphins move in water that is from two to six meters deep, and they spread out in a line that is longer than it is wide, with every animal at essentially the same depth and as close to the shore as possible. In this formation individual dolphins nose their way among the rocks and poke into crevices as they search for their prey.

Although the average size of subgroups was 15 dolphins, the average was

lower (14) during the summer than it was during the winter (20). It can be argued that fewer dolphins are needed for successful individual feeding near the shore. Indeed, it is possible that the resource—large fishes inhabiting crevices—is limited and that a group of about 14 dolphins in one area at one time represents the upper limit of the area's carrying capacity. The limit is higher when dolphins are feeding cooperatively on schooling fish.

Other reasons for the seasonal fluctuation in the size of dolphin subgroups must also be considered. For example, one could argue that the size of subgroups increases because dolphins are more susceptible to killer-whale attacks when they feed farther from shore and that with more animals per subgroup they can protect themselves better. Perhaps the fluctuations reflect seasonal peaks in mating and calving. On the basis of the information now available, however, food seems to be an important determinant of the size (and presumably the composition) of subgroups of bottlenose dolphins.

What can be said about the internal structure of subgroups? The fact that

individual dolphins shift about from one subgroup to another suggests that the animals have what is known among mammalogists as an open society. Among terrestrial mammals the African chimpanzee (*Pan troglodytes*) exhibits a quite similar system. Certain African ungulates (hoofed mammals) also have a degree of openness, with individuals moving frequently from subgroup to subgroup within a more rigidly defined herd. Among chimpanzees the variations appear to be in response to the availability of food; the animals search for food in small units but aggregate into larger units when the food has been found. Melany Würsig and I have found a similar situation among dusky dolphins, but it is not yet clear to what extent a patchy distribution of food might govern variations in the size of schools of bottlenose dolphins.

Notwithstanding the tendency to openness, groups of five or six dolphins were almost always observed traveling together. One such group included a notably large animal and a second adult that traveled with a calf during the entire 21-month study. I therefore believe the second adult was a female. The large

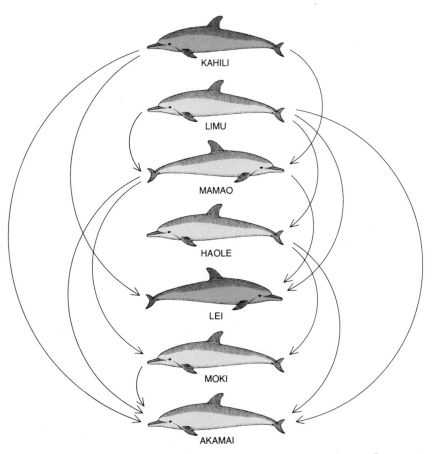

DOMINANCE HIERARCHY in a group of captive dolphins studied by Gregory Bateson at the Oceanic Institute of Hawaii is depicted. The group consisted of two spotted dolphins (*dark gray*) and five spinner dolphins (*light gray*). In this drawing males face to the left, females to the right. The arrows indicate threats (threatening entails showing teeth or lunging) made by one animal to another. The largest male, Kahili, and the male Limu were never threatened. Although Akamai, the smaller male dolphin at the bottom of the hierarchy, never threatened any other member of the group, he copulated with the female Lei on at least one occasion.

animal was the only one that consistently slapped its tail against the surface of the water when a boat approached. Experienced dolphin trainers say that tail slapping can signify anger, aggression or warning. It is tempting to say that the large dolphin was in some sense the leader of the group, but I do not have enough information about the social system and behavior of dolphins to substantiate the hypothesis.

Two other groups of five and six dolphins that we always found together consisted of adults of approximately the same size. It is possible (again I conjecture) that the groups were made up of nonbreeding members of the population, much as bachelor herds of elephants travel together.

The most detailed findings on the age and sex composition of dolphin schools have come from the work of Ir-

vine, Wells and Scott on Florida bottlenose dolphins. They captured 47 dolphins and put tags and other identifying marks on them in order to recognize them later in their natural habitat. As they tagged the animals they were able to determine the sex and size of the individuals.

Once the dolphins had been released Irvine, Wells and Scott found that the home range of the resident herd covered about 85 square kilometers. Females and calves often traveled in groups that included only a few adult males or none. Such males tended to associate more often with calfless females than with mothers and young, and they rarely associated with subadult males. The subadult males were at times found in bachelor groups far from the other dolphins. Several females were seen with their calves for as long as 15 months. Hence a

strong social bond exists between mother and calf, probably continuing long after weaning. No such long-term association between a male and a calf has been observed.

It should be emphasized that the social relations of dolphins are not clearcut or immutable. They are highly variable. Nevertheless, a few major features are apparent. The bonds between mother and calf are strong; the bonds between male and female and male and calf appear to be less so. This comparison suggests that mating is somewhat promiscuous. Subadult males may be excluded from the normal social routine but subadult females are not, which suggests that adult males may copulate more with various females than females do with different males. Such a relation is indicative of a polygynous mating system, which is also common among terrestrial mammals.

MEANS OF IDENTIFICATION of individual dolphins is provided by the pattern of nicks and scars on the trailing edge of the dorsal fin. These photographs represent a sampling of 12 bottlenose dolphins from a group of about 50. All individuals in the group could be identified.

Less can be said about the behavior of dolphins than about their social relations, even though a considerable body of accounts of behavior has been built up from observations of dolphins in nature and in captivity. I believe not enough is yet known to support any firm and broad statements. Still, a few major examples of behavior can be cited. Bottlenose dolphins (and other species) appear to engage in courtship and copulation throughout the year, as is often indicated in the wild by belly-to-belly swimming. Yet bottlenose dolphins and some other species have a definite yearly calving peak (sometimes two peaks). Among the dolphins we observed off Argentina all the calves were born in the summer. This finding indicates that a physiological change in the male or the female causes conception to occur in a limited period. Such a change has been documented in seasonal increases in the weight of the testes in the males of several dolphin species.

Yearlong mating also implies that courtship may have more than a sexual connotation. Several investigators have suggested that such interactions may also serve to define and strengthen social hierarchies and bonds. The argument is reinforced by the frequent homosexual activity seen at least among captive dolphins. Future studies may show that "homosocial" might be a better term. A carry-over of sexual signals to dominance hierarchies is seen in many other mammalian groups.

A second behavior found in almost all dolphin species is leaping. I have mentioned that it tends to occur most often when animals are widely separated and so may have a communicative function. Bottlenose dolphins off Argentina leaped far less than dusky dolphins in the same vicinity did, even when both species were hunting fish in essentially the same manner. The bottlenose dolphins, however, moved in one school

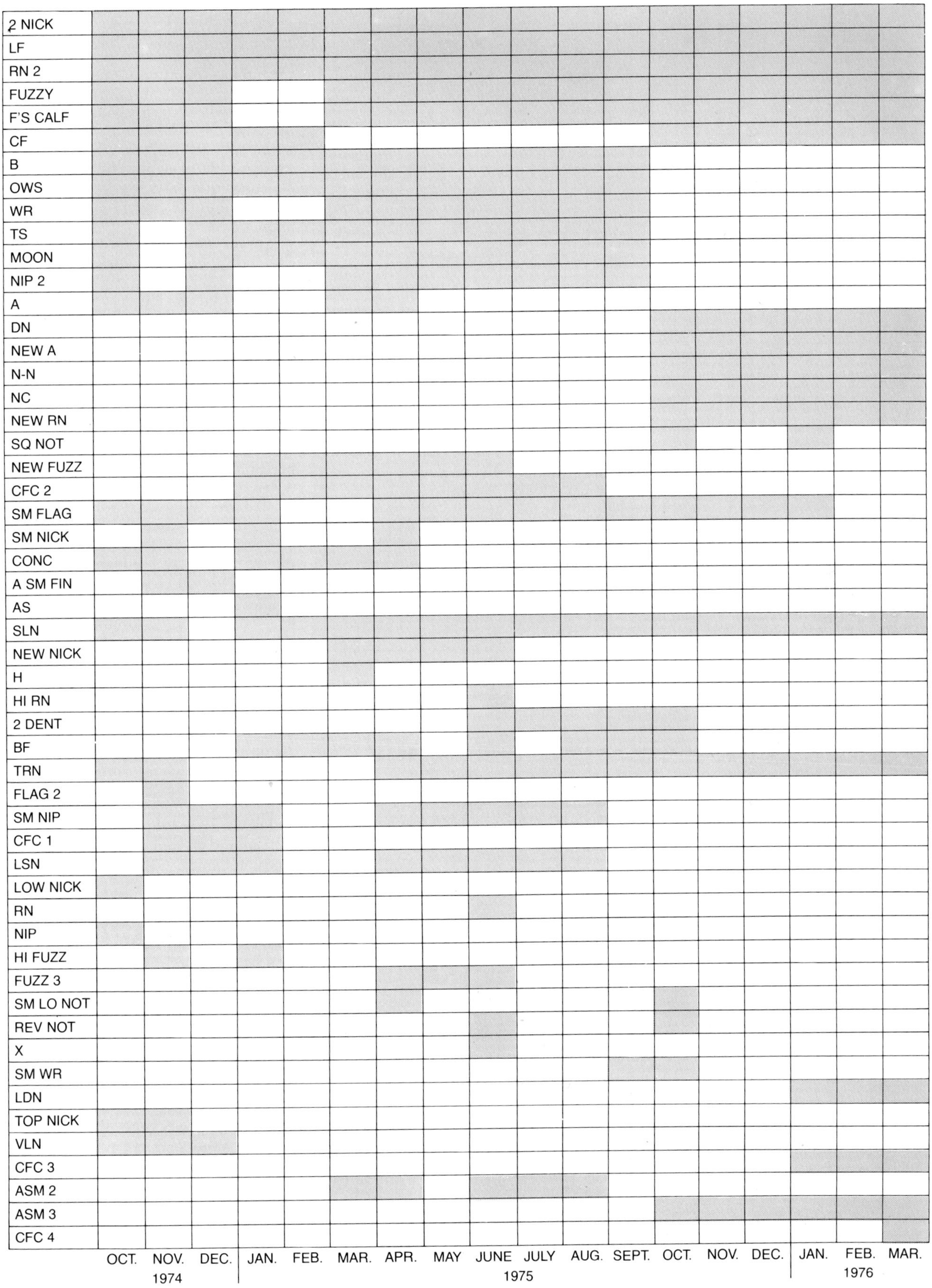

2 NICK
LF
RN 2
FUZZY
F'S CALF
CF
B
OWS
WR
TS
MOON
NIP 2
A
DN
NEW A
N-N
NC
NEW RN
SQ NOT
NEW FUZZ
CFC 2
SM FLAG
SM NICK
CONC
A SM FIN
AS
SLN
NEW NICK
H
HI RN
2 DENT
BF
TRN
FLAG 2
SM NIP
CFC 1
LSN
LOW NICK
RN
NIP
HI FUZZ
FUZZ 3
SM LO NOT
REV NOT
X
SM WR
LDN
TOP NICK
VLN
CFC 3
ASM 2
ASM 3
CFC 4

OCT.	NOV.	DEC.	JAN.	FEB.	MAR.	APR.	MAY	JUNE	JULY	AUG.	SEPT.	OCT.	NOV.	DEC.	JAN.	FEB.	MAR.
1974								1975							1976		

GROUP COHESIVENESS of dolphins is indicated by this chart of the presence and absence of 53 known bottlenose dolphins off the coast of Argentina over a period of 18 months. The designations at the left are abbreviated names the observers gave the dolphins. A bar opposite a designation indicates that the individual was seen near shore at least once during the corresponding month; a blank space means the animal was not seen during that month. A dolphin shown as having been seen may have been seen more than once in the month.

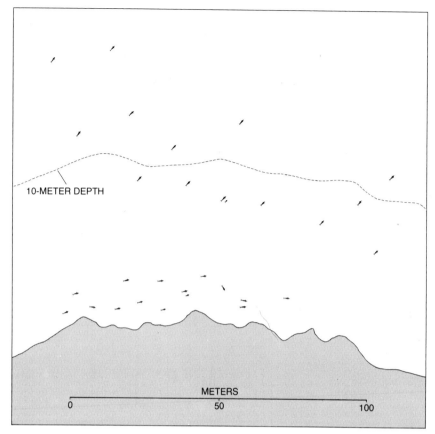

GROUP FORMATIONS of bottlenose dolphins off the coast of Argentina varied according to whether the animals were in shallow water close to shore or in deeper water farther out. In shallow water the dolphins were individually hunting rock-dwelling fishes, whereas in deep water they functioned as a group to find schools of anchovies and to herd them to the surface.

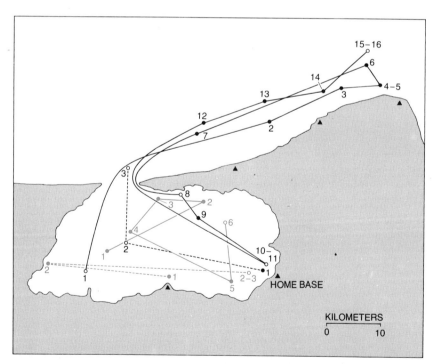

RADIO TRACKING of four dolphins off Argentina (in the Gulf of San José and the adjoining open ocean) produced these patterns of movement. Each dolphin is represented by a different track, and the number beside each circle shows the animal's position that number of days after it was fitted with a radio. Solid circles represent known positions, obtained either by triangulation from the shore points indicated by triangles or by approaching the dolphin in a boat, and open circles represent estimated positions. The animals' daily locations are given as of midday.

and so would have little cause to communicate with other members of the school by leaping. The dusky dolphins moved in as many as 30 small schools in one vicinity, and we often saw schools coalesce when leaping began. Leaps undoubtedly have other functions, such as helping to herd or catch prey, but the function of communication among members of a particular species may be important.

On the basis of the rather modest body of knowledge about dolphins, what can be said of their intelligence? Dolphins are certainly adept at learning complex tasks, as they demonstrate in their tricks in oceanariums, and they remember the tasks for years. They also have been shown to be capable of relatively abstract thinking. For example, Karen Pryor, working at the Oceanic Institute in Hawaii, trained rough-toothed dolphins (*Steno bredanensis*) to perform a new trick for a reward of fish. After several days of training they exhibited ever-different types of leaps and contortions, apparently "realizing" that the forms of behavior they had displayed previously would not be rewarded. Still, various trainers have pointed out that the same thing happens among dogs and other mammals and possibly even among pigeons, which implies that one does not have to invoke superintelligence to explain what the dolphins are doing.

Edward O. Wilson of Harvard University has suggested that the brain of the dolphin may be larger in relation to the size and weight of the body than the brains of most other mammals because of the same reputed imitative abilities that have made dolphins such a favorite with animal trainers and the public. The question then is: Why should an animal be such a superb imitator? R. J. Andrew of the University of Sussex has noted that vocal mimicry may be important to animals that often travel out of sight of one another. Individuals of a widely spaced group could then recognize other members because of an elaborate convergence of signals among the animals of one group or herd. This system, in the form of dialects, has been shown to operate among some primates and birds. It is plausible that the system evolved to an even greater degree among dolphins, which rely heavily on sound.

What about mimicry of movement? Wilson suggests that individual dolphins may imitate the members of the group that are most successful at catching fish and avoiding predation. Furthermore, it is advantageous for animals in social societies that cooperate to hunt food (as has been shown in at least some dolphin species) to know one another's movements well and for individuals to be able to take several roles in the herding of a school of fish. Wilson argues that imitation alone is enough to explain the size of the dolphin's brain and that the social signals of dolphins are probably no more sophisticated than those of most mammals and birds. In my opinion not enough is known about the social signals of dolphins to provide a basis for such a statement. Norris views the imitative powers of dolphins as not being necessarily better than those of many other mammals with brains that are smaller and physically less complex. It seems futile at present to compare the intelligence of dolphins with that of other mammals simply because of the lack of appropriate information about dolphins and the great differences between their environments and those of terrestrial mammals.

The need for better information about dolphins turns one's mind to better means of obtaining it. One possibility is to try to habituate dolphins to observers to such an extent that they will go about their daily activities as if the observers were not present. George Schaller studied mountain gorillas in this way, and Jane Goodall similarly opened a new era of work on chimpanzees. They moved with the animals and sat patiently until the animals either accepted them or simply ignored them.

How might one follow a group of dolphins in the ocean? Perhaps it is not necessary. Jody Solow of the University of California at Santa Cruz recently learned to make a sound underwater that at times called individuals of a group of nearby Hawaiian spinner dolphins to her. Her achievement opens the possibility that an investigator could eventually recognize all the members of a group, learn their social patterns and interactions and gain a better idea of their natural behavior.

VARIETY OF LEAPS performed by dolphins is suggested by these photographs of three dusky dolphins (*Lagenorhynchus obscurus*) in the open ocean. Dolphins in captivity have been observed to increase the variation in their leaps when they are receiving rewards of food.

V

THE HUMAN TOUCH

V THE HUMAN TOUCH

INTRODUCTION

Amidst uncertainties about the future, one long-standing prediction becomes steadily more sure: the sea simply is not the cornucopia that will feed a human population that has swelled beyond the nutritional capacity of the land. Where it can be farmed, as Gifford B. Pinchot describes in "Marine Farming," the sea can often be made astonishingly fertile, to the immense benefit of nearby regions and the relief of often severe malnutrition there. Some efforts in this endeavor are ages old, while others are increasingly feasible with new methods, materials, and ingenuity. But these undertakings are coastal, local ones that touch merely the rim and shallow margins of the marine realm. The great fisheries of the high seas remain the domain of hunters, however sophisticated their weapons.

Since the articles collected here were first published in *Scientific American*, several developments have occurred in world fisheries that warrant mention. Particularly, as the United Nations Food and Agriculture Organization has tabulated, the world's catch of marine fish stabilized in the 1970's at about 60 to 64 million metric tons. It is no longer rising with the tremendous momentum that characterized the two decades after World War II. Some fisheries have developed impressively—for example, jacks and mullets, some tunas, shrimp and prawns, some molluscs such as scallops, and a variety of minor fisheries that lend themselves to the initiatives in aquaculture that Pinchot describes. Increases in high-seas fisheries reflect the use of radically new ways of taking the crop (such as seining for tuna) or the use of new fishing grounds (to the far north and in the south Pacific, especially) or the turn to previously unexploited creatures (such as krill). The past decade has seen no dramatic increase of any traditionally fished stock in any traditional grounds. What increase there has been, rather, has been largely in what D. H. Cushing has called "the byways of the marine ecosystem."

This overall stabilization of the worldwide catch masks the dwindling of some famous fisheries, much as S. J. Holt anticipates in his article, "The Food Resources of the Ocean." A few of these declines have been really calamitous. The most stunning one has been the collapse of the Peruvian anchovy industry. What C. P. Idyll describes in "The Anchovy Crisis" has become chronic. Whether the grounds were simply fished beyond recovery by their intense exploitation through 1971, or whether El Niño or other environmental perturbations have scattered or destroyed the anchoveta in these waters, catches since 1971 have rarely exceeded four million tons annually. Since 1976 matters have even worsened, and each subsequent season has been abruptly closed to try to save what little appeared to be left. As this

Reader is compiled in 1980–81, the Peruvian anchovy fishery is entirely shut down, the boom utterly bust.

The North Atlantic and North Sea herring fisheries also have suffered declines that, though less awesome in their proportions than those of the Peruvian anchovy fisheries, are even more disturbing in their implications for the success of "management." In the northwest Atlantic, catches are down by half in a decade. In the North Sea and adjacent waters, region after region has failed and finally the whole fishery has collapsed. These old grounds, once one of northern Europe's treasures, are closed to the hunt for herring as this is written. The debacle in the North Sea should give us particular pause. How could such long study, such forewarning, such regulation and monitoring and attempted adjustment of pressure to circumstances—how could such careful management be to so little avail? The European herring fishery has failed disastrously before, in the Baltic in the fifteenth century after three centuries of exploitation by the Hanseatic League. Surely, we can expect that our knowledge has grown enough since then to give us a better chance to control the magnitude of such catastrophes. But events have not borne out this expectation.

With this dramatic reversal of major fisheries, there has been a bewildering increase in legal havoc. Nations have sought to protect their coastal grounds—those richest and most accessible areas of the sea—by extending maritime boundaries generally to 200 miles. These extensions have come in a disorderly way and have provoked several "fish wars" and widespread confusion and hostility. The United States and Canada have yet to work out who can take what from the Gulf of Maine. Great Britain's famous "cod war" with Iceland in 1976 has been echoed in the "tuna wars" that the United States currently fights against Canada, Ecuador, and even Fiji. Norwegian–Soviet tangles in the Barents Sea, U.S.–Mexican arguments about Gulf shrimp fisheries—the depressing list grows. And now the enormous risks of marine oil drilling and of supertankers ineluctably lead to actual spills ("dispersed," but to where?) that are often spread upon the very waters that these competing national claims have sought to protect. These risks and the unclear liabilities for damage further complicate every aspect of the matter, dimming the optimism engendered by Willard Bascom's sound assessment in "The Disposal of Waste in the Ocean." His realism aside, it *is* hard to escape the impression that there eventually will be more lawyers than fishermen at sea.

The dwindling of food species and fishing grounds despite our best efforts at control demonstrates not only that these efforts are belated but also that we apparently do not understand ecological factors that are crucial to sustaining these fisheries. For all we do know about the biology of the herring and the ecology of the North Sea, for example, we evidently do not know enough to explain beyond conjecture the bases of the fishery's collapse there. Conjecture may stir vigorous faultfinding and blame-laying, but it scarcely leads to the comprehension that might help prevent such disasters elsewhere. Beyond this ignorance lies impotence in the face of the factors that will transform marine ecosystems anyway. Thus, the Peruvian anchovy fishery surely was hit hard by man, but this destruction just as surely did not alter the course of the ocean currents there. Similarly, the disappearance of the California sardine stock may have been hastened by the efforts of Cannery Row, but we now know that sardines have thrived and disappeared in these waters repeatedly over millennia without any human touch—that we harvested them by coincidence with their abundance and were largely helpless witnesses to another such crash. Clearly, then, even though overfishing is at fault in the demise of many fisheries—as glum predictions come true for populations that are fished too hard to recover between seasons—circumstances utterly beyond our control or even our recognition also intervene naturally.

By pressing the catch toward maximum efficiency, of course, we really are permitting these unknown natural phenomena minimal leeway. Regulatory adjustments must be commensurately fine tuned. But how? Most likely, the fertile regions of the sea do suffer natural disturbances analogous, for example, to droughts on land. But the causes of these marine catastrophes are still largely mysterious to us even when, as with El Niño, we can describe, like reporters, the events themselves as they occur. In confronting this dilemma, we must accept the very youth of marine biology as a science, the patchiness of any new knowledge. Given the urgency of sustaining our fisheries, the magnitude and expanse of our perplexity are dismaying. On the other hand, one must feel a certain buoyancy about a science in which solid effort holds at least the hope of benefit for so many people. What we learn about the sea assuages more than curiosity; it can ease the awfulness of hunger, as well. This is a rare vindication of one's work.

What is not so apparent is how scientists can mesh their knowledge with the intimate understanding, the craftsmanship, and the absolute feel for their work that mark veteran fishermen. It is tempting to dictate with the heady confidence of "new knowledge." But regulations, even those set with the best of intentions, should be as cautious in their application as they are severe in the restraints they seek to impose. Arbitrarily posted by authorities, even good rules invite rebuke. Thus, *The Wall Street Journal* recently quoted a New Bedford fisherman's comment that he didn't like being told how to fish by people who "can't catch anything [but] a train or a cold." Some fisheries biologists, in turn, denounce fishermen as "the last of the buffalo-hunters." Such exchanges may be pithy, but they don't brighten the prospects of an uncertain future.

The Food Resources
of the Ocean

by S. J. Holt
September 1969

*The present harvest of the oceans is roughly 55
million tons a year half of which is consumed directly
and half converted into fish meal. A well-managed
world fishery could yield more than 200 million tons*

I suppose we shall never know what was man's first use of the ocean. It may have been as a medium of transport or as a source of food. It is certain, however, that from early times up to the present the most important human uses of the ocean have been these same two: shipping and fishing. Today, when so much is being said and written about our new interests in the ocean, it is particularly important to retain our perspective. The annual income to the world's fishermen from marine catches is now roughly $8 billion. The world ocean-freight bill is nearly twice that. In contrast, the wellhead value of oil and gas from the seabed is barely half the value of the fish catch, and all the other ocean mineral production adds little more than another $250 million.

Of course, the present pattern is likely to change, although how rapidly or dramatically we do not know. What is certain is that we shall use the ocean more intensively and in a greater variety of ways. Our greatest need is to use it wisely. This necessarily means that we use it in a regulated way, so that each ocean resource, according to its nature, is efficiently exploited but also conserved. Such regulation must be in large measure of an international kind, particularly insofar as living resources are concerned. This will be so whatever may be the eventual legal regime of the high seas and the underlying bed. The obvious fact about most of the ocean's living resources is their mobility. For the most part they are lively animals, caring nothing about the lines we draw on charts.

The general goal of ecological research, to which marine biology makes an important contribution, is to achieve an understanding of and to turn to our advantage all the biological processes that give our planet its special character. Marine biology is focused on the prob-

lems of biological production, which are closely related to the problems of production in the economic sense. Our most compelling interest is narrower. It lies in ocean life as a renewable resource: primarily of protein-rich foods and food supplements for ourselves and our domestic animals, and secondarily of materials and drugs. I hope to show how in this field science, industry and government need each other now and will do so even more in the future. First, however, let me establish some facts about present fishing industries, the state of the art governing them and the state of the relevant science.

The present ocean harvest is about 55 million metric tons per year. More than 90 percent of this harvest is finfish; the rest consists of whales, crustaceans and mollusks and some other invertebrates. Although significant catches are reported by virtually all coastal countries, three-quarters of the total harvest is taken by only 14 countries, each of which produces more than a million tons annually and some much more. In the century from 1850 to 1950 the world catch increased tenfold—an average rate of about 25 percent per decade. In the next decade it nearly doubled, and this rapid growth is continuing [*see illustration on page 206*]. It is now a commonplace that fish is one of the few major foodstuffs showing an increase in global production that continues to exceed the growth rate of the human population.

This increase has been accompanied by a changing pattern of use. Although some products of high unit value as luxury foods, such as shellfish, have maintained or even enhanced their relative economic importance, the trend has been for less of the catch to be used directly as human food and for more to be reduced to meal for animal feed. Just be-

fore World War II less than 10 percent of the world catch was turned into meal; by 1967 half of it was so used. Over the same period the proportion of the catch preserved by drying or smoking declined from 28 to 13 percent and the proportion sold fresh from 53 to 31 percent. The relative consumption of canned fish has hardly changed but that of frozen fish has grown from practically nothing to 12 percent.

While we are comparing the prewar or immediate postwar situation with the present, we might take a look at the composition of the catch by groups of species. In 1948 the clupeoid fishes (herrings, pilchards, anchovies and so on), which live mainly in the upper levels of the ocean, already dominated the scene (33 percent of the total by weight) and provided most of the material for fish meal. Today they bulk even larger (45 percent) in spite of the decline of several great stocks of them (in the North Sea and off California, for example). The next most important group, the gadoid fishes (cod, haddock, hake and so on), which live mainly on or near the bottom, comprised a quarter of the total in 1948. Although the catch of these fishes has continued to increase absolutely, the proportion is now reduced to 15 percent. The flounders and other flatfishes, the rosefish and other sea perches and the mullets and jacks have collectively stayed at about 15 percent; the tunas and mackerels, at 7 percent. Nearly a fifth of the total catch continues to be recorded in statistics as "Unsorted and other"—a vast number of species and groups, each contributing a small amount to a considerable whole.

The rise of shrimp and fish meal production together account for another major trend in the pattern of fisheries development. A fifth of the 1957 catch was sold in foreign markets; by 1967, two-fifths were entering international trade and export values totaled $2.5 billion. Furthermore, during this same period the participation of the less developed countries in the export trade grew from a sixth to well over 25 percent. Most of these shipments were destined for markets in the richer countries, particularly shrimp for North America and fish meal for North America, Europe and Japan. More recently several of the less developed countries have also become importers of fish meal, for example Mexico and Venezuela, South Korea and the Republic of China.

The U.S. catch has stayed for many years in the region of two million tons, a low figure considering the size of the country, the length of the coastline and the ready accessibility of large resources on the Atlantic, Gulf and Pacific seaboards. The high level of consumption in the U.S. (about 70 pounds per capita) has been achieved through a steady growth in imports of fish and fish meal: from 25 percent of the total in 1950 to more than 70 percent in 1967. In North America 6 percent of the world's human population uses 12 percent of the world's catch, yet fishermen other than Americans take nearly twice the amount of fish that Americans take from the waters most readily accessible to the U.S.

There has not been a marked change in the broad geography of fishing [see illustration on these two pages]. The Pacific Ocean provides the biggest share (53 percent) but the Atlantic (40 percent, to which we may add 2 percent for the Mediterranean) is yielding considerably more per unit area. The Indian Ocean is still the source of less than 5 percent of the catch, and since it is not a biologically poor ocean it is an obvious target for future development. Within the major ocean areas, however, there have been significant changes. In the Pacific particular areas such as the waters off Peru and Chile and the Gulf of Thailand have rapidly acquired importance. The central and southern parts of the Atlantic, both east and west, are of growing interest to more nations. Al-

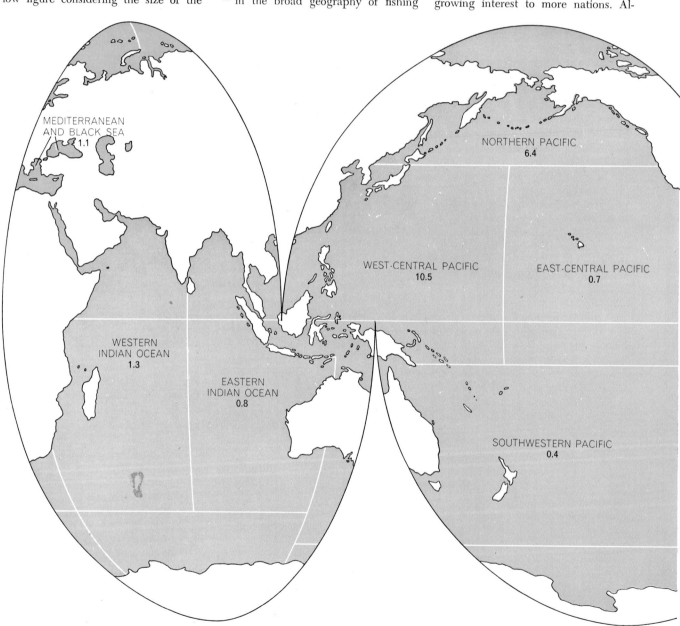

MAJOR MARINE FISHERY AREAS are 14 in number: two in the Indian Ocean (*left*), five in the Pacific Ocean (*center*) and six in the Atlantic (*right*). Due to the phenomenal expansion of the Peru fishery, the total Pacific yield is now a third larger than the Atlantic total. The bulk of Atlantic and Pacific catches, however, is still taken well north of the Equator. The Indian Ocean, with a

though, with certain exceptions, the traditional fisheries in the colder waters of the Northern Hemisphere still dominate the statistics, the emergence of some of the less developed countries as modern fishing nations and the introduction of long-range fleets mean that tropical and subtropical waters are beginning to contribute significantly to world production.

Finally, in this brief review of the trends of the past decade or so we must mention the changing importance of countries as fishing powers. Peru has become the leading country in terms of sheer magnitude of catch (although not of value or diversity) through the development of the world's greatest one-species fishery: 10 million tons of anchovies

per year, almost all of which is reduced to meal [see illustration on page 208]. The U.S.S.R. has also emerged as a fishing power of global dimension, fishing for a large variety of products throughout the oceans of the world, particularly with large factory ships and freezer-trawlers.

At this point it is time to inquire about the future expectations of the ocean as a major source of protein. In spite of the growth I have described, fisheries still contribute only a tenth of the animal protein in our diet, although this proportion varies considerably from one part of the world to another. Before such an inquiry can be pursued, however, it is necessary to say something

about the problem of overfishing.

A stock of fish is, generally speaking, at its most abundant when it is not being exploited; in that virgin state it will include a relatively high proportion of the larger and older individuals of the species. Every year a number of young recruits enter the stock, and all the fish—but particularly the younger ones—put on weight. This overall growth is balanced by the natural death of fish of all ages from disease, predation and perhaps senility. When fishing begins, the large stock yields large catches to each fishing vessel, but because the pioneering vessels are few, the total catch is small.

Increased fishing tends to reduce the level of abundance of the stock progressively. At these reduced levels the losses accountable to "natural" death will be less than the gains accountable to recruitment and individual growth. If, then, the catch is less than the difference between natural gains and losses, the stock will tend to increase again; if the catch is more, the stock will decrease. When the stock neither decreases nor increases, we achieve a sustained yield. This sustained yield is small when the stock is large and also when the stock is small; it is at its greatest when the stock is at an intermediate level—somewhere between two-thirds and one-third of the virgin abundance. In this intermediate stage the average size of the individuals will be smaller and the age will be younger than in the unfished condition, and individual growth will be highest in relation to the natural mortality.

The largest catch that on the average can be taken year after year without causing a shift in abundance, either up or down, is called the maximum sustainable yield. It can best be obtained by leaving the younger fish alone and fishing the older ones heavily, but we can also get near to it by fishing moderately, taking fish of all sizes and ages. This phenomenon—catches that first increase and then decrease as the intensity of fishing increases—does not depend on any correlation between the number of parent fish and the number of recruits they produce for the following generation. In fact, many kinds of fish lay so many eggs, and the factors governing survival of the eggs to the recruit stage are so many and so complex, that it is not easy to observe any dependence of the number of recruits on the number of their parents over a wide range of stock levels.

Only when fishing is intense, and the stock is accordingly reduced to a small

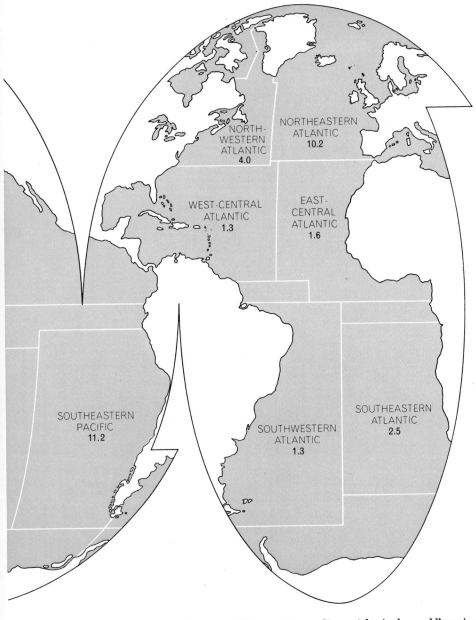

total catch of little more than two million metric tons, live weight, is the world's major underexploited region. The number below each area name shows the millions of metric tons landed during 1967, as reported by the UN Food and Agriculture Organization.

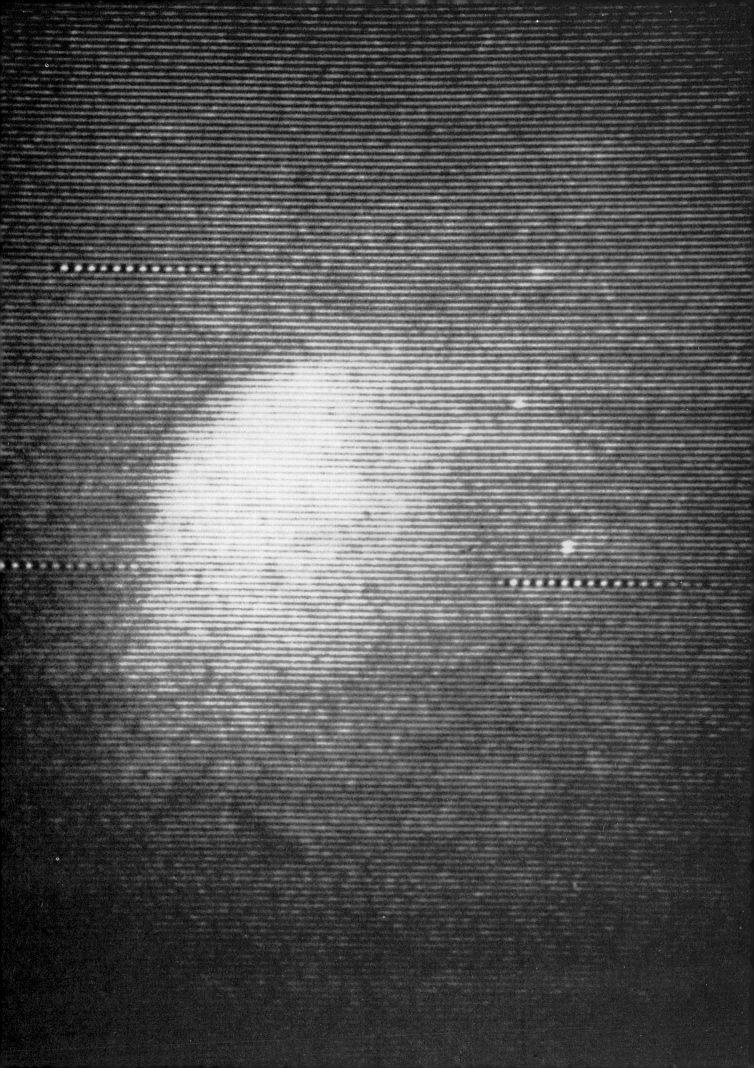

SCHOOL OF FISH is spotted from the air at night by detecting the bioluminescent glow caused by the school's movement through the water. As the survey aircraft flew over the Gulf of Mexico at an altitude of 3,500 feet, the faint illumination in the water was amplified some 55,000 times by an image intensifier before appearing on the television screen seen in the photograph on the opposite page. The fish are Atlantic thread herring. Detection of fish from the air is one of several means of increasing fishery efficiency being tested at the Pascagoula, Miss., research base of the U.S. Bureau of Commercial Fisheries.

fraction of its virgin size, do we see a decline in the number of recruits coming in each year. Even then there is often a wide annual fluctuation in this number. Indeed, such fluctuation, which causes the stock as a whole to vary greatly in abundance from year to year, is one of the most significant characteristics of living marine resources. Fluctuation in number, together with the considerable variation in "availability" (the change in the geographic location of the fish with respect to the normal fishing area), largely account for the notorious riskiness of fishing as an industry.

For some species the characteristics of growth, natural mortality and recruit-

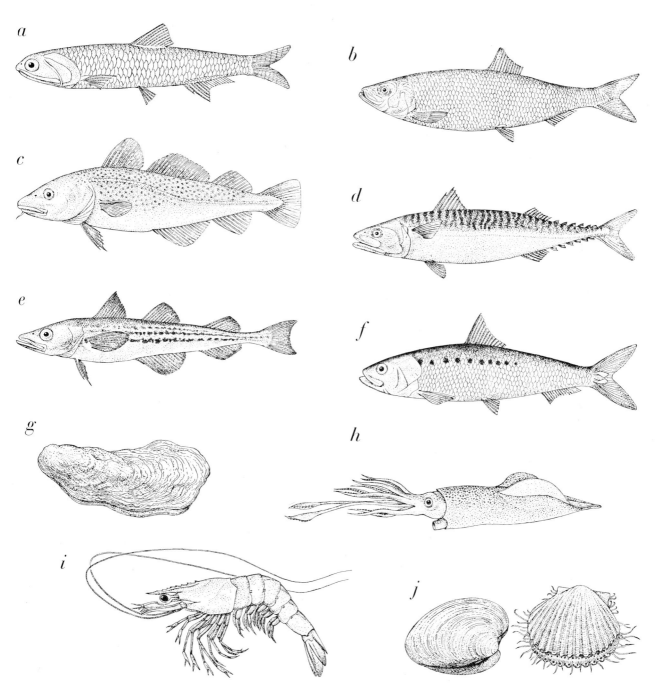

LARGEST CATCHES of individual fish species include the five fishes shown here (left). They are, according to the most recent detailed FAO fishery statistics (1967), the Peruvian anchoveta (a), with a catch of more than 10.5 million metric tons; the Atlantic herring (b), with a catch of more than 3.8 million tons; the Atlantic cod (c), with a catch of 3.1 million tons; the Alaska walleye pollack (d), with a catch of 1.7 million metric tons, and the South African pilchard (e), with a catch of 1.1 million tons. No single invertebrate species (right) is harvested in similar quantities. Taken as a group, however, various oyster species (f) totaled .83 million tons in 1967; squids (g), .75 million tons; shrimps and prawns (h), .69 million tons; clams and cockles (i), .48 million tons.

ment are such that the maximum sustainable yield is sharply defined. The catch will decline quite steeply with a change in the amount of fishing (measured in terms of the number of vessels, the tonnage of the fleet, the days spent at sea or other appropriate index) to either below or above an optimum. In other species the maximum is not so sharply defined; as fishing intensifies above an optimum level the sustained catch will not significantly decline, but it will not rise much either.

Such differences in the dynamics of different types of fish stock contribute to the differences in the historical development of various fisheries. If it is unregulated, however, each fishery tends to expand beyond its optimum point unless something such as inadequate demand hinders its expansion. The reason is painfully simple. It will usually still be profitable for an individual fisherman or ship to continue fishing after the *total* catch from the stock is no longer increasing or is declining, and even though his own rate of catch may also be declining. By the same token, it may continue to be profitable for the individual fisherman to use a small-meshed net and thereby catch young as well as older fish, but in doing so he will reduce both his own possible catch and that of others in future years. Naturally if the total catch is declining, or not increasing much, as the amount of fishing continues to increase, the net economic yield from the fishery—that is, the difference between the total costs of fishing and the value of the entire catch—will be well past its maximum. The well-known case of the decline of the Antarctic baleen whales provides a dramatic example of overfishing and, one would hope, a strong incentive for the more rational conduct of ocean fisheries in the future.

There is, then, a limit to the amount that can be taken year after year from each natural stock of fish. The extent to which we can expect to increase our fish catches in the future will depend on three considerations. First, how many as yet unfished stocks await exploitation, and how big are they in terms of potential sustainable yield? Second, how many of the stocks on which the existing fisheries are based are already reaching or have passed their limit of yield? Third, how successful will we be in managing our fisheries to ensure maximum sustainable yields from the stocks?

The first major conference to examine the state of marine fish stocks on a global basis was the United Nations Scientific Conference on the Conservation and

Utilization of Resources, held in 1949 at Lake Success, N.Y. The small group of fishery scientists gathered there concluded that the only overfished stocks at that time were those of a few high-priced species in the North Atlantic and North Pacific, particularly plaice, halibut and salmon. They produced a map showing 30 other known major stocks they believed to be underfished. The situation was reexamined in 1968. Fishing on half of those 30 stocks is now close to or beyond that required for maximum yield. The fully fished or overfished stocks include some tunas in most ocean areas, the herring, the cod and ocean perch in the North Atlantic and the anchovy

in the southeastern Pacific. The point is that the history of development of a fishery from small beginnings to the stage of full utilization or overutilization can, in the modern world, be compressed into a very few years. This happened with the anchovy off Peru, as a result of a massive local fishery growth, and it has happened to some demersal, or bottom-dwelling, fishes elsewhere through the large-scale redeployment of long-distance trawlers from one ocean area to another.

It is clear that the classical process of fleets moving from an overfished area to another area, usually more distant and less fished, cannot continue indefinitely. It is true that since the Lake Success

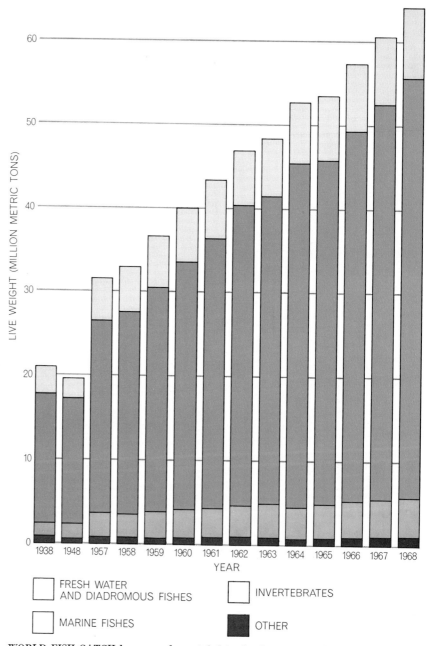

WORLD FISH CATCH has more than tripled in the three decades since 1938; the FAO estimate of the 1968 total is 64 million metric tons. The largest part consists of marine fishes. Humans directly consume only half of the catch; the rest becomes livestock feed.

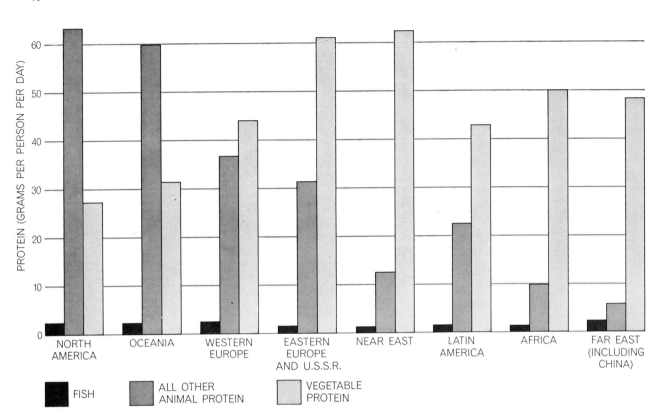

RELATIVELY MINOR ROLE played by fish in the world's total consumption of protein is apparent when the grams of fish eaten per person per day in various parts of the world (*left column in* *each group*) is compared with the consumption of other animal protein (*middle column*) and vegetable protein (*right column*). The supply is nonetheless growing more rapidly than world population.

meeting several other large resources have been discovered, mostly in the Indian Ocean and the eastern Pacific, and additional stocks have been utilized in fishing areas with a long history of intensive fishing, such as the North Sea. In another 20 years, however, very few substantial stocks of fish of the kinds and sizes of commercial interest and accessible to the fishing methods we know now will remain underexploited.

The Food and Agriculture Organization of the UN is now in the later stages of preparing what is known as its Indicative World Plan (IWP) for agricultural development. Under this plan an attempt is being made to forecast the production of foodstuffs in the years 1975 and 1985. For fisheries this involves appraising resource potential, envisioning technological changes and their consequences, and predicting demand. The latter predictions are not yet available, but the resource appraisals are well advanced. With the cooperation of a large number of scientists and organizations estimates are being prepared in great detail on an area basis. They deal with the potential of known stocks, both those fished actively at present and those

exploited little or not at all. Some of these estimates are reliable; others are naturally little more than reasonable guesses. One fact is abundantly clear: We still have very scrappy knowledge, in quantitative terms, of the living resources of the ocean. We can, however, check orders of magnitude by comparing the results of different methods of appraisal. Thus where there is good information on the growth and mortality rates of fishes and measures of their numbers in absolute terms, quite good projections can be made. Most types of fish can now in fact virtually be counted individually by the use of specially calibrated echo sounders for area surveys, although this technique is not yet widely applied. The size of fish populations can also be deduced from catch statistics, from measurements of age based on growth rings in fish scales or bands in fish ear stones, and from tagging experiments. Counts and maps of the distribution of fish eggs in the plankton can in some cases give us a fair idea of fish abundance in relative terms. We can try to predict the future catch in an area little fished at present by comparing the present catch with the catch in another area that has similar oceanographic char-

acteristics and basic biological productivity and that is already yielding near its maximum. Finally, we have estimates of the food supply available to the fish in a particular area, or of the primary production there, and from what we know about metabolic and ecological efficiency we can try to deduce fish production.

So far as the data permit these methods are being applied to major groups of fishes area by area. Although individual area and group predictions will not all be reliable, the global totals and subtotals may be. The best figure seems to be that the potential catch is about three times the present one; it might be as little as twice or as much as four times. A similar range has been given in estimates of the potential yield from waters adjacent to the U.S.: 20 million tons compared with the present catch of rather less than six million tons. This is more than enough to meet the U.S. demand, which is expected to reach 10 million tons by 1975 and 12 million by 1985.

Judging from the rate of fishery development in the recent past, it would be entirely reasonable to suppose that the maximum sustainable world catch of between 100 and 200 million tons could be reached by the second IWP target date, 1985, or at least by the end of the century. The real question is whether or not this will be economically worth the effort. Here any forecast is, in my view, on soft ground. First, to double the catch we have to more than double the amount of fishing, because the stocks decline in abundance as they are exploited. Moreover, as we approach the global maximum more of the stocks that are lightly fished at present will be brought down to intermediate levels. Second, fishing will become even more competitive and costly if the nations fail to agree, and agree soon, on regulations to cure overfishing situations. Third, it is quite uncertain what will happen in the long run to the costs of production and the price of protein of marine origin in relation to other protein sources, particularly from mineral or vegetable bases.

In putting forward these arguments I am not trying to damp enthusiasm for the sea as a major source of food for coming generations; quite the contrary. I do insist, however, that it would be dangerous for those of us who are interested in such development to assume that past growth will be maintained along familiar lines. We need to rationalize present types of fishing while preparing ourselves actively for a "great leap forward." Fishing as we now know it will need to be made even more efficient; we shall need to consider the direct use of the smaller organisms in the ocean that mostly constitute the diet of the fish we now catch; we shall need to try harder to improve on nature by breeding, rearing and husbanding useful marine animals and cultivating their pasture. To achieve this will require a much larger scale and range of scientific research, wedded to engineering progress; expansion by perhaps an order of magnitude in investment and in the employment of highly skilled labor, and a modified legal regime for the ocean and its bed not only to protect the investments but also to ensure orderly development and provide for the safety of men and their installations.

To many people the improvement of present fishing activities will mean increasing the efficiency of fishing gear and ships. There is surely much that could be done to this end. We are only just beginning to understand how trawls, traps, lines and seines really work. For example, every few years someone tries a new design or rigging for a deep-sea trawl, often based on sound engineering and hydrodynamic studies. Rarely do these "improved" rigs catch more than the old ones; sometimes they catch much

less. The error has been in thinking that the trawl is simply a bag, collecting more or less passive fish, or at least predictably active ones. This is not so at all. We really have to deal with a complex, dynamic relation between the lively animals and their environment, which includes in addition to the physical and biological environment the fishing gear itself. We can expect success in understanding and exploiting this relation now that we can telemeter the fishing gear, study its hydrodynamics at full scale as well as with models in towing tanks, monitor it (and the fish) by means of underwater television, acoustic equipment and divers, and observe and experiment with fish behavior both in the sea and in large tanks. We also probably have something to learn from studying, before they become extinct, some kinds of traditional "primitive" fishing gear still used in Asia, South America and elsewhere—mainly traps that take advantage of subtleties of fish behavior observed over many centuries.

Successful fishing depends not so much on the size of fish stocks as on their concentration in space and time. All fishermen use knowledge of such concentrations; they catch fish where they have

gathered to feed or to reproduce, or where they are on the move in streams or schools. Future fishing methods will surely involve a more active role for the fishermen in causing the fish to congregate. In many parts of the world lights or sound are already used to attract fish. We can expect more sophistication in the employment of these and other stimuli, alone and in combination.

Fishing operations as a whole also depend on locating areas of concentration and on the efficient prediction, or at least the prompt observation, of changes in these areas. The large stocks of pelagic, or open-sea, fishes are produced mainly in areas of "divergencies," where water is rising from deeper levels toward the surface and hence where surface waters are flowing outward. Many such areas are the "upwellings" off the western coasts of continental masses, for example off western and southwestern Africa, western India and western South America. Here seasonal winds, currents and continental configurations combine to cause a periodic enrichment of the surface waters.

Divergencies are also associated with

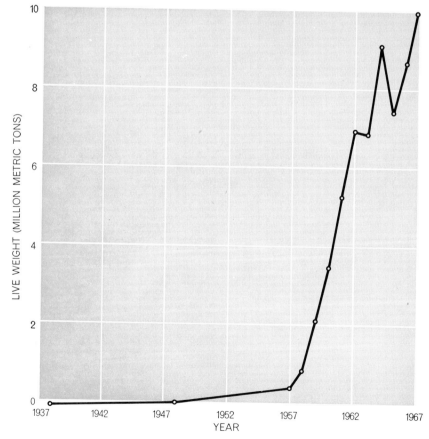

EXPLOSIVE GROWTH of the Peruvian anchoveta fishery is seen in rising number of fish taken between 1938 and 1967. Until 1958 the catch remained below half a million tons. By 1967, with more than 10.5 million tons taken, the fishery sorely needed management.

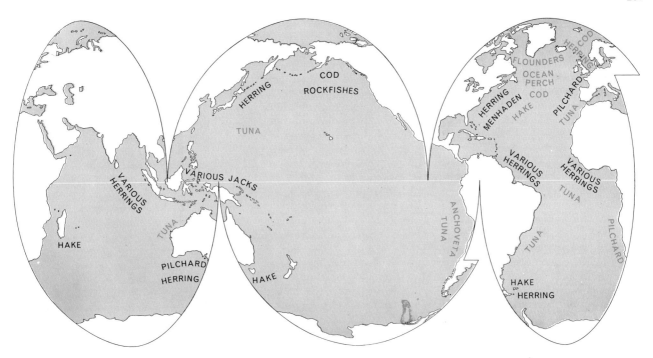

EXPLOITATION OF FISHERIES during the past 20 years is evident from this map, which locates 30 major fish stocks that were thought to be underfished in 1949. Today 14 of the stocks (*color*) are probably fully exploited or in danger of being overfished.

certain current systems in the open sea. The classical notion is that biological production is high in such areas because nutrient salts, needed for plant growth and in limited supply, are thereby renewed in the surface layers of the water. On the other hand, there is a view that the blooming of the phytoplankton is associated more with the fact that the water coming to the surface is cooler than it is associated with its richness in nutrients. A cool-water regime is characterized by seasonal peaks of primary production; the phytoplankton blooms are followed, after a time lag, by an abundance of herbivorous zooplankton that provides concentrations of food for large schools of fish. Fish, like fishermen, thrive best not so much where their prey are abundant as where they are most aggregated. In any event, the times and places of aggregation vary from year to year. The size of the herbivore crop also varies according to the success of synchronization with the primary production cycle.

There would be great practical advantage to our being able to predict these variations. Since the weather regime plays such a large part in creating the physical conditions for high biological production, the World Weather Watch, under the auspices of the World Meteorological Organization, should contribute much to fishery operations through both long-range forecasting and better short-term forecasting. Of course our interest is not merely in atmospheric forecasts,

nor in the state of the sea surface, but in the deeper interaction of atmosphere and ocean. Thus, from the point of view of fisheries, an equal and complementary partner in the World Weather Watch will be the Integrated Global Ocean Station System (IGOSS) now being developed by the Intergovernmental Oceanographic Commission. The IGOSS will give us the physical data, from networks of satellite-interrogated automatic buoys and other advanced ocean data acquisition systems (collectively called ODAS), by which the ocean circulation can be observed in "real time" and the parameters relevant to fisheries forecast. A last and much more difficult link will be the observation and prediction of the basic biological processes.

So far we have been considering mainly the stocks of pelagic fishes in the upper layers of the open ocean and the shallower waters over the continental shelves. There are also large aggregations of pelagic animals that live farther down and are associated particularly with the "deep scattering layer," the sound-reflecting stratum observed in all oceans. The more widespread use of submersible research vessels will tell us more about the layer's biological nature, but the exploitation of deep pelagic resources awaits the development of suitable fishing apparatus for this purpose.

Important advances have been made in recent years in the design of pelagic trawls and in means of guiding them in

three dimensions and "locking" them onto fish concentrations. We shall perhaps have such gear not only for fishing much more deeply than at present but also for automatically homing on deep-dwelling concentrations of fishes, squids and so on, using acoustic links for the purpose. The Indian Ocean might become the part of the world where such methods are first deployed on a large scale; certainly there is evidence of a great but scarcely utilized pelagic resource in that ocean, and around its edge are human populations sorely in need of protein. The Gulf of Guinea is another place where oceanographic knowledge and new fishing methods should make accessible more of the large sardine stock that is now effectively exploited only during the short season of upwelling off Ghana and nearby countries, when the schools come near the surface and can be taken by purse seines.

The bottom-living fishes and the shellfishes (both mollusks and crustaceans) are already more fully utilized than the smaller pelagic fishes. On the whole they are the species to which man attaches a particularly high value, but they cannot have as high a global abundance as the pelagic fishes. The reason is that they are living at the end of a longer food chain. All the rest of ocean life depends on an annual primary production of 150 billion tons of phytoplankton in the 2 to 3 percent of the water mass into which light penetrates and photosynthesis can occur. Below this "photic" zone dead and dying organisms sink as a continual rain of organic matter and are eaten or decompose. Out in the deep ocean little, if any, of this organic matter reaches the bottom, but nearer land a substantial quantity does; it nourishes an entire community of marine life, the benthos, which itself provides food for animals such as cod, ocean perch, flounder and shrimp that dwell or visit there.

Thus virtually everywhere on the bed of the continental shelf there is a thriving demersal resource, but it does not end there. Where the shelf is narrow but primary production above is high, as in the upwelling areas, or where the zone of high primary production stretches well away from the coast, we may find considerable demersal resources on the continental slopes beyond the shelf, far deeper than the 200 meters that is the average limiting depth of the shelf itself. Present bottom-trawling methods will work down to 1,000 meters or more, and it seems that, at least on some slopes, useful resources of shrimps and bottom-dwelling fishes will be found even down to 1,500 meters. We still know very little about the nature and abundance of these resources, and current techniques of acoustic surveying are not of much use in evaluating them. The total area of the continental slope from, say, 200 to 1,500 meters is roughly the same as that of the entire continental shelf, so that when we have extended our preliminary surveys there we might need to revise our IWP ceiling upward somewhat.

Another problem is posed for us by the way that, as fishing is intensified throughout the world, it becomes at the same time less selective. This may not apply to a particular type of fishing operation, which may be highly selective with regard to the species captured. Partly as a result of the developments in processing and trade, and partly because of the decline of some species, however, we are using more and more of the species that abound. This holds particularly for species in warmer waters, and also for some species previously neglected in cool waters, such as the sand eel in the North Sea. This means that it is no longer so reasonable to calculate the potential of each important species stock separately, as we used to do. Instead we need new theoretical models for that part of the marine ecosystem which consists of animals in the wide range of sizes we now utilize: from an inch or so up to several feet. As we move toward fuller utilization of all these animals we shall need to take proper account of the interactions among them. This will mean devising quantitative methods for evaluat-

ing the competition among them for a common food supply and also examining the dynamic relations between the predators and the prey among them.

These changes in the degree and quality of exploitation will add one more dimension to the problems we already face in creating an effective international system of management of fishing activities, particularly on the high seas. This system consists at present of a large number—more than 20—of regional or specialized intergovernmental organizations established under bilateral or multilateral treaties, or under the constitution of the FAO. The purpose of each is to conduct and coordinate research leading to resource assessments, or to promulgate regulations for the better conduct of the fisheries, or both. The organizations are supplemented by the 1958 Geneva Convention on Fishing and Conservation of the Living Resources of the High Seas. The oldest of them, the International Council for the Exploration of the Sea, based in Copenhagen and concerned particularly with fishery research in the northeastern Atlantic and the Arctic, has had more than half a century of activity. The youngest is the International Commission for the Conservation of Atlantic Tunas; the convention that establishes it comes into force this year.

For the past two decades many have hoped that such treaty bodies would en-

sure a smooth and reasonably rapid approach to an international regime for ocean fisheries. Indeed, a few of the organizations have fair successes to their credit. The fact is, however, that the fisheries have been changing faster than the international machinery to deal with them. National fishery research budgets and organizational arrangements for guiding research, collecting proper statistics and so on have been largely inadequate to the task of assessing resources. Nations have given, and continue to give, ludicrously low-level support to the bodies of which they are members, and the bodies themselves do not have the powers they need properly to manage the fisheries and conserve the resources. Add to this the trend to high mobility and range of today's fishing fleets, the problems of species interaction and the growing number of nations at various stages of economic development participating in international fisheries, and the regional bodies are indeed in trouble! There is some awareness of this, yet the FAO, having for years been unable to give adequate financial support to the fishery bodies it set up years ago in the Indo-Pacific area, the Mediterranean and the southwestern Atlantic, has been pushed, mainly through the enthusiasm of its new intergovernmental Committee on Fisheries, to establish still other bodies (in the Indian Ocean and in the east-central and southeastern Atlantic) that will be no better supported than the ex-

RUSSIAN FACTORY SHIP *Polar Star* lies hove to in the Barents Sea in June, 1968, as two vessels from its fleet of trawlers unload their catch for processing. The worldwide activities of the Russian fishing fleet have made the U.S.S.R. the third-largest fishing nation.

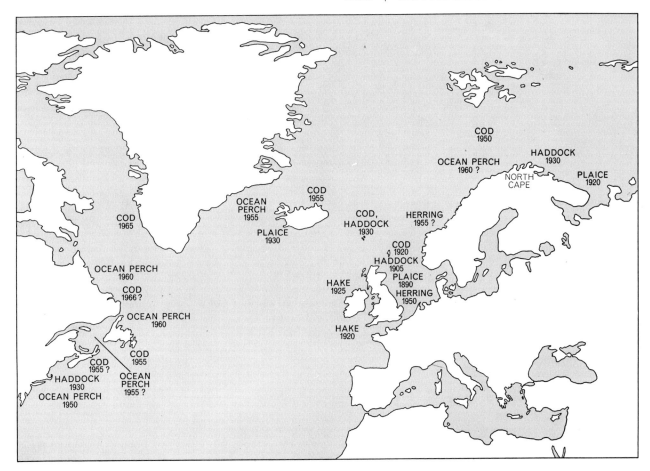

OVERFISHING in the North Atlantic and adjacent waters began some 80 years ago in the North Sea, when further increases in fishing the plaice stock no longer produced an increase in the catch of that fish. By 1950 the same was true of North Sea cod, haddock and herring, of cod, haddock and plaice off the North Cape and in the Barents Sea, of plaice, haddock and cod south and east of Iceland and of the ocean perch and haddock in the Gulf of Maine. In the period between 1956 and 1966 the same became true of ocean perch off Newfoundland and off Labrador and of cod west of Greenland. It may also be true of North Cape ocean perch and Labrador cod.

isting ones. A grand plan to double the finance and staff of the FAO's Department of Fisheries (including the secretariats and working budgets of the associated regional fishery bodies) over the six-year period 1966–1971, which member nations endorsed in principle in 1965, will be barely half-fulfilled in that time, and the various nations concerned are meanwhile being equally parsimonious in financing the other international fishery bodies.

Several of these bodies are now facing a crucial, and essentially political, problem: How are sustainable yields to be shared equitably among participating nations? It is now quite evident that there is really no escape from the paramount need, if high yields are to be sustained; this is to limit the fishing effort deployed in the intensive fisheries. This could be achieved by setting total quotas for each species in each type of fishery, but this only leads to an unseemly scramble by each nation for as large a share as possible of the quota. This can only be avoided by agreement on national al-

locations of the quotas. On what basis can such agreement be reached? On the historical trends of national participation? If so, over what period: the past two years, the past five, the past 20? On the need for protein, on the size or wealth of the population or on the proximity of coasts to fishing grounds? Might we try to devise a system for maximizing economic efficiency in securing an optimum net economic yield? How can this be measured in an international fishery? Would some form of license auction be equitable, or inevitably loaded in favor of wealthy nations? The total number or tonnage of fishing vessels might be fixed, as the United Kingdom suggested in 1946 should be done in the North Sea, but what flags should the ships fly and in what proportion? Might we even consider "internationalizing" the resources, granting fishing concessions and using at least a part of the economic yield from the concessions to finance marine research, develop fish-farming, police the seas and aid the participation of less developed nations?

Some of my scientific colleagues are optimistic about the outcome of current negotiations on these questions, and indeed when the countries participating are a handful of nations at a similar stage of economic and technical development, as was the case for Antarctic whaling, agreement can sometimes be reached by hard bargaining. What happens, however, when the participating countries are numerous, widely varying in their interests and ranging from the most powerful nations on earth to states most newly emerged to independence? I must confess that many of us were optimistic when 20 years ago we began proposing quite reasonable net-mesh regulations to conserve the young of certain fish stocks. Then we saw these simple—I suppose oversimple—ideas bog down in consideration of precisely how to measure a mesh of a particular kind of twine, and how to take account of the innumerable special situations that countries pleaded for, so that fishery research sometimes seemed to be becoming perverted from its earlier clarity and broad perspective.

Apprehension and doubt about the ultimate value of the concept of regulation through regional commissions of the present type have, I think, contributed to the interest in recent years in alternative regimes: either the "appropriation" of high-seas resources to some form of international "ownership" instead of today's condition of no ownership or, at the other extreme, the appropriation of increasingly wide ocean areas to national ownership by coastal states. As is well known, a similar dialectic is in progress in connection with the seabed and its mineral resources. Either solution would have both advantages and disadvantages, depending on one's viewpoint, on the time scale considered and on political philosophy. I do not propose to discuss these matters here, although personally I am increasingly firm in the conclusion that mankind has much more to gain in the long run from the "international" solution, with both seabed and fishery resources being considered as our common heritage. We now at least have a fair idea of what is economically at stake.

Here are some examples. The wasted effort in capture of cod alone in the northeastern Atlantic and salmon alone in the northern Pacific could, if rationally deployed elsewhere, increase the total world catch by 5 percent. The present catch of cod, valued at $350 million per year, could be taken with only half the effort currently expended, and the annual saving in fishing effort would amount to $150 million or more. The cost of harvesting salmon off the West Coast of North America could be reduced by three-quarters if management policy permitted use of the most efficient fishing gear; the introduction of such a policy would increase net economic returns by $750,000 annually.

The annual benefit that would accrue from the introduction and enforcement of mesh regulations in the demersal fishery—mainly the hake fishery—in the east-central Atlantic off West Africa is of the order of $1 million. Failure to regulate the Antarctic whaling industry effectively in earlier years, when stocks of blue whales and fin whales were near their optimum size, is now costing us tens of millions of dollars annually in loss of this valuable but only slowly renewable resource. Even under stringent regulation this loss will continue for the decades these stocks will need to recover. Yellowfin tuna in the eastern tropical Pacific are almost fully exploited. There is an annual catch quota, but it is not allocated to nations or ships, with the classic inevitable results: an increase in the catching capacity of fleets, their use in shorter and

shorter "open" seasons and an annual waste of perhaps 30 percent of the net value of this important fishery.

Such regulations as exist are extremely difficult to enforce (or to be seen to be enforced, which is almost as important). The tighter the squeeze on the natural resources, the greater the suspicion of fishermen that "the others" are not abiding by the regulations, and the greater the incentive to flout the regulations oneself. There has been occasional provision in treaties, or in *ad hoc* arrangements, to place neutral inspectors or internationally accredited observers aboard fishing vessels and mother ships (as in Antarctic whaling, where arrangements were completed but never implemented!). Such arrangements are exceptional. In point of fact the effective supervision of a fishing fleet is an enormously difficult undertaking. Even to know where the vessels are going, let alone what they are catching, is quite a problem. Perhaps one day artificial satellites will monitor sealed transmitters compulsorily carried on each vessel. But how to ensure compliance with minimum landing-size regulations when increasing quantities of the catch are being processed at sea? With factory ships roaming the entire ocean, even the statistics reporting catches by species and area can become more rather than less difficult to obtain.

Some of these considerations and pessimism about their early solution have, I think, played their part in stimulating other approaches to harvesting the sea.

One of these is the theory of "working back down the food chain." For every ton of fish we catch, the theory goes, we might instead catch say 10 tons of the organisms on which those fish feed. Thus by harvesting the smaller organisms we could move away from the fish ceiling of 100 million or 200 million tons and closer to the 150 billion tons of primary production. The snag is the question of concentration. The billion tons or so of "fish food" is neither in a form of direct interest to man nor is it so concentrated in space as the animals it nourishes. In fact, the 10-to-one ratio of fish food to fish represents a use of energy—perhaps a rather efficient use—by which biomass is concentrated; if the fish did not expend this energy in feeding, man might have to expend a similar amount of energy—in fuel, for example—in order to collect the dispersed fish food. I am sure the technological problems of our using fish food will be solved, but only careful analysis will reveal whether or not it is better to turn fish food, by way of fish meal, into chickens or rainbow trout than to harvest the marine fish instead.

There are a few situations, however, where the concentration, abundance and homogeneity of fish food are sufficient to be of interest in the near future. The best-known of these is the euphausiid "krill" in Antarctic waters: small shrimp-like crustaceans that form the main food of the baleen whales. Russian investigators and some others are seriously charting krill distribution and production, relating them to the oceanographic features of the Southern Ocean, experiment-

JAPANESE MARICULTURE includes the raising of several kinds of marine algae. This array of posts and netting in the Inland Sea supports a crop of an edible seaweed, *Porphyra*.

AUSTRALIAN MARICULTURE includes the production of some 60 million oysters per year in the brackish estuaries of New South Wales. The long racks in the photograph have been exposed by low tide; they support thousands of sticks covered with maturing oysters.

ing with special gear for catching the krill (something between a mid-water trawl and a magnified plankton net) and developing methods for turning them into meal and acceptable pastes. The krill alone could produce a weight of yield, although surely not a value, at least as great as the present world fish catch, but we might have to forgo the whales. Similarly, the deep scattering layers in other oceans might provide very large quantities of smaller marine animals in harvestable concentration.

An approach opposite to working down the food chain is to look to the improvement of the natural fish resources, and particularly to the cultivation of highly valued species. Schemes for transplanting young fish to good high-seas feeding areas, or for increasing recruitment by rearing young fish to viable size, are hampered by the problem of protecting what would need to be quite large investments. What farmer would bother to breed domestic animals if he were not assured by the law of the land that others would not come and take them as soon as they were nicely fattened? Thus mariculture in the open sea awaits a regime of law there, and effective management as well as more research.

Meanwhile attention is increasingly given to the possibilities of raising more fish and shellfish in coastal waters, where the effort would at least have the protection of national law. Old traditions of shellfish culture are being reexamined,

and one can be confident that scientific bases for further growth will be found. All such activities depend ultimately on what I call "productivity traps": the utilization of natural or artificially modified features of the marine environment to trap biological production originating in a wider area, and by such a biological route that more of the production is embodied in organisms of direct interest to man. In this way we open the immense possibilities of using mangrove swamps and productive estuarine areas, building artificial reefs, breeding even more efficient homing species such as the salmon, enhancing natural production with nutrients or warm water from coastal power stations, controlling predators and competitors, shortening food chains and so on. Progress in such endeavors will require a better predictive ecology than we now have, and also many pilot experiments with corresponding risks of failure as well as chances of success.

The greatest threat to mariculture is perhaps the growing pollution of the sea. This is becoming a real problem for fisheries generally, particularly coastal ones, and mariculture would thrive best in just those regions that are most threatened by pollution, namely the ones near large coastal populations and technological centers. We should not expect, I think, that the ocean would not be used at all as a receptacle for waste—it is in some ways so good for such a purpose: its large volume, its deep holes, the hydrolyzing, corrosive and biologically degrading

properties of seawater and the microbes in it. We should expect, however, that this use will not be an indiscriminate one, that this use of the ocean would be internationally registered, controlled and monitored, and that there would be strict regulation of any dumping of noxious substances (obsolete weapons of chemical and biological warfare, for example), including the injection of such substances by pipelines extending from the coast. There are signs that nations are becoming ready to accept such responsibilities, and to act in concert to overcome the problems. Let us hope that progress in this respect will be faster than it has been in arranging for the management of some fisheries, or in a few decades there may be few coastal fisheries left worth managing.

I have stressed the need for scientific research to ensure the future use of the sea as a source of food. This need seems to me self-evident, but it is undervalued by many persons and organizations concerned with economic development. It is relatively easy to secure a million dollars of international development funds for the worthy purpose of assisting a country to participate in an international fishery or to set up a training school for its fishermen and explore the country's continental shelf for fish or shrimps. It is more difficult to justify a similar or lesser expenditure on the scientific assessment of the new fishery's resources and the investigation of its ocean environment. It is much more difficult to secure even quite limited support for international measures that might ensure the continued profitability of the new fishery for all participants.

Looking back a decade instead of forward, we recall that Lionel A. Walford of the U.S. Fish and Wildlife Service wrote, in a study he made for the Conservation Foundation: "The sea is a mysterious wilderness, full of secrets. It is inhabited only by wild animals and, with the exception of a few special situations, is uncultivated. Most of what we know about it we have had to learn indirectly with mechanical contrivances to probe, feel, sample, fish." There are presumably fewer wild animals now than there were then—at least fewer useful ones—but there seems to be a good chance that by the turn of the century the sea will be less a wilderness and more cultivated. Much remains for us and our children to do to make sure that by then it is not a contaminated wilderness or a battlefield for ever sharper clashes between nations and between the different users of its resources.

The Anchovy Crisis

by C. P. Idyll

June 1973

Over the past decade the world's largest fishery has been in the Peru Current. A periodic ecological disturbance, combined with the heavy fishing, now threatens to destroy the industry

The world of the Peruvian anchovy is the sweep of a great cold ocean current. In a slow northward drift the current carries the little fishes along in company with countless tiny plants and animals that the anchovies avidly devour and with larger fishes, squids and a host of other marine animals. The anchovies form thronging legions that wheel and dart in the current. Their world is often entered by aliens: birds that plunge from above, snatching up the anchovies by the hundreds of thousands, and men who cast great net enclosures around the fishes, carrying them off to shore by the millions.

In the brief life-span of the anchovy, rarely longer than three years, its cold-current environment usually changes only within narrow limits. During the lifetime of some generations, however, their world may be put out of joint. The slow northward drift of the current, only two-tenths to three-tenths of a knot (compared with the six knots of the Gulf Stream off Florida), becomes still slower and may even reverse itself. The water grows warmer and less salty; the makeup of its populations changes and many of the usually abundant microscopic plants and animals dwindle in number. Finding their world poorer, the little anchovies scatter; many may die prematurely and many in the successor generation may not be born at all. The sea change known as El Niño has arrived.

Coastal Peru is normally a cool and misty land, quite unlike the steamy Tropics that occupy the same latitudes on the eastern coast of South America. The Peru coast is kept that way by the temperature of the ocean current, which in the south can be as cool as 10 degrees Celsius (50 degrees Fahrenheit) and only reaches about 22 degrees C. (71.6 degrees F.) in the north. Although the northernmost part of Peru is a mere three degrees of latitude below the Equator, the average air temperature is a moderate 18 to 22 degrees C. (64.4 to 71.6 degrees F.). Along the 1,475 miles of Peruvian coastline, a distance 100 miles longer than the Pacific coast of the U.S., there are no marshes, mud flats and estuaries—only arid desert, most of it a treeless, monotonous, barren brown. This bleak strip of sand extends a short distance inland, rarely more than 40 miles or so, to the upthrust Andes, one of the most awesome mountain ranges in the world. The high rain shadow of the Andes robs the prevailing southeast winds of their moisture and keeps the coastal region arid. A few streams that rise in the mountains cross the desert; their narrow valleys support what little agriculture exists along the coast.

The Peru Current consists of four components, the interaction of which creates, molds and changes the world of the anchovy. Two of the components travel in a northerly direction: the Coastal Current, flowing next to the shore, and the Oceanic Current, located farther out to sea. Between the two, on and near the surface, runs the Peru Countercurrent. Beneath all three runs the Peru Undercurrent. Both the Countercurrent and the Undercurrent flow southward.

The Coastal Current runs deep and hugs the land from about Valparaiso in Chile in the south to north of Chimbote in Peru, a stretch of some 2,000 miles. The anchovies live mostly in the northern part of this great band of water, which constantly changes shape and size, becoming wider or narrower, deeper or shallower, altering and twisting like an elongated amoeba. The Oceanic Current is longer than the Coastal Current. It is often several hundred miles wide, and it runs as deep as 700 meters. It flows north to a point about opposite the Gulf of Guayaquil before bending west.

The great northward sweep of water is often called the Humboldt Current, after the German naturalist who described the phenomenon following a visit to South America in 1803. Humboldt thought that the cold of the water was the chill of the Antarctic. His conjecture was partly right: the current does include subantarctic water. Much of the cold, however, is the cold of subsurface water. As the water on the ocean surface is swept away by the prevailing winds, deeper low-temperature water wells up slowly to replace it. The trade winds in this part of the world, channeled and bent by the Andes, blow from the south and southeast, mostly parallel to the shore. This prevailing wind urges the surface water northward at the same time that another influence, the Coriolis force, deflects it to the west. As the resulting steady offshore drift skims off the surface layer the cold subsurface water rises with stately slowness to replace it, traveling vertically at a rate ranging from 20 to 100 meters per month, depending on the location and the season.

The biological effect of the upwelling is enormous. That stretch of water, only a tiny fraction of the ocean surface, produces fully 22 percent of all the fish caught throughout the world. Its richness springs from a constantly renewed supply of the chemical nutrients—principally phosphates and nitrates—that stimulate plant growth. Accumulated gradually in the deep layers of the ocean as the debris of dead marine plants and animals sinks to the bottom, the nutrients travel with the upwelling water to the top levels. There the light is sufficient to drive photosynthesis, and the nutrients help the marine plants to flourish. The concentration of nutrients in the Peru upwelling is many times greater than that in the open ocean. In terms of

ANCHOVY TRAWLERS by the score lie at anchor in the fishing port of Pisco in Peru. Structures in the foreground are part of a fish-meal factory. Tonnage of anchovies alone landed in Peru in a normal year outweighs the combined fish catch of any other nation.

CASCADE OF ANCHOVIES spills from a draining grate into the cargo hold of a trawler as crewmen keep watch. The fish gather by the billions near shore and close to the surface, particularly in summer months; trawlers can then net anchovies 100 tons at a time.

the amount of carbon fixed photosynthetically per cubic meter of water per day, the range in the upwelling region is from 45 to 200 milligrams, compared with less than 15 milligrams in the waters immediately adjacent. Perhaps only one other part of the world ocean is richer: the Benguela Current off the southwestern coast of Africa.

The Peru upwelling sustains an enormous flow of living matter. The food chain begins with the microscopic diatoms and other members of the phytoplankton that comprise the pasturage of the sea. The plants absorb the nutrients and grow in rich profusion, providing fodder for billions of grazing animals, principally minute crustaceans such as copepods but including arrowworms and a wide variety of other small marine herbivores. The food chain can then go on to several more links, progressing from the small fishes that eat the herbivores to the larger fishes and squids that prey on the small fishes and perhaps continuing to include one or more further advanced levels of marine predation. The food chain in the Peru Current does go on in this fashion to some degree, but most of its energy flow stops with the anchovies. This single fish species has succeeded in capturing an extraordinarily high proportion of the total energy available in the ecosystem and in converting it into enormous quantities of living matter. At the height of the anchovies' annual cycle the total bulk of the species is probably of the order of 15 to 20 million metric tons.

The Peruvian anchovy belongs to the same genus (*Engraulis*) as the common anchovy of the eastern Atlantic and the Mediterranean (*E. encrasicolus*), but it comprises a separate species (*E. ringens*). Its life begins in the form of an egg, a tiny oval spot of nearly transparent protoplasm adrift in the sea. Eggs can be spawned at almost any time of the year but there are two periods when the anchovies' reproductive activity is highest. The major spawning occurs in August and September, during the southern winter, and it is repeated on a lesser scale in January and February. Anchovies are precocious: most females are capable of spawning when they are a year old. By then each female, a little over four inches long, may produce 10,000 eggs. If she survives to the age of two and reaches a length of six inches, her output increases to some 20,000 eggs.

The delicate larvae that hatch from the eggs lead a perilous existence. Many species of fish produce eggs that contain a considerable store of yolk; the reserve

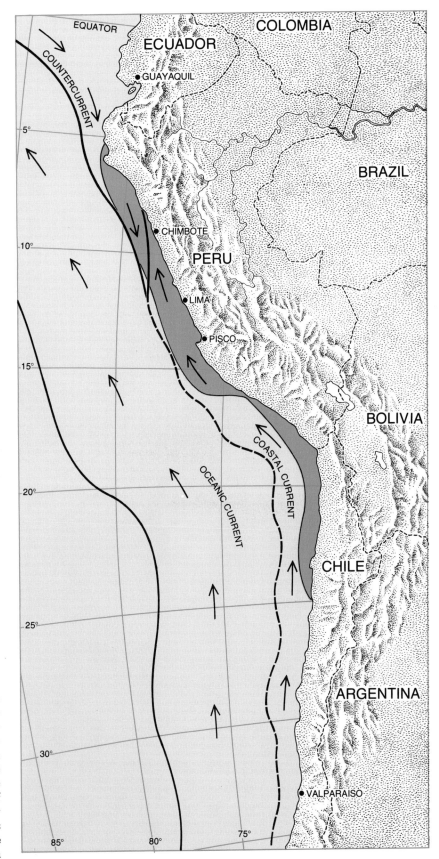

TWO NORTH-FLOWING COMPONENTS of the Peru Current are the deep, narrow Coastal Current that hugs the land from Valparaiso in Chile to north of Chimbote in Peru and the deeper and wider Oceanic Current that reaches the latitude of the Gulf of Guayaquil. The anchovies (*dark color*) are normally found in the Coastal Current between 25 and five degrees south latitude; they may consist of a northern and a southern population.

of nutrient helps to sustain the newly hatched young until they adjust to finding their own food. The anchovy egg has a negligible yolk store, and so the larva must locate food quickly or starve. To make matters worse, the larva has limited swimming powers and a high rate of metabolism. If more than a few wiggles are required to obtain the food it needs, it will not survive. Because every larva consumes plankton in substantial amounts and because the peak hatches produce larvae numbering in multiples of billions, only enormous swarms of microscopic plants and larval crustaceans can sustain the anchovy stock. Nor is starvation the only peril: the larval anchovies feed swarms of predators. They are eaten by the same copepods that will, if the little fish survive, be the an-

chovies' own main sustenance. Arrowworms also devour them, and so do their own parents.

One month after the time the anchovy larvae are hatched more than 99 percent of them have perished. Even with such a high mortality rate, a process that begins with billions of spawning fish, each casting 10,000 to 20,000 eggs, produces enormous quantities of anchovy larvae. The little fish grow rapidly; in the course of the first year they attain a length of 4.2 to 4.3 inches. The year-old fish are so slender, however, that they weigh a scant third of an ounce.

The anchovy schools do not move at random; apparently because of a strong preference for the cold water of the Coastal Current they remain within a comparatively restricted zone. The

Coastal Current is at its narrowest during the southern summer, running close to shore and seldom exceeding 200 meters in depth. Within this shrunken world the anchovies press together in enormous concentrations near the shore and close to the surface. It is now that predators fare best. The larger fishes and the squids feed well; the several species of guano birds have to fly only short distances from their island nesting grounds and need not dive deep to reach their prey. The greatest of the predators, the fisherman, finds summer work the easiest. He can often set his purse seine within sight of port and gather in anchovies 100 tons at a time.

El Niño occurs at irregular intervals. There is said to be a seven-year cycle, but in actuality the phenomenon is far

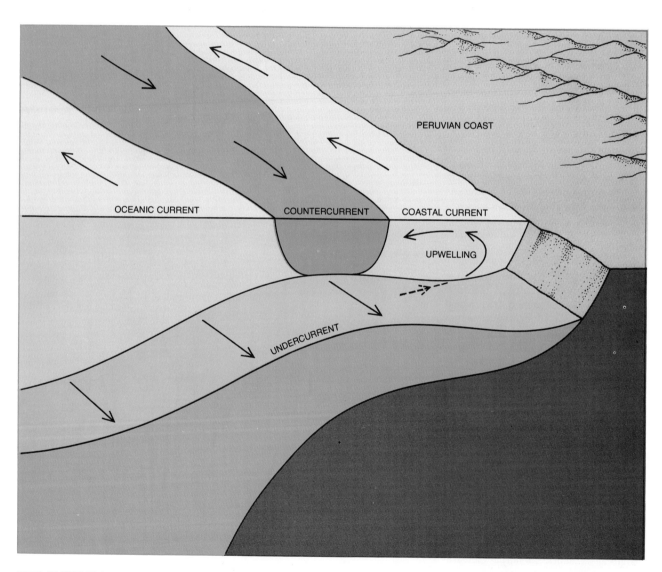

TWO SOUTH-FLOWING COMPONENTS of the Peru Current are seen schematically in relation to the two north-flowing components in this diagram. The Countercurrent (*dark color*), a surface or near-surface stream of water, intrudes between the north-flowing cold Coastal and Oceanic currents. Normally the Countercurrent does not extend much south of the Equator, but when the wind that moves the north-flowing currents along falters or changes direction, its warm water pushes far to the south with disastrous biological consequences. Deep below all three currents is the second, far larger south-flowing component, the Undercurrent (*light color*).

less precise in its appearance. The severe environmental dislocations may be repeated for two or more years in a row or may not recur for a decade or longer. Another kind of regularity, however, has given El Niño its name. The change usually begins around Christmastime and so is given the Spanish name for the Christ Child. The complicated chain of events in a year of El Niño disturbs the anchovies, sometimes profoundly. The wind now comes from the west rather than from the southeast, and it is laden with moisture from the Pacific. With no mountains to rob the air of its burden of water the arid coast is often subjected to torrential rains and severe windstorms. In a desert region where even a heavy mist can cause problems the floods of El Niño are often devastating. Oddly enough, however, in some Niño years no rain falls.

A warning that the sea change may be on its way is given when the temperature of the coastal water begins to rise. If the increase in temperature persists and spreads, the delicately adjusted world of the anchovy tilts. With the warm water come unfamiliar inhabitants of the northern Tropics: the yellowfin tuna, the dolphinfish, the manta ray and the hammerhead shark. Some of them

feed on the anchovies. A greater threat to the anchovies' survival, however, is a slowing of the northbound Coastal Current and a decline or even a halt in the usual upwelling of subsurface waters. As the supply of nutrients diminishes, the planktonic plant life that provides the base of the ocean food chain becomes less abundant. As a result herbivorous planktonic animals become scarcer, and so it goes link by link up the chain. Furthermore, the water temperature is now too high to suit the anchovies themselves. Even if the shortage of food has not yet greatly reduced their numbers, the fish scatter, no longer forming the enormous schools that normally afford the guano birds and fishermen such rewarding targets.

The effect on the guano birds and marine animals is among the most serious of the changes wrought by El Niño. The birds starve or fly away, deserting their nestlings. Fishes, squids and even turtles and small sea mammals die. Their decaying bodies release evil-smelling hydrogen sulfide that bubbles up through the water and blackens the paint on the boats in the harbors. This unpleasant phenomenon is called El Pintor (The Painter). Patches of reddish, brownish or yellow water similar to the "red tides" that upset the Florida tourist industry

become relatively common. They are caused by prodigious blooms of dinoflagellates: microscopic planktonic plants that are toxic in high concentrations. The greater frequency of the blooms during Niño years may be because the nutrient composition of the seawater suits the organisms better then or because the less vigorous currents fail to disperse accumulating clusters of dinoflagellates as quickly as usual.

The causes of El Niño are wind changes and sea changes on a very large scale. When the steady southeast trade winds weaken or when the wind blows from the west, the ocean currents that run to the northwest are no longer pushed along with the same vigor. Under normal circumstances the south-flowing Peru Countercurrent is weak, but when the prevailing winds fail or are reversed, the Countercurrent thrusts a tongue of warm water into the cleft between the now less vigorous Coastal and Oceanic currents. As it meets less resistance the Countercurrent penetrates farther south, pushing the weak north-flowing currents aside and covering their cold waters with a 30-meter layer of warm tropical water. The water may have come from as far away as the Panama Bight, north of the Equator, and part of it may even have originated as land runoff in Central America. It can be as much as seven degrees C. warmer than the north-flowing Coastal Current and is lower in salinity, deficient in oxygen and poor in nutrients. In some Niño years the Countercurrent pushes the tropical water as far as 600 miles south of the Equator.

The organisms most obviously affected by a Niño year are the guano birds, various species that are colloquially lumped together under the same name that is applied to the droppings that accumulate in large quantities on the rocky islands where they nest. There are three principal species: the guanay, or cormorant (*Phalacrocorax bougainvillii*); the piquero, or booby (*Sula variegata*), and the alcatras, or pelican (*Pelicanus thagus*). Over the millenniums the bird droppings have accumulated in piles as high as 150 feet on some islands. Because guano is perhaps the finest natural fertilizer known, the guano islands have provided the foundation for a valuable industry. Of the guano birds' diet between 80 and 95 percent is made up of anchovies, and the coastal waters of Peru support what is probably the largest population of oceanic birds anywhere in the world. In recent years estimates of the birds' total number have gone as

ANCHOVY CATCH remained below two million metric tons until the 1960's. Exploitation of the fishery skyrocketed thereafter; the annual rate approached or exceeded nine million tons for six of eight consecutive years. The peak year, 1970, saw a catch of 12.3 million tons.

SACKS OF FISH MEAL awaiting shipment abroad line a wharf at a Peruvian port. The meal is used to enrich feeds for poultry and other livestock. In 1970 the export of fish meal and fish oil earned some $340 million, a third of Peru's total export revenue.

high as 30 million. The five million individuals that inhabited one particular guano island are believed to have consumed 1,000 tons of anchovies a day. The guano birds' annual anchovy catch in recent years is calculated to average 2.5 million metric tons, or between a fourth and a fifth of the commercial-fishery catch.

Following every Niño year of any consequence the bird population declines just as the anchovy population does. When the warmer water scatters the dense surface schools of fish, the birds find it harder to feed themselves, let alone their nestlings. Adult birds fly to other areas. Juvenile birds, less efficient fishers than their parents, perish in large numbers. The deserted nestlings are doomed to starvation. After the severe Niño year of 1957 the guano-bird population, then estimated to be 27 million, plummeted to six million and dropped to a low of 5.5 million the following year. Numbers slowly increased thereafter, so that there were 17 million birds when the Niño year of 1965 arrived. That year the population fell to 4.3 million.

Since then the guano birds have failed to recover at the normal rate. There is concern that the commercial anchovy fishery, which has expanded greatly in the same period, is depriving the birds of so much food that their numbers may fall below the level that is critical to their survival as social species. The late Robert Cushman Murphy of the American Museum of Natural History devoted some years to the study of these populations, and it was his opinion that the birds and the fishermen were essentially incompatible. It seems likely, however, that in spite of Murphy's contrary view the two competitors will be able to coexist at some suitable level of commercial fishing. At the same time it may well be that the size of the commercial catch in recent years has prevented the bird population from regaining its former numbers.

It is not commonly known that in the past few years the anchovy fishery has made Peru the world's leading fish-producing nation. Until recently Peru was harvesting anchovies at a rate of 10 million metric tons or more per year. This is a greater weight than that of all the species of fish being caught by any one nation in the Old World, and is twice the tonnage of the combined all-species catch of all the nations of North and Central America. The fish meal made from the Peruvian catch is sold around the world to enrich feeds for poultry and other livestock; the fish oil goes into margarine, paint, lipstick and a score of other products. In 1970 the export of fishery products brought Peru some $340 million, nearly a third of the nation's foreign-exchange earnings. In addition to this the tax revenues from the industry and the domestic employment it provides have become major elements in Peru's economy.

The anchovy industry began in earnest in 1957. Within 10 years the profits that could be made from catching and processing the fish attracted hundreds of fishing boats and led to the construction of dozens of fish-meal factories. No fish stock, however, can stand unchecked exploitation. Government authorities and fishery biologists became concerned about the future of the resource. Peru was a newcomer to large-scale commercial fishing and had neither fishery scientists nor administrators with experience in the complexities of fishery research and management. The Peruvian government turned to the United Nations for help.

In 1960, with a grant from the UN Development Programme and a matching amount in Peruvian funds, the Insti-

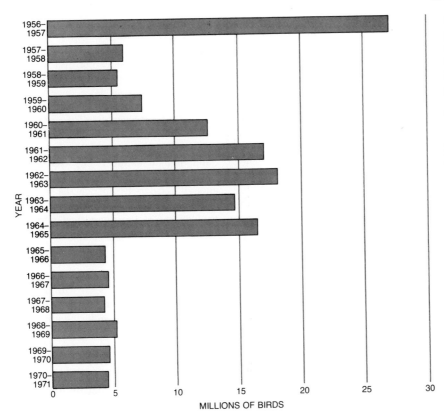

GUANO-BIRD POPULATION, hard hit by El Niño in 1957, had regained more than half its former numbers when El Niño reappeared in 1965. The population did not recover at the normal rate thereafter, possibly because the large commercial anchovy catches cut down the birds' food supply. Only a million or so guano birds may have survived the 1972 Niño.

ond, "recruitment" (the numbers of fish grown big enough to enter the commercial fishery in any year) has been by far the smallest ever observed. It is scarcely 13 percent of the recruitment in a normal year.

What is causing the trouble? Very possibly El Niño has been a major factor, but some puzzling circumstances have made the scientists closest to the subject uncertain about the extent of the relation between the sea change and the reduction in the anchovy stock. It is clear, however, that the Niño year of 1972 was one of the most severe ever observed. Instead of remaining at the normal level of 22 degrees C. the surface temperature of the Coastal Current rose to 30.3 degrees in February. Although the temperature fluctuated thereafter, it remained high for the rest of the year and was still above normal in January, 1973. It did not fall to near-normal temperature until March of this year. At the same time that the temperature rose the salinity of the water declined from a normal 35 or 36 parts per 1,000 to 32.7 parts per 1,000.

Tropical and subtropical marine plants and animals began to appear far south of their usual limits: dolphinfish, skipjack tuna and the tropical crab *Euphilax*. The guano birds fled from their nesting islands, abandoning their young; their population may now number no more than a million. The warm Countercurrent forced the cold-seeking anchovies so close to the shore that the fishermen often found the water too shallow for their nets. Moreover, the crowded fish did not spawn as abundantly as usual and the eggs and larvae, already reduced in numbers, did not survive at the usual rate because of greater predation by their own close-packed parents.

Biologists are nonetheless unwilling to blame El Niño for all these occurrences. For example, with respect to the anchovies they note that recruitment of young fish to the adult population was observed to fall below normal levels before it became obvious that the surface temperature of the Coastal Current had risen. In seeming contradiction to this, however, the tropical crabs too made their appearance before the surface temperature rose, apparently indicating that a body of tropical water had by then already invaded the world of the anchovies. In the light of such oddities the experts frankly admit that they do not know how much influence El Niño exerted on the anchovies in 1972.

tuto del Mar del Peru was set up to conduct research on the anchovy stocks and to advise the government on management of the fishery. The Food and Agriculture Organization of the UN (FAO) recruited experienced fishery scientists from around the world to work at the institute, conduct research on the anchovy stocks and train a Peruvian staff. Located in Lima, the institute is now a firmly established fishery-research center where more than 50 young Peruvian scientists are conducting the studies needed to establish conservation regulations for the anchovy fishery. The Peruvian staff is advised by a few resident FAO scientists and by a panel of distinguished experts from around the world, organized by the FAO, that meets twice a year.

It has taken biologists nearly a century to unravel the intricacies of fish populations and the complexities of their response to the dual stresses of environmental change and human exploitation. Until last year fishery biologists could point to the management of the Peruvian fishery as an exemplary application of this hard-won knowledge. A major stock had been put under rational control before exploitation had depleted it, and conservation measures seemed to be en-

suring an enormously high yield at the limit of the biological capacity of the Peru Current ecosystem. Then in 1972 such pride was chastened, if not utterly humbled.

After the Niño year of 1965 the fishery had enjoyed several very successful seasons, culminating in 1970 with an anchovy catch of 12.3 million metric tons. Then, toward the end of April, 1972, and only a few weeks after the start of the season, fishing suddenly faltered. By the end of June catches had dwindled to almost nothing, and at the close of the 1972 season only 4.5 million tons of anchovies had been harvested. The catch this year threatens to be even poorer. Indeed, there is some reason to fear that the world's greatest stock of fish may have been irreversibly damaged, in which case the Peru fishery would be destined to collapse altogether.

That is the gloomiest outlook. It is based on two disturbing circumstances. First, the size of the "standing stock" (the total anchovy population) now appears to be far smaller than normal. It may be as low as one or two million metric tons, compared with an average of 15 million tons in recent years and an estimated 20 million tons in 1971. Sec-

Quite apart from the sea change, however, the Peruvian commercial fishery must accept a share of the blame. The 1970 catch of 12.3 million tons considerably exceeded the 10-million-ton level that fishery biologists had estimated to be the maximum sustainable yield of the Peruvian stock. Several economic and political stresses were responsible for the excessive catch. Foremost among these harsh realities is that there are many more fishing boats and fish-meal factories in Peru than are needed to harvest and process the catch. The anchovy fleet is so large that it could harvest the equivalent of the annual U.S. catch of yellowfin tuna in a single day or the annual U.S. salmon catch in two and a half days. The fleet could be reduced by more than 25 percent and still comfortably harvest a rational quota of 10 million tons of anchovies a year. Moreover, the record 1970 catch figure does not measure the full toll the fishery took of the anchovy stock that year. Conservative estimates of losses from spoilage at sea and unloading and processing ashore raise the commercial total to some 13 or 14 million tons.

At this writing the future of the Peruvian anchovy is uncertain. The gloomy forecasts based on biological sampling in 1972 have been confirmed by additional observations early this year and by the results of trial fishing allowed at that time. For three weeks in March the anchovy fleet went to sea and about one million tons of fish were caught. During that brief period the catch per unit of fishing effort (a statistic that provides a measure of the size of the stock) declined rapidly. This suggests that the fleet had caught a significant proportion of all the fish that were available. Most of the catch consisted of fish recruited since July, 1972. There will not be any substantial additions to the stock until this coming October, when the progeny of the present population, much reduced by the fishing in March, have grown big enough to enter the fishery. Even so, fishing was authorized again in April with the quota set at 800,000 tons. Only 400,000 tons were taken. At present the 1973 catch is forecast at no more than three million tons.

If things are as bad as the worst prognostications indicate, the anchovy fishery may, like the California sardine industry and the Hokkaido herring industry, collapse forever. Many aspects of the history of the Peru fishery bear a disturbing resemblance to the events that brought about these earlier disasters.

Nature being what it is, the Peruvian coast will sooner or later once again have normal winds and ocean currents. If the anchovy population has not been too severely reduced, the fishery will then begin to recover. On the other hand, human nature being what it is, difficulties may arise in enforcing soon enough and strictly enough the moderate catch quotas required to avoid overexploitation of the diminished population. Unless such a policy of moderation is achieved, not only will the fish stock suffer but also the world of the Peruvian anchovy will be permanently changed. The guano-bird population will be further reduced and perhaps even eliminated. There will also be enormously complex effects among the many other animals that depend to a greater or lesser degree on the presence of the little fish. Finally, if the anchovies' world is allowed to go awry, the biggest loser will be man. He will have lost not only a rich natural resource but also some of the quality of his own world.

GUANO ISLAND off the coast of Peru is a nesting ground for a vast population of piqueros, or boobies. Cormorants, pelicans and boobies are the principal guano birds. Although now reduced in numbers, they once ate some 2.5 million tons of anchovies per year.

Marine Farming

23

by Gifford B. Pinchot
December 1970

*Man gets food from the sea essentially by hunting
and gathering. Yet the farming of fish and shellfish
has been pursued for some 2,000 years, and its
potentialities are far from being exhausted*

A major concern of modern man is the possibility that the earth will not be able to produce enough food to nourish its expanding population. A particularly controversial issue is the question of how much food can ultimately be obtained from the sea. It is argued on the one hand that, on the basis of area, the oceans receive more than twice as much solar energy—the prime source of all biological productivity—as the land. This suggests that the oceans' potential productivity should greatly exceed the land's. On the other hand, most of the sea is biologically a desert. Its fertile areas are found where runoff from the land or the upwelling of nutrient-rich deep water fertilizes the surface water and stimulates the growth of marine plants, the photosynthetic organisms on which all other marine life depends. Even at today's high level of exploitation the fisheries of the world provide only a small fraction of human food needs, and there is some danger that they may supply even less in the future because of overfishing.

Does this mean that there is no hope of increasing our yield of food from the sea? I do not think so. It does mean, however, that instead of concentrating exclusively on more efficient means of fishing we must also learn to develop the potential of the oceans by farming them, just as early man learned that farming rather than hunting was the more effective method of feeding a human population. The purpose of this article is to

examine briefly the contribution marine farming now makes to our food supply, and to consider some possibilities for its future role.

Marine farming has a long history. The earliest type of farming was the raising of oysters. Laws concerning oyster-raising in Japan go back to well before the time of Christ. Aristotle discusses the cultivation of oysters in Greece, and Pliny gives details of Roman oyster-farming in the early decades of the Christian Era. By the 18th century the natural oyster beds in France were beginning to be overexploited and were saved only by extensive developments in rearing practices.

Carp (*Cyprinus carpio*) were commonly raised in European freshwater ponds in both Roman and medieval times. Records concerning the regulation of salt or brackish ponds for raising milkfish (*Chanos chanos*) in Java date back to the 15th century. Carp and milkfish are both herbivores that thrive on a diet of aquatic plants. Oysters, as filter-feeders, can also be loosely classified as herbivores.

Oysters are particularly appropriate for marine farming because their spawn can be collected and used for "seeding" new areas of cultivation. An oyster produces more than 100 million eggs at a single spawning. The egg soon develops into a free-swimming larval form, known as a veliger, which settles to the bottom after two or three weeks. Veligers attach themselves to any clean surface and de-

velop into miniature adult oysters, called "spat" because oystermen once believed the adult oysters spat them out. At this point the oyster farmer enters the picture. He distributes a supply of "cultch": clean material with a smooth, hard surface, such as old oystershell or ceramic tile. The cultch receives a "set" of spat and is then used to seed new oyster beds.

The bottom is prepared for seeding by removing as many natural enemies of the oyster as possible. In the eastern U.S. this is usually done by dragging a rope mat along the bottom to sweep the area clear of starfish, one of the major predators. In France, where more intensive labor is employed, the spat are usually planted on the exposed bottom of an estuary at low tide. The predators are removed by hand and the oyster bed is fenced to prevent their return. The oysters are moved after a few years to *claires*, special fattening areas where the water is rich in diatoms. This produces oysters of improved taste and color. When the oysters have reached marketable size, they are moved again to shallower water, where they must stay closed for longer periods at low tide. The French oystermen believe this treatment prepares the oysters for their trip to the market.

A significant advance in oyster-farming is the use of suspension cultures. This method, pioneered in Japan, is now spreading to the rest of the world. The spat are collected on shells that are

NUTRIENT-RICH WATER from the depths of the Pacific appears dark blue in the photograph on the following page. Land (*left*) is the coast of Taiwan as seen from *Gemini X*. Natural upwellings such as this are caused by winds or currents. Areas of continuous upwelling are extremely productive fishing grounds because the organisms eaten by fish flourish in them.

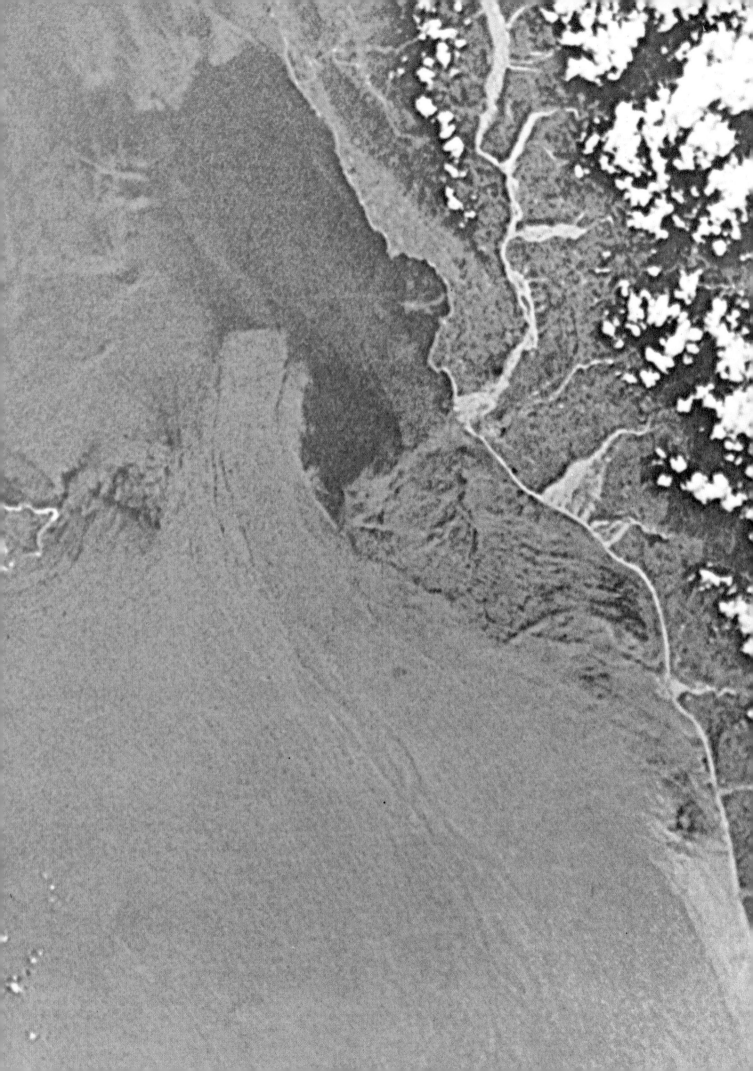

strung in long bundles and immersed in tidal water. The strings, which do not touch the bottom, are sometimes attached to stakes but more generally are attached to rafts. The suspension method has a number of advantages over growth on the bottom. The oysters are protected from predators and from silting, and they feed on the suspended food in the entire column of water rather than being limited to what reaches the bottom. The result is faster growth, rounder shape and superior flavor.

In small areas of Japan's Inland Sea suspension cultures of oysters annually yield 46,000 pounds of shucked meats per acre of cultivated area. This does not mean that one can multiply the total acreage of the Inland Sea by this figure to estimate the potential productivity of the area. Tidal flow allows the anchored oysters to filter much larger volumes of water than surround them at any given time. In addition, inshore waters are generally more productive than those farther from land. The figure does illustrate, however, the production of meat that is

possible with our present farming practices in inshore waters.

Luther Blount of Warren, R.I., has tested oyster suspension cultures in Rhode Island waters over the past several years, using spat set on scallop shells. Blount spaces seven scallop shells well apart on each suspension string. At the end of seven months' growth he harvested one group of suspended oysters from 3,200 square feet of float area. The oysters weighed nearly 40,000 pounds and yielded 2,500 pounds of oyster meat. His experience suggests that the coastal waters of the eastern U.S. might yield more than 16,000 tons of meat per square mile of float per year.

Although the farming of oysters in suspension cultures is a comparatively recent development, the same technique has long been used in Europe to raise mussels. The Bay of Vigo is one of the many Spanish ports where acres of mussel floats are a common sight. French and Italian mussel growers are less inclined to use rafts. Their mussel strings are usually suspended from stakes set in

the estuary bottom.

John H. Ryther and G. C. Matthiessen of the Woods Hole Oceanographic Institution have studied the yields obtained by the mussel farmers of Vigo. The annual harvest produces an average of 240,000 pounds of mussel meat per acre. This is equivalent to 70,000 tons of meat per square mile of float, or better than four times the yield of oysters in suspension cultures in the U.S. and Japan.

The farming of fish is more difficult than the farming of bivalves for at least two reasons. First, the fish, being motile, must be held in ponds. Second, the saltwater species that are most commonly farmed—milkfish and mullet (*Mugil*)—breed only at sea. This means that the fry have to be caught where and when they occur naturally, and in some years the supply is not adequate. Furthermore, unwanted species and predators have to be sorted out by hand, with the inevitable result that some of both are introduced into the ponds along with the desired species.

In spite of such handicaps pond farm-

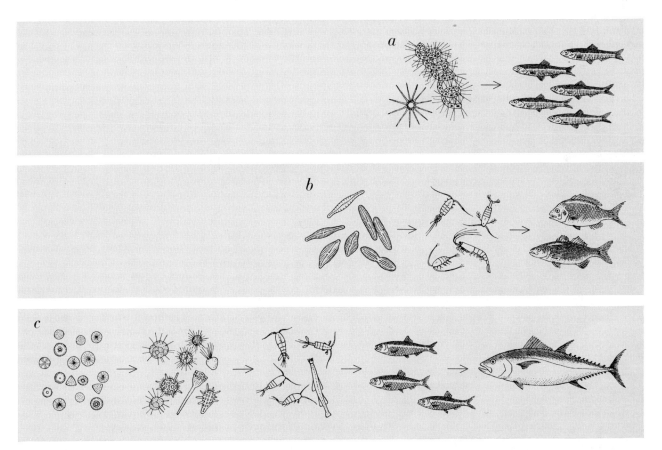

HIGH PRODUCTIVITY of upwelling areas and coastal waters, in contrast to the low productivity of the open sea, is not due only to greater mineral enrichment. In upwelling areas (*a*) the phytoplankton at the bottom of the food chain are usually aggregates of colonial diatoms that are large enough to feed fish of exploitable size. As a result the food chain is very short, with an average of 1.5 steps. The food chain in coastal water (*b*) is longer, averaging 3.5 steps. In the open sea (*c*), where phytoplankton at the bottom of the chain are widely scattered, single-celled diatoms, five steps are needed to produce exploitable fish and the energy transfer at each step is low in efficiency. The length of the chains was calculated by John H. Ryther of the Woods Hole Oceanographic Institution.

OYSTER FARM at Port Stephens in New South Wales in Australia is seen at low tide. The rack-and-stick cultivation system was adopted after depletion of oyster beds in the 1870's.

SUSPENSION CULTURE of young oysters dangles from a float in a Rhode Island oyster pond. Each string supports a series of scallop shells set with oyster spat, a product used commercially for seeding conventional oyster beds. Luther Blount, who is farming the oyster suspension cultures experimentally, has recorded four-year weight gains of 1,000 percent.

ing is remarkably productive. In the Philippine Republic, for example, the annual milkfish harvest is estimated at some 21,000 tons and the productivity of the ponds averages 78 tons per square mile. A comparable estimate for the annual productivity of free-swimming fish in coastal waters, as calculated by Ryther and Matthiessen, falls between six and 17 tons per square mile. In the Philippines, moreover, it is not customary to enrich the pond waters artificially, a process that accelerates the growth of the fishes' plant food. In Taiwan, where milkfish ponds are fertilized, the average annual yield is 520 tons per square mile, and in Indonesia, where sewage is diverted into the ponds in place of commercial fertilizer, the annual yield reaches 1,300 tons per square mile.

Fish farming in Asia is still a long way from reaching its maximum potential. The United Nations Food and Agriculture Organization has calculated that more than 140,000 square miles of land in southern and eastern Asia could be added to the area already devoted to milkfish husbandry. Even if this additional area were no more productive than the ponds of Taiwan, its yield would be more than today's total catch from all the world's oceans. Assuming an adequate supply of milkfish fry, such an increase could be achieved without any technological advance over present methods of pond farming. Even the fry problem may be close to solution. Mullet, a largely herbivorous fish, is now extensively farmed not only in Hawaii and China but also in India and even in Israel. Recently it has proved possible to breed mullet in the laboratory, which brings closer the prospect of mullet hatcheries and a steady supply of mullet fry.

In looking for ways to increase the potential yield of fishponds throughout the world, we are faced with two problems. The first is whether or not we can overcome the sanitary and aesthetic objections to using sewage as a growth stimulant. This is a complex question, but it is worth noting that some practical progress is being made by transferring shellfish from polluted areas to unpolluted ones for a period of "cleaning" before shipment to market.

An equally important question is to what degree commercial fertilizer could increase productivity. Oysters or mussels suspended from rafts in small ponds should provide a simple test organism for such experiments, and they are particularly appropriate because of their high natural yields.

The effect of adding commercial fertilizer to Long Island Sound water has been studied by Victor L. Loosanoff of the U.S. Fish and Wildlife Service. He wanted to produce large amounts of marine plants as food for experiments in rearing oysters and clams. He found marked stimulation of plant growth, but the zooplankton—the marine animals in the water—also grew and ate the plants, thus competing with the shellfish for food. After trying various methods of inhibiting the zooplankton's growth, Loosanoff finally came to the use of pure cultures of the plants, but this would be a very expensive practice on a commercial scale.

The growing of both marine plants and marine animals in a pond could be rewarding, and the zooplankton could be converted from a pest to an asset by adding an organism that feeds on them. Rainbow trout might fill this requirement: they are carnivorous, adapt readily to salt water and are said to grow faster and have a better flavor than when they live in fresh water. In addition they are readily available from hatcheries and

have a good market value. To dispose of the inevitable organic debris sinking to the bottom of the pond one might add clams and a few lobsters, since both are in demand and their young are being reared in hatcheries and could be obtained.

It seems to me of the utmost importance that we follow the principles of ecology in our efforts to develop marine farming, by working with nature to establish balanced, stable communities rather than by supporting large single crops artificially, as we do on land, with what are now becoming recognized as disastrous side effects. Perhaps the single most exciting challenge we have in marine farming is this opportunity to make a new start in the production of food, utilizing the ecological knowledge now available.

If the results of the pond experiments are satisfactory, it is technically feasible to consider applying fertilizer to estuaries or even to the open ocean. The mechanical problem here is that the applied fertilizer sinks to the bottom in estuaries and tends to become absorbed by mud, and in the open ocean it simply sinks below the zone where the marine

plant life grows. A solution for this problem would be to combine the fertilizer with some floating material that would disintegrate and liberate it slowly. The political and legal problems of controlling the harvest of the crop seem more difficult than the technical one of developing floating fertilizer.

Beyond the continental shelf in the open ocean the surface water is normally poor in nutrients and as barren as any desert on land. As irrigation projects have frequently shown, the addition of water makes the desert bloom. Adding nutrients to the ocean has much the same effect. There is, however, a significant difference between the two measures. The availability of fresh water may ultimately limit our agricultural output on land, but the deep ocean holds an immense supply of nutrients that is constantly being renewed.

The concentration of nitrogen and phosphorus compounds in the ocean reaches its maximum value at depths of from 2,000 to 3,000 feet below the surface. That is well below the region penetrated by sunlight, making the nutrients unavailable for plant growth. What then

ARTIFICIAL PONDS are widely used in Asia to raise fish fry, netted at sea, until they reach edible size. The ponds seen here are in Indonesia, where the use of sewage as fertilizer for pond algae brings a harvest that is equal to 1,300 tons of fish per square mile.

is the practical possibility of bringing these nutrients to the surface? If we were able to do it, would the number of fish increase? The answer to the second question can be found in nature. The upwelling of nutrient-rich deep water occurs naturally in some parts of the ocean, and the world's most productive fisheries are found in these areas. One of the best-known of these is the Peru Current on the west coast of South America. Along the shores of Chile and Peru the southeast trade winds blow the surface water away from the land, with the result that it is replaced by deeper water containing the nutrients needed for plant growth. In 1968, 10.5 million tons of fish—mostly anchovies—were harvested in an area 800 miles long and 30 miles wide along this coast. That is a yield of 440 tons per square mile of ocean surface. If, as seems likely, an equal quantity of fish was taken by predators, it means that this area of natural upwelling approaches the productivity of the heavily fertilized Asiatic fishponds. Incidentally, it far surpasses the production of protein by the raising of cattle on pastureland, which Ryther and Matthiessen give as between 1.5 and 80 tons per square mile.

The Peru Current demonstrates that bringing nutrient-rich water to the surface leads to an enormous increase in fish growth. In fact, the areas of natural upwelling, which comprise only .1 percent of the oceans' surface, supply almost half of the total fish catch, whereas the open oceans, where upwelling does not occur, account for 90 percent of the surface and yield only about 1 percent of the catch. In other words, natural upwelling increases the productivity of the open ocean almost 50,000-fold in terms of fish actually landed.

It would obviously be worthwhile to stimulate upwelling artificially, not only because of the probability of high fish yields but also because the stable ecological communities that inhabit the natural areas of ocean upwelling are models of efficient food production for man, with none of the drawbacks—such as herbicides, pesticides, pollution and excessive human intervention—that such highly productive systems usually entail ashore.

To achieve artificial upwelling we need first some kind of container. Deep water is cold and therefore dense, and without a container it would sink again,

taking its nutrients with it. We also need to surround the fish we hope to grow, not only to protect them from predators and to simplify harvesting, but even more important to keep them from being caught by fishermen who have not paid for the upwelling. We also need a land area where the pumping and processing activities can be located and of course a supply of deep water nearby.

There are hundreds of coral atolls in the Pacific and the Indian Ocean that meet these specifications. Rings of coral reef surrounding shallow lagoons, atolls vary in area from less than a square mile to more than 800 square miles. Low islands are often found on the encircling reef, and since atolls are the remains of sunken volcanic peaks topped with coral they are steepsided, with deep water generally less than a mile from the reef itself. The trade winds blow over many of these islands and carry energy enough to bring deep water to the surface; that, after all, is the mechanism in many natural upwelling areas. There is even a built-in stirring system: the trade winds produce a downwind current on the lagoon surface that is matched by a return current near the bottom of the lagoon.

FRESHWATER FISH, the herbivorous carp, has been reared in ponds around the world for more than two millenniums. The carp in the photograph are from a pond in Burma. The 1968 crop of carp and carplike fishes in neighboring China totaled 1.5 million tons.

Given an atoll lagoon filled with nutrient-rich water from the deeps, what kinds of marine plants and animals should be grown in it? Perhaps the simplest procedure would be to leave the passages between lagoon and ocean open and allow nature to take its course in introducing new species. A balanced and stable community such as the one found in the Peru Current might establish itself. If it did not, colonies of plants and fishes from the Peru Current could be introduced in the hope that they could establish themselves in the new location.

Perhaps a crop of suitably large zooplankton such as krill—the shrimplike animals that are the principal food of the baleen whales in the Antarctic—could be raised in a fertilized lagoon. In that case another particularly interesting experiment might be possible. This would be to determine whether or not baleen whales, particularly the now almost vanished blue whales, could adapt to such a restricted environment. A school of blue whales raised in captivity could be regularly culled for a significant yield in meat and edible oil, and at the same time its existence would protect the species from what now seems to be certain extinction.

We know that blue whales migrate into the tropical Pacific to bear their young. The migrants could be followed by attaching radio transmitters to them in Antarctica. Techniques for capturing, transporting and keeping smaller whales have already been worked out. Humpback whales, which are about half the length of blue whales, have been captured at sea by investigators at the Sea Life Park in Hawaii by dropping a net over the whale's head. There is a real possibility that whale farms could be started by capturing pregnant female blue whales and confining them in fertilized atolls.

Artificial upwelling, on a small scale at least, has already been achieved by Oswald A. Roels, Robert D. Gerard and J. Lamar Worzel of the Lamont-Doherty Geological Observatory of Columbia University. They have installed on St. Croix in the Virgin Islands a 3½-inch plastic pipe that extends nearly a mile into the Caribbean, enabling them to pump deep water with a temperature of 40 degrees Fahrenheit into small ponds on shore. They find that selected plant life from the seawater off St. Croix grows 27 times faster in water from the pipe than in water from the surface. They are now exploring the possibilities of feeding a variety of marine herbivores on these artificial blooms.

The Lamont-Doherty group has also

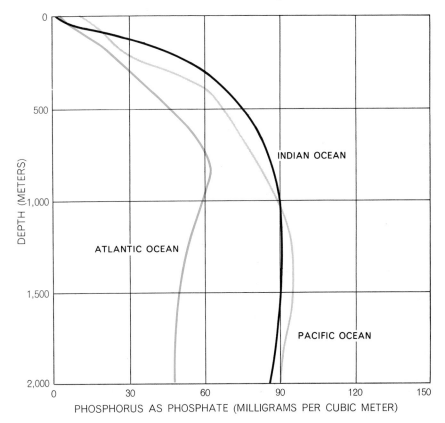

PHOSPHATE IN SEAWATER, present only in small amounts at the surface, increases in the deeper zones and reaches a near-maximum concentration of 90 milligrams per cubic meter in the Pacific and Indian Ocean and about 60 milligrams in the Atlantic at a depth of 1,000 meters. Data are from a study by Lela M. Jeffrey of the University of Nottingham.

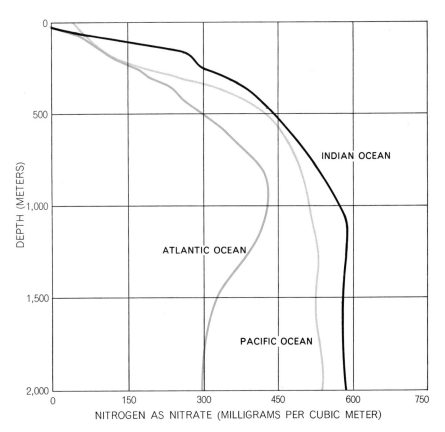

NITRATE IN SEAWATER is also scarce at the surface but approaches its maximum concentration at 1,000 meters. Again the Atlantic has the least; data are from Miss Jeffrey.

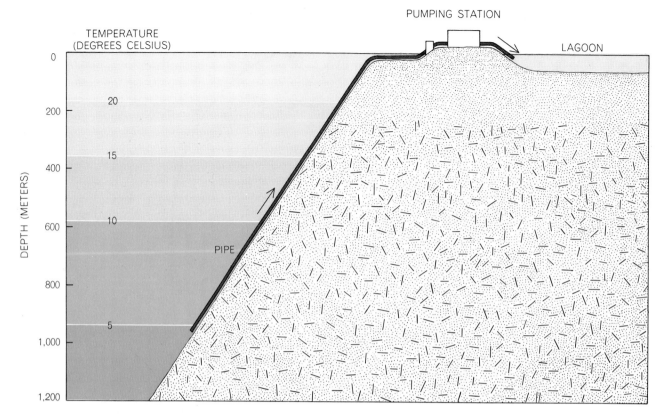

ARTIFICIAL UPWELLING of deep water might be contrived in an atoll setting, as this diagram suggests. The atoll's steep drop to seaward means that the wanted water would be pumped the least possible distance. The central lagoon would provide a catchment basin for the pumped water, retaining its nutrients at or near the surface. The difference in temperature between the surface and the deep water might be used to generate more power than is required for the pumps. A pilot version of this experiment is being conducted by workers in the Virgin Islands, who are pumping deep-sea water ashore and accelerating the growth of phytoplankton in ponds.

pointed out in a recent report that the energy represented by the nearly 40-degree difference in temperature between deep water and surface water can be used for air conditioning, the generation of power and the condensation of fresh water from the atmosphere. (The last idea emerged after observation of the condensation of atmospheric moisture on a Martini glass in a St. Croix bar.) In addition the low temperature of deep water offers the possibility of using the water to cool power plants, including nuclear reactors, without causing thermal pollution.

These fringe benefits, particularly the possibility of producing more than enough power to pump the deep water to the surface, may at first seem to suggest the dream of getting something for nothing. No physical laws would be violated, however; the water-temperature gradient is simply another product of solar-energy input, just as the energy fixed by photosynthetic plants is. From the standpoint of practical economics ar-

tificial upwelling may be too expensive to be feasible exclusively for fish farming at the present time. The system seems entirely practical, however, if its cost can be shared with some additional service such as air conditioning or the cooling of power plants.

A less elegant but much cheaper means of enriching the lagoons of atolls would be the addition of commercial fertilizer. To bring an atoll one square mile in area and 30 feet deep to a level of phosphate concentration equal to the level of nutrient-rich deep water would require only about 10 tons of fertilizer and might cost less than $500. In principle, if the lagoon were entirely enclosed, fertilizer would be removed only as the end product of the farming operation. In actual practice, of course, there would be other losses. Even assuming that one recovered only 10 percent of the fertilizer in marketable fish, however, the cost would be only half a cent per pound of fish produced. From the economic point of view this would seem to be a highly

practical experiment.

Advances in technology frequently generate further threats to the quality of our already overburdened environment. It is encouraging to realize that the use of deep water from the sea both to stimulate food production and to obtain power or fresh water is a pollution-free process. The deep water returns to the sea at the same temperature and with about the same nutrient concentration as the waters that receive it, without having an adverse effect on either the atmosphere or the ocean. The same is true of the use of commercial fertilizer in atoll lagoons, since the fertilizer is almost wholly consumed in the process. Yet at the same time animal-protein production could be stimulated to a level not yet approached by conventional agriculture. Large areas of our planet could be developed into highly productive marine farms. The time seems ripe for applying the fundamental knowledge we already possess to the practical problems of developing them.

The Disposal Of Waste in the Ocean

by Willard Bascom
August 1974

*Contrary to some widely held views, the ocean is the
plausible place for man to dispose of some of his
wastes. If the process is thoughtfully controlled, it will
do no damage to marine life*

No one would dispute the wisdom of protecting the sea and its life against harm from man's wastes. An argument can be made, however, that some of the laws the U.S. and the coastal states have adopted in recent years to regulate the wastes that can be put into the oceans are based on inadequate knowledge of the sea. It is possible that a great effort will be made to comply with laws that will do little to make the ocean cleaner.

This discussion of waste disposal will be limited to disposal in the ocean; it will not take up disposal in lakes, rivers, estuaries, harbors and landlocked bays. Indeed, part of the problem is that insufficient distinction has been drawn between the ocean and the other bodies of water, whose chemistry, circulation, biota and utilization differ from those of the ocean in many ways. It is not sensible to try to write one set of water-quality specifications that will cover all bodies of water. My concern here is only with the quality of ocean water and marine life along the U.S. Atlantic, Gulf and Pacific coasts. The scientific findings in those areas apply, however, to nearly any other coastal waters that are exposed to ocean waves and currents.

Some of the changes that human activities have wrought in the ocean environment are already irreversible. For example, rivers have been dammed, so that they release much smaller quantities of fresh water and sediment. Ports have been built at the mouth of estuaries, changing patterns of flow and altering habitats. On the other hand, certain abuses of the ocean have already been stopped almost completely by the U.S. Nuclear tests are no longer conducted in the atmosphere, so that radioactive material is no longer distributed over the land and the sea; the massive dumping of DDT has been halted, and the reckless development of coastal lands has been restrained by laws calling for detailed consideration of the impact on the environment.

Between these extremes is a broad realm of uncertainty. Exactly how clean should the ocean be? How unchanged should man try to keep an environment that nature is changing anyway? The problem is to decide what is in the best interest of the community and to achieve the objective at some acceptable cost. At the same time it is necessary to guard against the danger that excessive demands made in the name of preserving ecosystems will lead to action that is both useless and expensive.

Waste disposal automatically suggests pollution, which is a highly charged word meaning different things to different people. A definition is needed for evaluating accidental and deliberate inputs into the ocean. Athelstan F. Spilhaus of the National Oceanic and Atmospheric Administration, who has written extensively on pollution, defines it as "anything animate or inanimate that by its excess reduces the quality of living." The key word is excess, because most of the substances that are called pollutants are already in the ocean in vast quantities: sediments, salts, dissolved metals and all kinds of organic material. The ocean can tolerate more of them; the question is how much more it can tolerate without damage.

One approach to the question was suggested by the National Water Commission in its report of June, 1973, to the President and Congress: "Water is polluted if it is not of sufficiently high quality to be suitable for the highest uses people wish to make of it at present or in the future." What are "the highest uses" that can be foreseen for ocean waters, particularly those near the shore? They are probably water-contact sports, the production of seafood and the preservation of marine life.

Water-contact sports are occasionally

MONTEREY BAY in California appears in a deliberately overexposed aerial photograph on the following page. The overexposure through a filter is a technique that shows details of turbidity in the water caused by mud, organisms or waste. At bottom right a harbor projects into the bay. The light spot of water along the shore above and slightly to the left of the harbor is a sewage outfall. The dark brown spots are beds of kelp, and the reddish purple is a "red tide" consisting of large numbers of the marine organism *Gonyaulax*.

inhibited by pollutants on the seacoasts of the U.S. Where such conditions exist they should be corrected at once. Even where coastal waters are clean the community must be alert to keep them so.

To maintain the ocean waters at an acceptable level of quality it is necessary to consider the main inputs of possible pollutants resulting from human activity. One of them is fecal waste (75 grams dry weight of solids per person per day), which after various degrees of treatment ends up in the ocean as "municipal effluent." Wastes also flow from a host of industrial activities. They are usually processed for the removal of the constituents that are most likely to be harmful, and the remaining effluent is discharged through pipes into the ocean. Dumping from barges into deep water offshore is a means of disposing of dredged materials, sewage sludge and chemical wastes. Thermal wastes include the warmed water from coastal power plants and cooled water from terminals where ships carrying liquid natural gas are berthed. In addition ships heave trash and garbage overboard and pump oily waste from their ballast tanks and bilges.

Such are the intentional discharges, but pollutants reach the ocean in other ways. Aerial fallout brings minute globules of pesticide sprayed on crops, particles of soot from chimneys and the residue of the exhaust of automobiles and airplanes. Painted boat bottoms exude small amounts of toxicants intended to discourage the growth of algae and barnacles. Forest fires put huge amounts of carbon and metallic oxides into the air and thence into the sea. Oil spills from ship collisions and blowouts during underwater drilling operations add an entire class of compounds.

Moreover, natural processes contribute things to the sea that would be called pollutants if man put them there. Streams add fresh water, which is damaging to marine organisms such as coral, and they also bring pollutants washed by rain from trees and land. Volcanic eruptions add large quantities of heavy metals, heat and new rock. Oil has seeped from the bottom since long before man arrived.

Finally, the ocean is neither "pure" nor the same everywhere. It already contains vast amounts of nearly everything, including a substantial burden of metals at low concentration and oxygen at relatively high concentration, plus all kinds of nutrients and chemicals. It has hot and cold layers, well stratified by the thermocline (the boundary between the warm, oxygen-rich upper layer and the

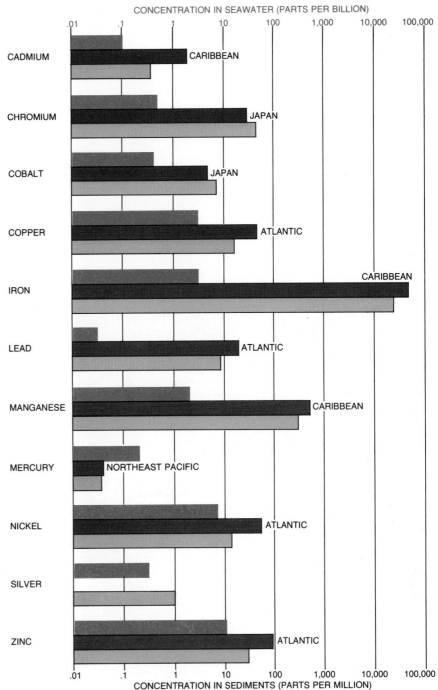

TRACE METALS in seawater (*color*) and top 10 centimeters of sediment (*gray*) are charted. Seawater figures are a worldwide average. The darker gray bars show concentrations at several sampling sites and the lighter bars the average from five sites along California coast.

cold, oxygen-poor depths). Waves and currents keep the water constantly in motion. It is against this complex background that man must measure the effects of his own discharges.

Even if there were no people living on seacoasts, it would be impossible to predict accurately the kind and quality of marine life because of the natural variability of the ocean. The biota shifts constantly because the temperature and the currents change. Great "blooms" of plankton develop rapidly when conditions have become exactly right and then die off in a few days, depleting the oxygen in the water on both occasions. Within a single year the population of such organisms as salps, copepods and euphausids can change by a factor of 10. When the waters off California become warmer as the current structure shifts, red "crabs" (which look more like small lobsters and are of the genus *Pleuroncodes*) float by in fantastic numbers, fol-

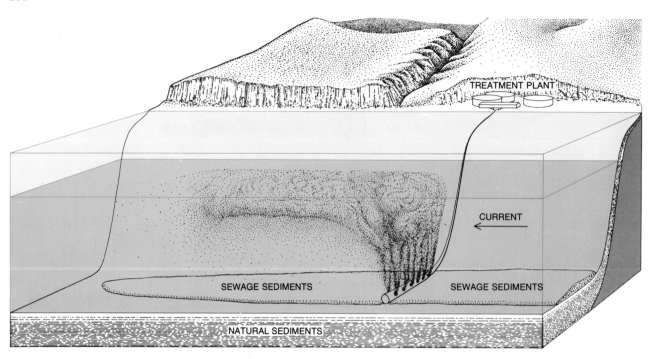

OCEANIC DISPOSAL of waste from a sewage-treatment plant is portrayed. The system is for a plant with a capacity of about 100 million gallons per day. Effluent from the plant flows a distance of from two to five miles through an outfall pipe that is from six to 12 feet in diameter. For about a quarter of a mile at its end the pipe has dozens of six-inch discharge ports. The mostly liquid material it discharges rises to the thermocline, which is the boundary between deep, cool water and the warmer surface layer. Prevailing current mixes material and moves it to one side or the other, depending on wind and tide. Some solid particles settle on the bottom.

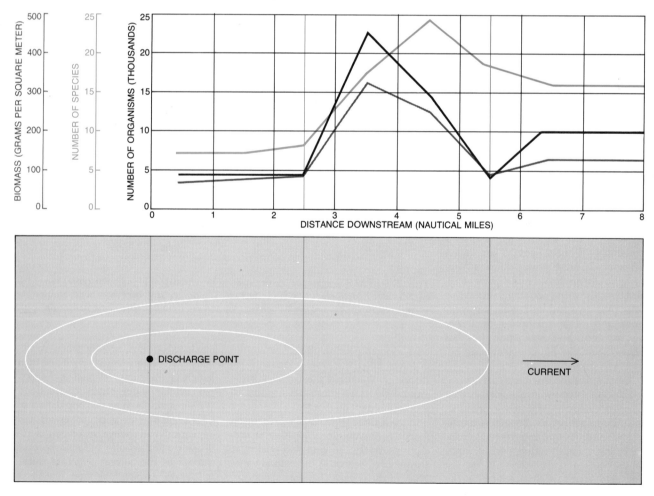

EFFECT OF OUTFALL on a community of polychaete worms is depicted. For about three miles downstream from the outfall the worm population is reduced. For the next three miles it is above normal. Thereafter it is about the same as in uninfluenced seabed.

lowed by large populations of bonito and swordfish. They came in 1973 as they did in 1958 and 1963, but the water soon turned cold again and the fish departed, leaving windrows of dead *Pleuroncodes* along the beaches.

The investigator's problem is to learn enough about the major natural changes so that he can tell whether or not human activities have any effect, either positive or negative. It is a signal-to-noise problem; here the changes one is trying to detect are often only a tenth of the natural biological and oceanic background variations. Both types of variation are hard to quantify.

In the case of the sardine, however, a record of the natural changes has been preserved below the floor of the Santa Barbara Basin off the coast of southern California. The bottom of the basin is anaerobic, that is, lacking in oxygen and so supporting little life. The particles that sift down to form sediments are undisturbed by burrowing creatures and therefore remain exactly as they land, in thin strata, layer on layer, one per year. The years can be counted backward, and the count can be confirmed with the lead-210 dating technique.

Some years ago John D. Isaacs of the Scripps Institution of Oceanography, who is also director of the California Marine Life Program, discovered in work with Andrew Soutar that each layer contains identifiable fish scales. Each layer showed a more or less constant number of anchovy and hake scales, but the sardine scales were present erratically, indicating major changes in the population. When the sardines disappeared about 1950, human activities were blamed. The geologic record clearly shows, however, that the sardines had come and gone many times before man arrived. Someday they will return.

It is obvious that some of man's wastes can be damaging to sea life; indeed, products such as DDT, chlorine and ship-bottom paint have been specifically designed to protect man against insects, bacteria and barnacles. Ionic solutions of certain metals are also known to be toxic at some level, as are numerous other substances. The problem is to determine what level is harmful, remembering that some of the substances are actually required for life processes. For example, copper is beneficial or essential for a number of organisms, including crabs, mollusks and oyster larvae. Other marine animals seem to require nickel, cobalt, vanadium and zinc.

Oceanographers would like to be able to demonstrate cause and effect in the ocean, that is, to show that some specific level of a metal does not harm marine life. Proving the absence of damage, however, including long-term and genetic effects, is difficult. Only on fairly rare occasions has it been possible to directly link a specific oceanic pollutant with biological damage. Examples include the finding by Robert Risebrough of the University of California at Berkeley that the decline of the brown pelican off California was attributable to DDT, which inhibits the metabolism of calcium and so makes the shells of the eggs so thin that the mother pelican breaks her own eggs by sitting on them. After patient scientific detective work the source was found to be a single chemical plant in the Los Angeles area. As a result of the work the plant was required to stop discharging DDT wastes into the ocean, and the brown pelican is now returning to California.

From what I have said so far it can be seen that the question of what is a pollutant or what amount of a substance represents pollution is not always easy to answer. Let me now try to put the main kinds of waste in proper perspective.

Municipal sewage containing human fecal material is the type of waste one usually thinks of first. It is certainly a natural substance; indeed, as "night soil" it has long been in demand as a fertilizer in many countries. Since it is not appreciably different from the fecal material discharged by marine animals, is there any reason to think it will be damaging to the ocean, even without treatment? Isaacs has pointed out that the six million metric tons of anchovies off southern California produce as much fecal material as 90 million people, that is, 10 times as much as the population of Los Angeles, and the anchovies of course comprise only one of hundreds of species of marine life.

Two aspects of municipal sewage do require attention. One of them is disease microorganisms. Human waste contains vast numbers of coliform bacteria; they are not themselves harmful, and they die rapidly in seawater (90 percent of them in the first two hours), but they are routinely sampled along public beaches because they indicate the level of disease microorganisms. When there are no endemic diseases in the city discharging the waste (the normal condition in the U.S.), there will be none in the water. It should be noted, however, that the assumption that disease microorganisms die off at the same rate as coliform bacteria is being questioned. It is necessary to guard against the possibility that such organisms will survive in bottom muds

long enough to be stirred up by a major storm.

The usual way of reducing the bacterial count is to add chlorine to waste water that is about to be discharged. This approach seems reasonable, since chlorine is commonly added to drinking water and swimming pools to kill bacteria and algae. The trick is to add just the right amount, so that the chlorine exactly neutralizes the bacteria and no excess of either enters the ocean.

The other problem with sewage is one of aesthetics. People do not like to look at discolored water or oily films. A greater effort to reduce effluent "floatables" (tiny particles of plastic, wood, wax and grease) will help to reduce such effects. It will also reduce the number of bacteria reaching the shore, since many of them are attached to the particles.

Petroleum products are perhaps the most controversial marine pollutants. They are seen as small, tarlike lumps far out to sea and on beaches, as great slicks and as brown froth. From two to five million metric tons of oil enter the ocean annually. At least half of it is from land-based sources such as petroleum-refinery wastes and flushings from service stations. Significant quantities of oil enter the marine environment from airborne hydrocarbons. A considerable amount of oil must enter the ocean as natural seepage from the bottom, but it is obviously difficult to estimate how much.

Oil pollution from ships is the most serious problem. Oceangoing vessels shed oil in three ways: by accidents such as collisions; during loading and unloading, and by intentional discharge, which includes the pumping of bilges, the discharge of ballast by tankers and the cleaning of oil tanks by tankers. The ballast component is the worst.

After a tanker unloads its cargo of oil it takes on seawater (about 40 percent of the full load of oil) so that it will not ride too high in the water and be unmanageable. Any oil that remains in the tanks mixes with the water and is discharged with it when the ballast is pumped out in preparation for reloading the vessel with oil. The discharge of oil can be reduced in two ways. One is to wash the tank with water and stow the water aboard in a "slop tank," where the oil slowly separates from it. Then the water is discharged and the next load of oil is put on top of the oil that remains in the slop tank. This practice, which is described as being 80 percent effective, is followed in tankers carrying about 80 percent of the oil now transported at sea. The other stratagem is to build segre-

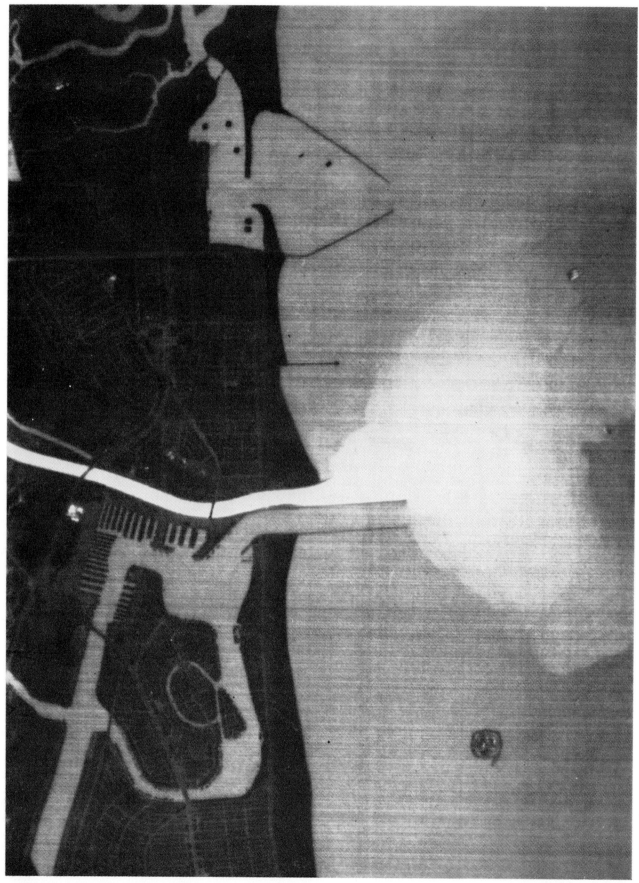

THERMAL EFFLUENT from generating plants of the Southern California Edison Co. and the Los Angeles Department of Water and Power appears in a thermograph of the San Gabriel River, which is the white strip at center, and San Pedro Bay. The intake temperature of cooling water from the bay was 18.9 degrees Celsius. On discharge into the river it was 26.7 degrees C. At the point where the river enters the bay the water temperature was 24.4 degrees C. The two plants jointly generate 900 megawatts of power.

gated ballast spaces into the double bottoms of new tankers, which reduces the discharge by 95 percent.

A system of international controls could virtually eliminate such discharges. There is an extra incentive for international controls because wherever oil is discharged, and by whatever ship, there is no telling to what shore it will be carried by winds, waves and currents. Substantial progress toward this kind of agreement has been made recently.

Ships are also responsible for most of the littering of the ocean and its shores. Waste consisting of paper, plastics, wood, metal, glass and garbage is customarily thrown overboard. The heavier material sinks quickly, littering the bottom; paper products disintegrate or become waterlogged and sink slowly, and the foods are soon consumed by marine scavengers. The wood, sealed containers and light plastics float ashore.

The estimated yearly litter from ships is about three million metric tons, much of which seems to come from the fishing fleets. The litter that Americans see and are annoyed by comes mostly from the land by way of streams or is thrown into harbors or tossed overboard from pleasure craft. Littering is an aesthetic problem rather than an ecological one, but it certainly reduces the quality of living. It can be curbed by the force of public opinion.

Dumping is a word with a specific meaning; it should not be confused with littering or with discharges from pipes. Dumping means carrying waste out to sea and discharging it at a designated site. Barges carrying solids simply open bottom doors and drop their cargo. Barges carrying liquids generally pump the material out through a submerged pipe into the turbulent wake of the vessel. Still other barges dump wastes enclosed in steel drums or other containers.

Much of the material is dredge spoil sucked up from harbor bottoms by hopper dredges to deepen ship channels. Some 28 million tons of this material were dumped into the Atlantic in 1968. Next in quantity in the New York area is relatively clean material removed from excavations for buildings; then comes sewage sludge, and finally industrial waste such as acids and other chemicals.

The amount of sludge dumped annually into New York Bight is about 4.6 million tons. Much of it is sewage sludge, which is a slurry of solid waste formed by sedimentation in primary sewage treatment or by secondary treatment in the activated-sludge process. For ocean discharge the material is thickened by settling or centrifuging to from 3 to 8 percent solids. Much of the solid material is silt, but complex organic materials and heavy metals are also present.

In some parts of the country the sludge is not dumped but is discharged into the ocean through special pipes. In others it is buried in landfill or spread as fertilizer, although the metals it contains may cause problems later. A broad spectrum of industrial effluents (solvents from pharmaceutical production, waste acid from the titanium-pigment industry, caustic solutions from oil refineries, metallic sodium and calcium, filter cake, salts and chlorinated hydrocarbons) are dumped intermittently at certain sites under Government license.

What damage is done to marine organisms by materials of this kind? The turbidity created by dumping is usually dispersed within a day. Dumped dredge spoil buries bottom-dwelling animals under a thin blanket of sediment, but many of them dig out and the others are replaced by recolonization in about a year. Sewage sludge is high in heavy metals, which may be toxic, particularly when they combine with organic materials to create a reducing (oxygen-poor) environment in which few animals can live. Sludge can also have a high bacterial count. It is clear that much industrial waste could be harmful to marine life and should not be dumped into the sea.

The entire matter of dumping needs more study. With reliable data it will be possible to retain the option of disposal at sea for some materials, such as dredge spoil, and to reject it for others, such as chemicals. Deep-water sites could be set aside for dumping on the same logic that applies to city dumps, namely that it is a suitable use for space of low value where few animals could be harmed.

Thermal waste is discharged into the sea by power plants because the sea is a convenient source of cooling water. The temperature of the water on discharge is typically 10 degrees Celsius higher than it was on intake. The difference is within the range of natural temperature variations and so is not harmful to most adult marine animals. The eggs, larvae and young animals that live in coastal waters, however, are sucked through the power plant with the cooling water. They are subjected to a sudden rise in temperature and decrease in pressure that is likely to be fatal. For this reason and others it would seem logical to put new power plants offshore. There they could draw deeper and cooler water from a level that is not rich in living organisms. For a nuclear plant the hazards of a nuclear accident would be reduced; for an oil- or coal-fired plant fuel could be delivered directly by ship, and the shoreline could be reserved for nonindustrial purposes.

Some industries discharge substantial quantities of heavy metals and complex organic compounds into municipal waste-water systems whose effluent reaches the ocean. Certain of the metals (mercury, chromium, lead, zinc, cadmium, copper, nickel and silver) are notably toxic and so are subject to stringent regulation. The most dangerous substances, however, are synthetic organic compounds such as DDT and polychlorinated biphenyls. The discharge of these substances as well as the heavy metals must be prevented. The best way to do so is by "source control," meaning the prevention of discharge into the sewer system. Each plant must be held responsible for removing and disposing of its own pollutants.

Other waste substances that generate controversy are those with nutrient value. Since they are decomposed by bacteria, oxygen is required. This biological oxygen demand is commonly measured in units that express how much oxygen (in milligrams per liter) will be required in a five-day period.

There is good reason to restrict the amount of nutrient material that is discharged into lakes and rivers, where oxygen is limited and a reducing environment can be created. The ocean is another matter. It is an essentially unlimited reservoir of dissolved oxygen, which is kept in motion by currents and is constantly being replenished by natural mechanisms.

It is nonetheless possible to overwhelm a local area of the ocean with a huge discharge of nutrient material that may form a deposit on the sea floor if the local conditions are not carefully considered. The materials must be presented to the ocean in the right places and at reasonable rates. Among the ways of achieving that objective are the use of discharge pipes that lead well offshore and have many small diffuser ports and, if the volume of discharge is exceptionally large, the distribution of the effluent through several widely dispersed pipes.

Problems caused by the addition of nitrogen and phosphate to inland waters, which they overfertilize, do not apply to the ocean. There they could be helpful by producing the equivalent of upwelling, the natural process that brings nutrients from deep water to the surface waters where most marine organisms are found. As Isaacs has pointed out, "the sea is *starved* for the basic plant nutrients, and it is a mystery to me why

238

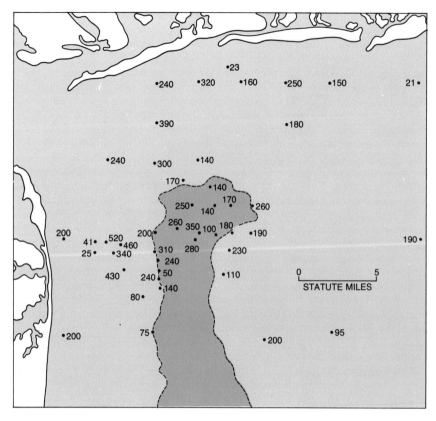

EFFECT OF DUMPING on sediments of the New York Bight is indicated by the concentration of chromium in parts per million in and around the dumping area. A bight is an open bay formed by a bend in a coast. Material transported to the bight to be dumped includes sewage sludge and rubble. The broken contour line represents a depth of 30 meters.

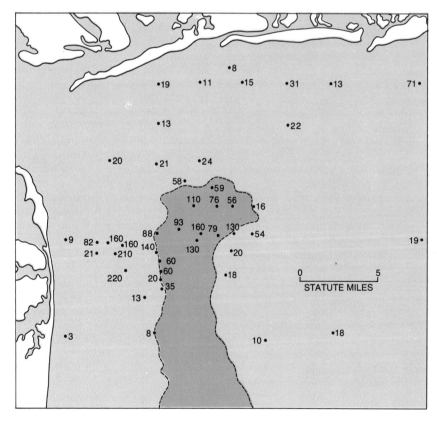

COPPER CONCENTRATION in sediments of the dumping area of the New York Bight is shown in parts per million. Data on chromium and copper in the New York Bight are based on work by Grant Gross of the oceanography section of the National Science Foundation.

we should be concerned with their thoughtful introduction into coastal seas in any quantity that man can generate in the foreseeable future."

Once possible pollutants reach the ocean it is necessary to keep track of where they go, the extent to which they are altered or diluted and what animals they affect. In order to obtain this information many of the techniques of oceanography are brought into play. Currents are measured above and below the thermocline; other instruments measure the temperature, salinity and dissolved oxygen. Water is sampled at various depths, and so are bottom sediments. It is also useful to directly monitor any changes in plant and animal communities with divers or television cameras.

A good indicator of change is the response of the polychaete worms in the bottom mud. Close to an effluent-discharge point the number of species may be as low as from four to 10 per sample and the total weight of worms as low as 50 grams per square meter. A short distance away the number of species may be 40 or more and the total weight 700 grams per square meter. At greater distances the figures drop off to normal: about 25 species and 300 grams per square meter. This local enrichment shows that worms thrive at some optimum level of organic material. Laboratory tests by Donald Reish and Jack Word of California State University at Long Beach show that worms have a similar optimum for toxic metals such as zinc and copper.

Man must do something with his wastes, and the ocean is a logical place for some of them. No single solution will be sensible for all kinds of waste or all locations, but the following suggestions may help to protect both the land and the sea in the long run. (1) Clearly define what is ocean, separating it from inland freshwaters and from harbors and shallow bays, and make laws that are appropriate for each environment. (2) Avoid the assumption that anything added to the ocean is necessarily harmful and consider instead what substances might cause damage and eliminate excesses of them. (3) Rigorously prohibit the disposal in the ocean of all man-made radioactive materials, halogenated hydrocarbons (such as DDT and polychlorinated biphenyls) and other synthetic organic materials that are toxic and against which marine organisms have no natural defenses. (4) Set standards based on water quality (after reasonable mixing) that are compatible with what is known about the threshold of

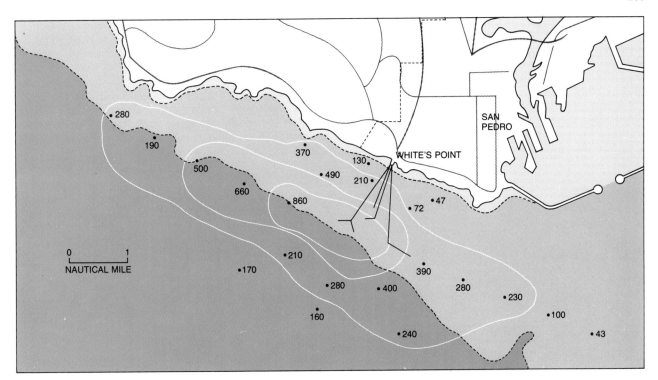

CONCENTRATION OF CHROMIUM in the upper sediments adjacent to a major industrial outfall off San Pedro, Calif., is charted in parts per million. Four outfall lines enter the sea from White's Point. The smallest area surrounded by an elliptical contour line is the area where the concentration is 800 parts per million or more, and the larger contoured areas have concentrations above 500 and 200 parts per million respectively. A depth of three fathoms is shown by the broken line close to shore. Farther out is a 50-fathom line. Other heavy metals discharged into the ocean make similar patterns based on current structure and the slope of the bottom.

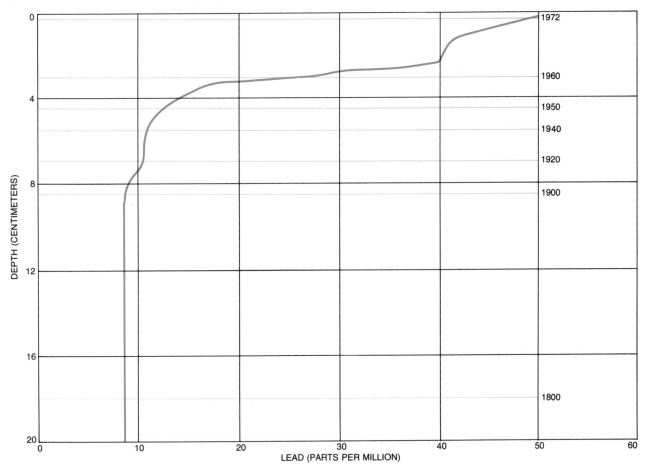

CONCENTRATION OF LEAD in the sediments of the San Pedro Basin shows an increase in recent years because of airborne lead that originates mainly from automobile emissions. The figures are based on samples of the sediments obtained by means of coring.

damage to marine life, providing a safety factor of at least 10. (5) Work to obtain international cooperation in prohibiting ships from disposing of litter or oil and from pumping bilges. (6) Set aside ocean areas of deep water and slow current where certain materials can be dumped with minimal damage. (7) Require each discharger to make studies to demonstrate how his specific effluent will influence the adjacent ocean. (8) Support additional research on the effect of pollutants on the ocean and its life. (9) Anticipate pollutants that may become serious as technology produces new chemical compounds in greater quantities.

A more rational basis is needed for making decisions about how to treat wastes and where to put them. No oceanographer wants damaging waste in the ocean where he works or on land where he lives. Since the waste must go to one place or the other, however, one would prefer the choice to be based on a knowledge of all the factors. Unemotional consideration of which materials can be introduced into the sea without serious damage to marine life will result in both an unpolluted ocean and a large saving of national resources.

SEWAGE-TREATMENT PLANT serves the City of Los Angeles, discharging into the Pacific Ocean about 235 million gallons per day of primary treated effluent and 100 million gallons per day of secondary effluent. The discharge pipe is 12 feet in diameter and nearly five miles long. At the discharge end it is in 197 feet of water. The plant separately discharges sludge, consisting of about 1 percent solids, through a seven-mile pipe to a depth of more than 300 feet. The sludge is discharged at the brink of a marine canyon.

MARINE ORGANISMS grow on the outfall pipe from the Los Angeles sewage-treatment plant. At left are anemones about three feet high and at right is a gorgonian. The location is near the discharge point of the outfall pipe at a depth of approximately 200 feet.

A BIBLIOGRAPHIC POSTSCRIPT

Even in a Reader as large as this, selecting the articles has called for hard decisions. From some two hundred articles initially considered, I have settled on a mere two dozen. To guide your further reading, I have listed here some of the *Scientific American* articles I have most reluctantly left out. Even this bibliography is selective. For example, it slights numerous articles in *Scientific American* during the 1960's that trace the transformation of perspective that plate tectonics brought to historical geology and paleoecology, although it does include those that have a distinctly marine cast. But I would direct the interested reader to the following articles next. Offprint numbers (where applicable) are given in parentheses.

Most pertinent are the many articles that were excluded from this Reader for, quite frankly, arbitrary and indefensible reasons, and by decisions that were sometimes reversed repeatedly as the balance of the volume was gradually struck. Go read these articles; they are part of this Reader even if they happen to remain outside its covers.

THE BIOSPHERE, by G. Evelyn Hutchinson, Sept. 1970 (and other articles in this issue of *Scientific American*) (#1188)

THE VOYAGE OF THE "CHALLENGER," by Herbert S. Bailey, Jr., May 1953

THE OCEANIC LIFE OF THE ANTARCTIC, by Robert Cushman Murphy, Sept. 1962

POISONOUS TIDES, by S. H. Hutner and John J. A. McLaughlin, Aug. 1958

MICROBIAL LIFE OF THE DEEP SEA, by Holger W. Jannasch and Carl O. Wirsen, June 1977 (#926)

THE PORTUGUESE MAN-OF-WAR, by Charles E. Lane, March 1960

OYSTERS, by Pieter Korringa, Nov. 1953

THE TEREDO, by Charles E. Lane, Feb. 1961

SALPA, by N. J. Berrill, Jan. 1961

THE HAGFISH, by David Jensen, Feb. 1966 (#1035)

SHARKS V. MEN, by George A. Llano, June 1957

The BEHAVIOR OF SHARKS, by Perry W. Gilbert, July 1962 (#127)

THE ICE FISH, by Johan T. Ruud, Nov. 1965

HOW FISHES SWIM, by Sir James Gray, Aug. 1957 (#1113)

REFLECTORS IN FISHES, by Eric Denton, Jan. 1971 (#1209)

"THE WONDERFUL NET," by P. F. Scholander, April 1957

FISHES WITH WARM BODIES, by Francis G. Carey, Feb. 1973 (#1266)

THE CHEMICAL LANGUAGES OF FISHES, by John H. Todd, May 1971 (#1222)

PENGUINS, by William J. L. Sladen, Dec. 1957

THE NAVIGATION OF PENGUINS, by John T. Emlen and Richard L. Penney, Oct. 1966

THE GREAT ALBATROSSES, by W. L. N. Tickell, Nov. 1970 (#1204)

SALT GLANDS, by Knut Schmidt-Nielsen, Jan. 1959

THE EVOLUTION OF BEHAVIOR IN GULLS, by N. Tinbergen, Dec. 1960 (#456)

THE NAVIGATION OF THE GREEN TURTLE, by Archie Carr, May 1965 (#1010)

THE WEDDELL SEAL, by Gerald L. Kooyman, Aug. 1969 (#1156)

THE BLUE WHALE, by Johan T. Ruud, Dec. 1956

THE LIFE OF AN ESTUARY, by Robert M. Ingle, May 1954

THE DIVING WOMEN OF KOREA AND JAPAN, by Suk Ki Hong and Hermann Rahn, May 1967 (#1072)

WHALES, PLANKTON AND MAN, by Willis E. Pequegnat, Jan. 1958 (#853)

THE DIMENSIONS OF HUMAN HUNGER, by Jean Mayer, Sept. 1976 (and other articles in this issue of *Scientific American*)

PELAGIC TAR, by James N. Butler, June

Many articles from *Scientific American* trace the advances of the marine sciences since World War II. Older ones become dated by virtue of changes in the very conditions they describe (for example, food resources or whale stocks) or the methods available to study the subject and consequently our knowledge of it (for example, the deep sea). However, the following articles are among those that will reward reading, albeit with this caveat of some datedness in mind:

THE BATHYSCAPH, by Robert S. Dietz, Russell V. Lewis, and Andreas S. Rechnitzer, April 1958

EXPLORING THE OCEAN FLOOR, by Hans Pettersson, Aug. 1950

THE DEEP-SEA LAYER OF LIFE, by Lionel A. Walford, Aug. 1951

ANIMALS OF THE ABYSS, by Anton F. Bruun, Nov. 1957

ANIMAL SOUNDS IN THE SEA, by Marie Poland Fish, April 1956

THE HOMING SALMON, by Arthur D. Hasler and James A. Larsen, Aug. 1955 (#411)

THE PHYSIOLOGY OF WHALES, by Cecil K. Drinker, July 1949

THE RETURN OF THE GRAY WHALE, by Raymond M. Gilmore, Jan. 1955

THE LAST OF THE GREAT WHALES, by Scott McVay, Aug. 1966

THE EELGRASS CATASTROPHE, by Lorus J. and Margery J. Milne, Jan. 1951

LIVING UNDER THE SEA, by Joseph B. MacInnes, March 1966 (#1036)

FOOD FROM THE SEA, by Gorden A. Riley, Oct. 1949 (Compare this article with S. J. Holt's 1969 essay, in this Reader, and Jean Mayer's article of September 1976.)

Finally, some of the oceanographic, geological, and paleontological articles that could well grace these pages (were there more) are these:

THE OCEAN, by Roger Revelle, Sept. 1969 (and other articles in this issue of *Scientific American*) (#879)

THE ANATOMY OF THE ATLANTIC, by Henry Stommel, Jan. 1955 (#810)

THE SARGASSO SEA, by John H. Ryther, Jan. 1956

THE ARCTIC OCEAN, by P. A. Gordienko, May 1961

THE ANTARCTIC OCEAN, by V. G. Kort, Sept. 1962 (#860)

THE EVOLUTION OF THE INDIAN OCEAN, by D. P. McKenzie and J. G. Sclater, May 1973 (#908)

THE EVOLUTION OF THE PACIFIC, by Bruce C. Heezen and Ian D. MacGregor, Nov. 1973 (#911)

THE MICROSTRUCTURE OF THE OCEAN, by Michael C. Gregg, Feb. 1973 (#905)

THE TOP MILLIMETER OF THE OCEAN, by Ferren MacIntyre, May 1974 (#913)

WHY THE SEA IS SALT, by Ferren MacIntyre, Nov. 1970 (#893)

THE CIRCULATION OF THE OCEANS, by Walter Munk, Sept. 1955 (#813)

THE PERU CURRENT, by Gerald S. Posner, March 1954

THE CROMWELL CURRENT, by John A. Knauss, April 1961

THE CIRCULATION OF THE ABYSS, by Henry Stommel, July 1958

OCEAN WAVES, by Willard Bascom, Aug. 1959 (#828)

BEACHES, by Willard Bascom, Aug. 1960 (#845)

THE CONTINENTAL SHELF, by Henry C. Stetson, March 1955 (#808)

THE CONTINENTAL SHELVES, by K. O. Emery, Sept. 1969 (#882)

THE DEEP-OCEAN FLOOR, by H. W. Menard, Sept. 1969 (#883)

WHEN THE MEDITERRANEAN DRIED UP, by Kenneth J. Hsü, Dec. 1972 (#904)

WHEN THE BLACK SEA WAS DRAINED, by Kenneth J. Hsü, May 1978 (#932)

THE HISTORY OF THE ATLANTIC, by John G. Sclater and Christopher Tapscott, June 1979 (#938)

PRE-CAMBRIAN ANIMALS, by Martin F. Glaessner, March 1961 (#837)

CORALS AS PALEONTOLOGICAL CLOCKS, by S. K. Runcorn, Oct. 1966 (#871)

THE EVOLUTION OF REEFS, by Norman D. Newell, June 1972 (#901)

INDEX